OBSERVATIONAL ASTRONOMY

Astronomy is fundamentally an observational science, and as such it is important for astronomers and astrophysicists to understand how their data are collected and analyzed. This book is a comprehensive review of current observational techniques and instruments.

Featuring instruments such as Spitzer, Herschel, Fermi, ALMA, Super-Kamiokande, SNO, IceCube, the Auger Observatory, LIGO, and LISA, the book discusses the capabilities and limitations of different types of instruments. It explores the sources and types of noise and provides statistical tools necessary for interpreting observational data. Due to the increasingly important role of statistical analysis, the techniques of Bayesian analysis are discussed, along with sampling techniques and model comparison.

With topics ranging from fundamental subjects such as optics, photometry, and spectroscopy, to neutrinos, cosmic rays, and gravitational waves, this book is essential for graduate students in astronomy and astrophysics.

EDMUND C. SUTTON is Associate Professor in the Astronomy Department at the University of Illinois. His research has been primarily in infrared and submillimeter astronomy with an emphasis on instrumentation.

OBSERVATIONAL ASTRONOMY

Techniques and Instrumentation

EDMUND C. SUTTON

University of Illinois

CAMBRIDGE
UNIVERSITY PRESS

CAMBRIDGE
UNIVERSITY PRESS

University Printing House, Cambridge CB2 8BS, United Kingdom

One Liberty Plaza, 20th Floor, New York, NY 10006, USA

477 Williamstown Road, Port Melbourne, VIC 3207, Australia

314-321, 3rd Floor, Plot 3, Splendor Forum, Jasola District Centre, New Delhi - 110025, India

79 Anson Road, #06-04/06, Singapore 079906

Cambridge University Press is part of the University of Cambridge.

It furthers the University's mission by disseminating knowledge in the pursuit of
education, learning and research at the highest international levels of excellence.

www.cambridge.org
Information on this title: www.cambridge.org/9781107010468

First published 2012

A catalogue record for this publication is available from the British Library

Library of Congress Cataloging in Publication data
Sutton, Edmund Charles.
Observational astronomy : techniques and instrumentation / Edmund C. Sutton.
p.   cm.
ISBN 978-1-107-01046-8 (hardback)
1. Astronomy – Textbooks.   2. Astronomy – Observations – Textbooks.   I. Title.
QB43.3.S88   2011
520–dc23
2011030682

ISBN 978-1-107-01046-8 Hardback

Additional resources for this publication at www.cambridge.org/9781107010468.

Contents

The color plates will be found between pages 232 and 233.

Illustrations

Tables

Preface

This book is based on a required course for graduate students in Astronomy which I taught for a number of years at the University of Illinois. The premise of the course is that both theoretical astronomers and observers should have a basic understanding of the techniques of observational astronomy. The emphasis is on the underlying physics of the methods of detection and analytical tools (statistical and otherwise) that astronomers find useful. The great variety of current instruments and the rapid introduction of new instruments preclude an in-depth treatment of the peculiarities and idiosyncrasies of many instruments. But every instrument has its own idiosyncrasies and its own ways of corrupting the data and deceiving the observer. The topics in this book, I believe, cover the minimum which is required of anyone attempting to understand or interpret observational astronomy data.

Throughout the book equations are given in mks (SI) units so that it is easy to relate the discussion to practical quantities such as volts and watts. This is true even in the chapter on gravitational waves, a subject for which many texts and references use geometrized units ($c = 1$, $G = 1$). I prefer to keep c and G around rather than having to figure out where to put them when I need to calculate power. I also like being able to check equations using dimensional analysis. In the text other units are freely worked in. Among astronomers, the gauss remains firmly fixed as the unit of magnetic flux density. And astronomers frequently use other cgs units. For example, cross sections are always in cm^2. And of course there is a plethora of astronomical units such as pc, AU(!), and $M_\odot$. An appendix is provided with physical constants in both mks and cgs units and with a list of other units used and their equivalents in mks and cgs units.

The reader will note that the chapters on neutrinos, cosmic rays, and gravitational waves are of a different nature than other parts of the book. These fields are sufficiently specialized that it is difficult to separate purely observational issues from the underlying science, Therefore, in these chapters I freely go back and forth between design and scientific goals.

In addition to the color plates, there are color versions of a large number of other figures. The complete set of color figures may be accessed and/or downloaded through this book's website: www.cambridge.org/9781107010468.

I am well aware of other topics that I could have included in this book. In particular, I regret not being able to include a thorough discussion of adaptive optics and not covering topics in astroparticle physics.

The outlook for possible future instruments has changed markedly since much of this text was written, largely due to budgetary constraints. A funding increment for DUSEL (Chapter 14) by the National Science Foundation was recently rejected by the US National Science Board. The fate of DUSEL currently rests with its remaining US sponsor, the Department of Energy. WFIRST (Chapter 5) remains a high priority project for NASA. If ESA assigns a similarly high priority to its Euclid mission, a merger of these projects is likely to be considered. The US commitments to IXO (Chapter 11) and LISA (Chapter 16) are very much in doubt. These international collaborations are expected to continue, but reduced financial support could lead to delays and reductions in scope. In any event, these instrument concepts are the current state of the art. Astronomers constantly need to readjust their plans in light of financial realities. If better ways can be found to pursue some of these scientific objectives, now is certainly the time for them.

Acknowledgements

I appreciate the willingness of my colleagues Brian Fields, Athol Kemball, Ben Wandelt, and Dick Crutcher to review limited sections of this text. Their comments have been very useful. Any remaining mistakes are, of course, solely my responsibility. I encourage anyone who discovers errors of any sort to communicate them to me at ecsutton@illinois.edu. The publisher and I will work together to maintain an online list of any errata.

The graduate students to whom I have had the pleasure to teach this material over the years are Scott Bain, Ian Barton, Yohann Beda, Jana Bilikova, Mark Butala, Karen Camarda, Christine Cecala, Nachiketa Chakraborty, Ray Chen, Rosie Chen, Yun Chen, Hsin-Fang Chiang, Samuel Crawford, Conley Ditsworth, Joshua Dolence, Bryan Dunne, Rich Frazin, Khurram Gillani, Daniel Goscha, Philip Grathoff, Michelle Griffin, Xiaoyue Guan, Troy Hacker, Thomasanna Hail, Nicholas Hakobian, Hassan Halataei, Brett Hayes, Nathan Hearn, Nicholas Indriolo, Rishi Khatri, Soyoung Kim, Robert Klinger, Scott Kruger, Hsin-Lun Kuo, Woojin Kwon, Shih-Ping Lai, I-Jen Lee, PoKin Leung, Amy Lien, Wen-Ching Lin, Jiayi Liu, Sheng-Yuan Liu, Justin Lowry, Zarija Lukic, Britt Lundgren, Patrick Lynch, Modhurita Mitra, Rosa Murphy, Erik Nelson, Christopher Neyman, Chenping Ni, Lisa Norton, Brian O'Neill, Kuo-Chuan Pan, Vasiliki Pavlidou, Tijana Prodanovic, Ramprasad Rao, David Rebolledo Lara, Ashley Ross, Jonathan Seale, Jerry Shaw, Hotaka Shiokawa, Jeeseon Song, Thomas Spinka, Ranjani Srinivasan, Ian Stephens, Shweta Sundararajan, Konstatinos Tassis, Daniel Thayer, Glenn Thurman, Toshiya Ueta, Scott Walker, Li-Bang Wang, Shiya Wang, Yiran Wang, Rui Xue, Amit Yadav, Chao-Chin Yang, Hsiang-Yi Yang, Jeong Yim, Alfredo Zenteno, and Jie Zou. They have done much to determine the direction my course has taken and have been invaluable at finding mistakes and inconsistencies in the notes. I am indebted to them.

I appreciate the help of the production staff at Cambridge University Press, especially Claire Poole, Vince Higgs, and Abigail Jones. Ms. Sehar Tahir was helpful with TeX support. Margaret Patterson was an excellent copy editor.

I am indebted to the developers of Inkscape, which made the production of the figures relatively painless. A portion of the author's proceeds from this publication has been donated in advance to the Software Freedom Conservancy to help defray further development costs for the Inkscape Project.

Every effort has been made to acknowledge and obtain permission for all figures used in this work. Figure 2.10 is included by kind permission of R. A. Jansen. Tables 4.1, 4.2, and 4.3 are reprinted with permission from Bracewell, *The Fourier transform and its applications*, 3rd edn. ©2000 The McGrawHill Companies, Inc. Figure 5.6 is reprinted with permission from Timothy (1983) *PASP*, **95**, 573, ©1983 University of Chicago Press. The image shown in Figure 10.10 was kindly obtained by D. Ketelsen, who has granted permission for its use here. Figures 11.6, 11.7, and 11.8 reproduced by permission of the AAS. The quotation from Bahcall & Ostriker, *Unsolved problems in astrophysics* ©1997 is used by permission of Princeton University Press. Figure 14.1 reproduced by permission of the AAS. Figures 14.4, 14.12, and 14.17 are reprinted with permission from Aharmim *et al.* (2005) *Phys. Rev. C*, **72**, 055502, Ashie *et al.* (2005) *Phys. Rev. D*, **71**, 112005, and Abbasi *et al.* (2008) *Phys. Rev. Lett.*, **100**, 101101, ©2005 and 2008 by the American Physical Society. Figures 14.9 and 14.10 reprinted with permission, ©Kamioka Observatory, ICRR, University of Tokyo. Figures 14.14, 14.15, and 14.6 are courtesy of SNO. Figures 14.18, 14.19, and 14.20 are based upon work supported by the National Science Foundation under Grant Nos. OPP-9980474 (AMANDA) and OPP-0236449 (IceCube), University of Wisconsin-Madison. Figure 15.1 reprinted with permission of Annual Reviews from Beatty & Westerhoff (2009) *ARNPS*, **59**, 319; permission conveyed through Copyright Clearance Center, Inc. Figure 15.3 reprinted from Gaisser & Stanev (2008) *Phys. Lett. B*, **667**, 254 with permission from Elsevier and D. Muller. Figure 15.7a reprinted from Stone *et al.* (1998) *Sp. Sci. Rev.*, **86**, 357 with permission from Elsevier. Figure 15.7b reprinted by permission of R. Ogliore. Figure 15.9 reproduced by permission of the AAS. Figure 15.12 reprinted from Boyer *et al.* (2002) *NIMPR A*, **482**, 457 with permission from Elsevier. Figure 15.13 reprinted by permission from Abbasi *et al.* (2008) *Phys. Rev. Lett.*, **100**, 101101 ©2008 by the American Physical Society. Figure 15.14 reprinted by permission of P. Mantsch on behalf of the Auger Project. Figure 16.6 reprinted from Abadie *et al.* (2010c) *NIMPR A*, **624**, 223 with permission from Elsevier. Figure 16.8 reprinted by permission, ©Science and Technology Facilities Council and Brett Shapiro/LIGO Laboratory. Figure 16.10 reprinted from Hild *et al.* (2010) *Class. Quantum Grav.*, **27**, 15003 with permission from IoP Publishing.

And finally I thank Jean for all of her support during the writing and production of this book.

1

Astrophysical information

In observational astronomy we study the processes by which Earth-bound astronomers obtain and interpret information about distant parts of the universe. Theoretical descriptions of the natural world and observational/experimental data are complementary, and their interplay is a fundamental feature of scientific inquiry. For progress in astronomy we need extensive, sensitive, and accurate observations. But such data do not come for free. They are not just lying around for anybody to pick up. Work is required. An observer who simply accepts data at face value is likely to encounter problems.

In studying the observational process it will be helpful to adopt the following point of view. There is something we will call *information* which is present in an astronomical source. This information leaves the source, perhaps in the form of electromagnetic radiation. As it travels from the source to the observer it passes through intervening regions, often being modified in the process. The information then reaches the detection system. This final stage inevitably involves significant modification of the information. Noise is added, much information is lost, and other changes occur. From this final state the astronomer attempts to infer characteristics of the original source.

1.1 Electromagnetic radiation

The most important carrier of astronomical information is electromagnetic radiation. The electromagnetic spectrum is commonly broken down into various wavelength bands, as indicated in Table 1.1. Each band may be roughly described by both a characteristic photon energy $h\nu$ and a characteristic temperature $h\nu/k$. Each band carries a different set of information, since radiation at different wavelengths is produced (and modified) by different physical processes. And in each band the information is carried in a variety of forms (spectral, temporal, spatial, polarization, intensity).

Table 1.1. *Characteristic photon
energies and temperatures*

Band	E_{typ} (hν)	T_{typ} (hν/k)
gamma ray	10^5 eV	10^9 K
x-ray	10^3 eV	10^7 K
ultraviolet	10 eV	10^5 K
visible	1 eV	10^4 K
infrared	0.1 eV	10^3 K
microwave	10^{-3} eV	10 K
radio	10^{-6} eV	0.01 K

The wide range of wavelengths implies, among other things, that a wide variety of detection mechanisms must be employed. We will focus on the basic methods of detection and then provide some detail about differences between the bands. We will also discuss limitations to sensitivity and spatial resolution and how these vary between bands. Similarly, each wavelength band will have its own characteristics associated with each type of analysis. Here again we will focus on the fundamentals of the various types of analysis (spectroscopy, high speed photometry, imaging, polarimetry, photometry), providing some detail about how these types of analysis vary between the different bands.

1.2 Other carriers of information

In addition to electromagnetic radiation we have information carried to us via neutrinos, cosmic rays, and gravitational waves. These will be discussed in Chapters 14, 15, and 16, respectively. The study of material such as meteorites, lunar rocks, and interplanetary dust particles, although important, is somewhat specialized and will not be discussed here. The possible future detection of exotic particles such as dark matter will also not be discussed.

Neutrinos have been detected both from the Sun (at characteristic energies of $\sim 10^5$ eV) and from supernovae ($\sim 10^7$ eV). They contain information in their flux, in their arrival times (in the case of supernovae), and in their spectra. The lower than expected flux of neutrinos from the solar core was a longstanding problem in astrophysics, now considered to be resolved. The discovery of neutrinos from SN 1987A provided important confirmation of our picture of core-collapse supernovae. Studies of neutrino energy spectra have been difficult due to the fact that detectors are typically sensitive to neutrinos of particular energies determined by the type of interaction material used (gallium, chlorine, etc.).

Cosmic rays consist of energetic electrons, protons, and heavy nuclei (out to Pb and beyond), reaching Earth from distant astrophysical sources. The lower energy particles may be detected directly from balloons and satellites. The higher energy particles range up to about 10^{20} eV. These create extensive air showers in Earth's atmosphere which can be detected with ground arrays sensitive to fluorescence from atmospheric nitrogen. Studies of the relative abundances of cosmic ray particles and their energy spectra reveal information about cosmic composition and the energetics of the generating sources (e.g. pulsars).

Gravitational waves from astrophysically interesting events are predicted with strains of

$$\delta L / L \lesssim 10^{-23}. \tag{1.1}$$

Current instruments such as LIGO and Virgo are approaching this level in the 100 Hz range.

1.3 Intervening regions

We speak of astronomy as an *observational* (not an experimental) science, implying that generally we do not have control over the conditions of our *experiment* and cannot directly probe or manipulate the object of interest. A corollary fact is that there exist intervening regions which can affect the flow of information from an astronomical source to the observer. These intervening regions include the intergalactic, interstellar, and interplanetary mediums and Earth's atmosphere. The effects of such regions on the flow of information are understood only *in part*. The intergalactic, interstellar, and interplanetary mediums are themselves astronomical entities about which we have limited observational information. And although we have abundant information about global properties of Earth's atmosphere, we generally lack sufficient detail down to the smallest relevant spatial scales (centimeters) and time scales (milliseconds).

There are situations in astrophysics in which the intervening region is itself the object of interest. Examples include galactic H I (21 cm) absorption, quasar (Ly α) absorption line systems, gravitational lensing and micro-lensing studies, and the Sunyaev–Zel'dovich effect. In these cases some knowledge about the background sources of radiation is required in order to study the effects produced by the intervening regions. For simplicity we will concentrate here on the more common case, in which one wishes to study a distant object and the intervening regions have the ability to modify the flow of radiation from the object to the observer.

1.3.1 Intergalactic/interstellar medium

The interstellar (and intergalactic) medium (ISM/IGM) contains gas and dust. Dust particles absorb and scatter light. If the size, shape, and composition of the dust grains were known, their effect on electromagnetic radiation could, in principle, be calculated (Mie, 1908; van de Hulst, 1957). But these properties are not readily determined and must be inferred from global characteristics such as the wavelength dependent coefficients of absorption and scattering, which may be different for different lines of sight through the Galaxy. The combined effect of scattering and absorption is referred to as *extinction*. The average value of extinction due to dust in the plane of the Galaxy is

$$\langle A_V \rangle / D \approx 2 \text{ mag kpc}^{-1}. \tag{1.2}$$

Interstellar extinction is dependent on wavelength. Since extinction is stronger in the blue portion of the visible spectrum, starlight is *reddened* as it passes through the ISM. A typical value for the reddening in the plane of the Galaxy (the differential extinction between the B and V photometric bands) is

$$\langle E_{B-V} \rangle / D \approx 0.6 \text{ mag kpc}^{-1}. \tag{1.3}$$

Interstellar reddening is likely to be highly dependent on the nature of the dust particles and thus variable from one region to another.

Atomic (neutral) gas in the ISM/IGM can produce various interstellar absorption lines. There is strong absorption shortwards of the Lyman limit (91 nm) due to neutral hydrogen (H I).

The ionized ISM/IGM (plasma) gives rise to effects such as pulsar dispersion, Faraday rotation, and radio scintillation. Pulsar dispersion is parameterized by the dispersion measure,

$$DM = \int_0^L n_e \, dl, \tag{1.4}$$

which is typically of order 10–100 parsec cm^{-3} for lines of sight towards nearby pulsars. Relative to a signal at sufficiently high frequency, a lower frequency ν will be delayed by a time

$$\Delta t = \frac{e^2}{2\pi m_e c} \frac{1}{\nu^2} DM. \tag{1.5}$$

Faraday rotation is parameterized by the rotation measure,

$$RM = \int_0^L n_e H_\parallel \, dl, \tag{1.6}$$

with the convention that RM and $H_\parallel$ are positive for magnetic fields pointing towards us. Typical values of the galactic magnetic field are of order a few μG. In traversing a region of rotation measure RM, a linearly polarized wave will have its plane of polarization rotated by an angle

$$\Delta\theta = \frac{e^3}{2\pi \, m_e^2 c^2} \frac{1}{v^2} \, \text{RM}, \tag{1.7}$$

counterclockwise as viewed by us for positive RM.

Cosmic ray particles are also influenced by the medium they pass through. The lower energy cosmic rays follow magnetic field lines, making it impossible to determine their point of origin. The higher energy cosmic rays interact with the photons of the microwave background radiation. Since they lose energy in these interactions, the highest energy cosmic rays that we see must have originated within about 50 Mpc of Earth.

1.3.2 Interplanetary medium

There also exists an interplanetary medium consisting of gas and dust, which can produce effects similar to the interstellar medium. The interplanetary medium is concentrated in the plane of the solar system. The zodiacal dust is evident by its scattering and absorption and, since it is warm, also by its thermal emission in the infrared. The plasma in the solar corona and solar wind influence the propagation of radio waves (Thompson *et al.*, 2001). Such effects vary with the 11-year solar activity cycle and on shorter time scales as well.

1.3.3 Earth's atmosphere

The Earth's atmosphere produces a multitude of effects, most of which interfere strongly with the free propagation of astronomical signals. Molecules and atoms in the atmosphere absorb radiation across almost the entire electromagnetic spectrum except in the visible and radio bands. Atmospheric dust produces scattering of visible light, the amount of which is very much dependent on such things as volcanic activity and wind patterns (e.g. dust from the Sahara). And of course there is variable cloud cover. The upper atmosphere emits radiation by a process known as airglow. The ionosphere cuts off the propagation of long wavelength radio waves. As with most ionospheric phenomena, this cutoff is dependent on the solar activity cycle. And finally, turbulence in the atmosphere gives rise to the effects known as seeing and scintillation.

Radio and microwave absorption

The radio, microwave, millimeter, and submillimeter wave bands extend out to frequencies of several hundreds of GHz, as shown in Figure 1.1. The atmospheric spectrum in this region contains many discrete rotational lines of molecules such as oxygen (O_2), water vapor (H_2O), and to a lesser extent ozone (O_3) and other

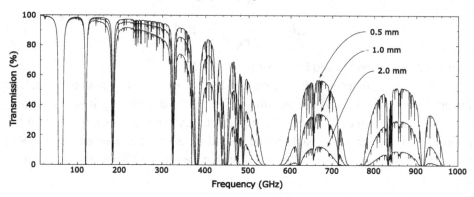

Figure 1.1 Atmospheric transmission from 10 to 1000 GHz for an elevation of 5108 m (Llano de Chajnantor) and precipitable water vapor content of 0.5, 1.0, and 2.0 mm. Except for the blended 60 GHz band of O_2 and isolated oxygen lines at 119, 368, 425, 487, 715, 774, and 834 GHz (Tretyakov *et al.*, 2005), the majority of strong lines are due to water vapor absorption. Courtesy of Atacama Pathfinder Experiment (APEX).

trace constituents. At the shorter wavelengths the water vapor lines blend into a quasi-continuous absorption.

The oxygen lines are magnetic dipole transitions and arise throughout the troposphere. The tropospheric pressure distribution is determined by hydrostatic equilibrium and is roughly exponential,

$$P(z) = P_0\, e^{-z/H}, \tag{1.8}$$

with a scale height

$$H = \frac{kT}{\langle \mu \rangle m_H g}, \tag{1.9}$$

which is of order 7 km for a typical temperature of $T = 250$ K and a mean molecular weight $\langle \mu \rangle = 29$. Most molecular species follow this exponential distribution.

Water vapor, on the other hand, is not well mixed. The saturation vapor pressure of water vapor is a strong function of temperature. In the troposphere, temperature drops with altitude, and colder air has a strongly reduced water vapor content. The distribution of water vapor is quasi-exponential, with a reduced scale height of

$$H_{H_2O} \approx 2\text{--}3 \text{ km}. \tag{1.10}$$

The profiles of tropospheric absorption lines are determined by pressure broadening. Therefore, most atmospheric line profiles are approximately Lorentzian in shape with very broad wings. Since the water vapor is preferentially present in the lower layers of the atmosphere, water vapor line widths are typically broader than

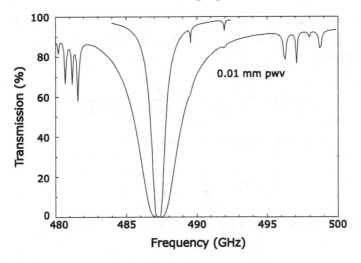

Figure 1.2 Detail of atmospheric transmission from 480 to 500 GHz for 0.01 mm precipitable water vapor, showing the presence of weak, narrow ozone lines. The lower curve (red in electronic edition) corresponds to an altitude of 5108 m. At an aircraft altitude of 12.5 km (upper, blue curve), the effect of pressure broadening is much less and atmospheric transmission can approach 98% near 492 GHz. Courtesy of Atacama Pathfinder Experiment (APEX).

those of other atmospheric constituents. By the same token, lines formed at high altitudes, such as those of ozone, are narrow, as shown in Figure 1.2. Water lines get considerably narrower wings when observed at high altitudes.

Lorentzian line shape

The quantum mechanical problem of the line shape for absorption or emission by an atom undergoing random collisions (Gross, 1955) is somewhat subtle. But for our present purposes we will treat it as analogous to a classical simple harmonic dipole oscillator with radiative damping, which exponentially decays (for $t \geqslant 0$),

$$x(t) = x_0 \, e^{-\Gamma t/2} \cos 2\pi \nu_0 t. \tag{1.11}$$

The $(-i)$ Fourier transform of $x(t)$ is

$$\hat{x}(\nu) = \frac{x_0}{2i} \left(\frac{1}{2\pi \nu - 2\pi \nu_0 - i\Gamma/2} + \frac{1}{2\pi \nu + 2\pi \nu_0 - i\Gamma/2)} \right), \tag{1.12}$$

the second term of which can be generally neglected. The total radiated power is given by the Larmor formula. In mks units

$$P = \frac{1}{4\pi \epsilon_0} \frac{2}{3} \frac{q^2 \ddot{x}^2}{c^3}. \tag{1.13}$$

According to Rayleigh's theorem the intensity profile $I(\nu)$ is proportional to $|\hat{x}(\nu)|^2$, so the radiated power is distributed in frequency according to the Lorentzian profile

$$\phi(\nu) \approx \frac{1}{4} \frac{\Gamma}{(2\pi\nu - 2\pi\nu_0)^2 + (\Gamma/2)^2},\qquad(1.14)$$

which for $\Gamma \ll \nu_0$ can be normalized for unit area as

$$\phi(\nu) \approx \frac{1}{\pi} \frac{\gamma}{(\nu - \nu_0)^2 + \gamma^2}.\qquad(1.15)$$

The HWHM line width γ is proportional to the rate of collisions.

Infrared absorption and background

The infrared spectrum is dominated by absorption from CO_2 and H_2O. Water vapor is a particular problem because the water molecule is an asymmetric top with a permanent dipole moment. These and other molecular lines leave the atmosphere totally opaque at many infrared wavelengths. Between the absorption bands there are a few discrete windows which are sufficiently transparent to allow ground-based observations. The best of these windows are centered near 1.2, 2.2, 3.4, 5.0, 10, and 20 μm. However, even in these bands ground-based observations are faced with strong atmospheric thermal background emission, especially near 5, 10, and 20 μm (near the peak of a blackbody spectrum for room temperature).

There is therefore a strong incentive for space-based observations throughout the infrared, but especially for those portions of the spectrum with significant atmospheric opacity. An extensive infrared satellite survey was conducted by the IRAS satellite, which measured long wavelength fluxes (12, 25, 60, 100 μm), albeit with low spatial resolution. More recent work was done by ISO, a satellite launched by the European Space Agency (ESA). The most important current infrared satellites are Spitzer (SIRTF) and Herschel (FIRST), which have both imaging and spectroscopic capabilities. SOFIA, an airborne telescope built jointly by Germany and the USA is also about to start producing data.

Low frequency EM wave propagation

Consider a free electron plasma of density N in a field $\vec{E}_0\, e^{-i2\pi\nu t}$. Neglecting collisions, the equation of motion for an electron displacement is

$$m\frac{d^2\vec{x}}{dt^2} = e\, \vec{E}_0\, e^{-i\omega t}.\qquad(1.16)$$

Solving for the electron velocity and current density, we get

$$\vec{v} = \frac{d\vec{x}}{dt} = \frac{-e}{im\omega} \vec{E}_0 \, e^{-i\omega t},$$ (1.17)

$$\vec{J} = Ne\vec{v} = -\frac{Ne^2}{im\omega} \vec{E}_0 \, e^{-i\omega t}.$$ (1.18)

Since the conductivity of a material is defined by $\vec{J} = \sigma \vec{E}$, the conductivity of the electron plasma is

$$\sigma = i \frac{Ne^2}{m\omega}.$$ (1.19)

Waves in a conducting medium propagate according to the damped wave equation

$$\nabla^2 \vec{E} - \epsilon\mu \frac{\partial^2 \vec{E}}{\partial t^2} - \sigma\mu \frac{\partial \vec{E}}{\partial t} = 0.$$ (1.20)

Adopting a trial solution of $\vec{E} = \vec{E}_0 \, e^{i(\tilde{k}z - \omega t)}$, we can get

$$\tilde{k}^2 = \frac{\omega^2}{c^2} \left(\frac{\epsilon}{\epsilon_0} + i \frac{\sigma}{\epsilon_0 \omega} \right)$$ (1.21)

$$\approx \frac{\omega^2}{c^2} \left(1 - \frac{\omega_p^2}{\omega^2} \right),$$ (1.22)

where $\omega_p^2 = Ne^2/m\epsilon_0$. For $\omega < \omega_p$, the wavenumber $\tilde{k}$ is imaginary, so waves are exponentially attenuated, not propagated. In the ionosphere electron densities are of order $10^6 \, \text{cm}^{-3}$, so frequencies below about 10 MHz are not propagated. The exact cutoff frequency varies with the day/night cycle and with solar activity.

Airglow

An additional factor to consider at visible and near-visible wavelengths is airglow, a form of fluorescent recombination which occurs in the upper atmosphere ($\sim$100 km). In the visible, strong airglow lines of O I occur at 558 and 630 nm, O_2 at 762 nm, etc. The intensity of telluric airglow is measured in units of rayleighs,

$$1 \text{ rayleigh} = 10^6/4\pi \quad \text{photons cm}^{-2} \, \text{s}^{-1} \, \text{sr}^{-1}$$ (1.23)

$$= 1.58 \times 10^{-11}/\lambda \quad \text{nm W cm}^{-2} \, \text{sr}^{-1}.$$ (1.24)

Rayleigh scattering

Atmospheric scattering (resonant, Rayleigh) is also a significant source of sky background in the ultraviolet, visible, and near-infrared. Consider a bound electron with a driving field. The equation of motion is

$$\ddot{x} + \Gamma \dot{x} + \omega_0^2 x = \frac{e}{m} E_0 \cos \omega t, \tag{1.25}$$

which, for a trial solution

$$x = x_0 \, e^{i\omega t}, \tag{1.26}$$

gives an amplitude displaying a resonant response where

$$x_0 = -\frac{e}{m} E_0 \frac{1}{\omega^2 - \omega_0^2 - i\omega_0 \Gamma}. \tag{1.27}$$

Calculating the radiated power by the Larmor formula gives

$$P = \frac{1}{4\pi\epsilon_0} \frac{2}{3} \frac{q^2}{c^3} \omega^4 |x_0|^2. \tag{1.28}$$

We get the interaction cross section by normalizing by the incident Poynting flux,

$$\sigma(\omega) = \sigma_T \frac{\omega^4}{(\omega^2 - \omega_0^2)^2 + (\omega_0 \Gamma)^2}, \tag{1.29}$$

$$\sigma_T = \frac{2}{3} \frac{q^4}{m^2 c^4} \frac{1}{4\pi\epsilon_0^2}. \tag{1.30}$$

At low frequencies this displays the ω^4 Rayleigh scattering and at high frequencies the constant Thomson cross section σ_T, with the resonant fluorescence peak in between, as shown in Figure 1.3.

Atmospheric turbulence

Atmospheric density inhomogeneities produce regions of different refractive indices which introduce wavefront corrugations, as shown in Figure 1.4. *Seeing*

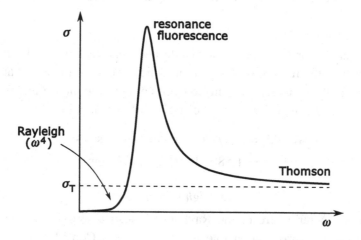

Figure 1.3 Scattering from bound electrons.

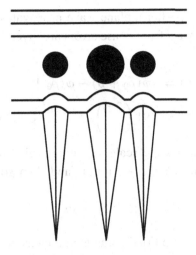

Figure 1.4 Dense pockets of air produce wavefront delays, leading to seeing and scintillation.

refers to a change in apparent stellar position or an increase in image size due to wavefront slope. *Scintillation* refers to intensity (brightness) fluctuations due to the focussing effects of curved wavefronts (this is weak focussing, therefore the fluctuations are apparent only after several kilometers).

If there exists *random turbulence*, an idealized theory in three spatial dimensions predicts that fluctuations over a wide range of spatial scales will follow the Kolmogorov spectrum, in which large scale sizes dominate. For turbulent mixing of air at different initial temperatures (densities), one can predict the following structure functions[1] for temperature,

$$D_T(\vec{r}) = \langle [T(\vec{r}_0 + \vec{r}) - T(\vec{r}_0)]^2 \rangle = C_T^2 r^{2/3}, \tag{1.31}$$

and for the index of refraction,

$$D_n(\vec{r}) = \langle [n(\vec{r}_0 + \vec{r}) - n(\vec{r}_0)]^2 \rangle = C_n^2 r^{2/3}, \tag{1.32}$$

where empirically the magnitudes of the fluctuations are of order

$$C_T \approx 10^{-2} \text{ K cm}^{-1/3}, \tag{1.33}$$

and

$$C_n \approx 10^{-8} \text{ cm}^{-1/3}, \tag{1.34}$$

---

[1] Some aspects of the theory of random processes are covered in Chapter 7. Here the structure function is introduced to allow us to consider what are known as *random processes with stationary first increments*, which in this case are *locally* homogeneous, isotropic random fields.

since $\partial n/\partial T \approx 10^{-6}$ K^{-1}. After a plane wave propagates through such an atmosphere, the accumulated wavefront phase errors are given by the structure function (Tatarskii, 1971)

$$D_S(\rho) = \langle [\phi(\vec{\rho}_0 + \vec{\rho}) - \phi(\vec{\rho}_0)]^2 \rangle \qquad (1.35)$$

$$= 2.91 \left(\frac{2\pi}{\lambda}\right)^2 \rho^{5/3} \int_0^L C_n^2(z) \, dz. \qquad (1.36)$$

It is convenient to discuss the scale size over which the wavefront is largely coherent. By convention, set $D_S \approx 1$ rad^2 and then solve for a characteristic separation,

$$\rho = r_0 \propto \lambda^{6/5}, \qquad (1.37)$$

a value which is known as the Fried parameter, a measure of the amount of turbulence. A typical value at 500 nm would be $r_0 \approx 10$ cm. Since the diffraction limit is proportional to λ/D, this implies that one can achieve better resolution (smaller image sizes) at long wavelengths (image size $\propto \lambda^{-1/5}$).

Exercises

1.1 Assume that the pressure of Earth's atmosphere falls off exponentially,

$$P(z) = P_0 \, e^{-z/H}, \qquad (1.38)$$

with $P_0 = 1013$ mb $= 101.3$ kPa (sea level) and a scale height $H = 7$ km. Ignoring temperature changes, this implies the atmospheric density also falls off exponentially. Assume that the ratio of water vapor density to air density (the "mixing ratio") also obeys an exponential law (a rather crude approximation to real life),

$$r(z) = \frac{\rho_{H_2O}}{\rho_{air}} = 10^{-2} \, e^{-z/H'}, \qquad (1.39)$$

where $H' = 2$ km.

a. Show that the density of water vapor also falls off exponentially and calculate its scale height.

b. Calculate the amount of water vapor present in a vertical column of air above a point at sea level ($z = 0$). Such a quantity is usually expressed in terms of the thickness of a layer of liquid water with an equivalent number of water molecules,

$$w = \frac{1}{\rho_{liquid}} \int_0^\infty \rho_{vapor}(z) \, dz. \qquad (1.40)$$

c. Do a reality check. Is your answer reasonable?

1.2 Go to www.apex-telescope.org/sites/chajnantor/atmosphere/transpwv/ and create a plot of the atmospheric transmission between 100 and 200 GHz for a precipitable water vapor content of 1 mm (a magnified version of part of Figure 1.1). Most atmospheric absorption lines have near-Lorentzian shapes,

$$\phi(v, v_0) = \frac{\Delta v / 2\pi}{(v - v_0)^2 + (\Delta v / 2)^2}. \tag{1.41}$$

a. For the water vapor line at 183 GHz, estimate the FWHM line width parameter Δv.

b. The line width is related to the mean time between collisions by the formula $\tau = 1/(\pi \Delta v)$. What is the mean collisional time?

c. Compare with the mean collisional time estimated from kinetic theory assuming a mean atmospheric pressure of 400 mb (at the characteristic height of water vapor) and a collisional cross section (for collisions between H_2O and N_2) of 3×10^{-19} m^{-2}.

d. Comment on the line at 119 GHz.

e. Make a magnified plot of the 160–170 GHz region. The line at 166 GHz is due to ozone. *Estimate* the height of the ozone layer.

2

Photometry

2.1 Specific intensity (brightness)

In studying radiometry and photometry we will make use of a fundamental quantity
known as the specific intensity (or brightness). This may not correspond with the
way you use the word brightness in everyday speech. We will define the *specific
intensity* as the rate of energy transport, *along a particular direction*, per unit area,
per unit solid angle, per unit frequency. We will always use the word *brightness* as
synonymous with specific intensity.

Consider a propagation direction $\hat{k}$, a differential solid angle $d\Omega$ around $\hat{k}$, and
a differential surface element dA normal to $\hat{k}$, as shown in Figure 2.1. The energy
crossing dA within $d\Omega$ of $\hat{k}$, within a frequency band of width $d\nu$ around ν, in a
time dt is

$$dE = I_\nu \, dA \, dt \, d\Omega \, d\nu. \tag{2.1}$$

This equation, in effect, defines the specific intensity I_ν, which has units of
$J \, s^{-1} m^{-2} \, sr^{-1} Hz^{-1}$ ($W \, m^{-2} \, sr^{-1} Hz^{-1}$). The specific intensity is sometimes also
called the spectral radiance. It is also closely related to the quantity in (non-
astrophysical) visible photometry known as the luminance, which has units of
lumen $m^{-2} \, sr^{-1}$ (candle m^{-2}).

2.2 Étendue

Jacquinot emphasized the importance of a quantity in optics named *étendue*.
This term has no satisfactory English translation, but it is sometimes rendered
as "throughput" or "area–solid angle product". Consider a bundle of rays passing through the areas dA_1 and dA_2, where dA_1 and dA_2 are separated by a distance
L, and where dA_1 and dA_2 are normal to the mean ray. For each point on dA_1,
the solid angle subtended by dA_2 is $d\Omega_1 = dA_2/L^2$ and similarly $d\Omega_2 = dA_1/L^2$,
therefore

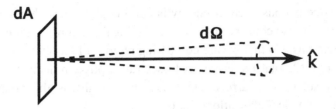

Figure 2.1 Geometry defining the specific intensity.

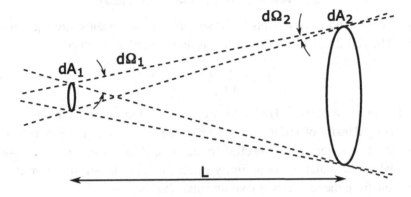

Figure 2.2 Conservation of étendue.

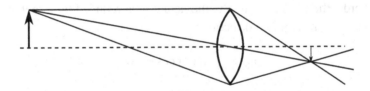

Figure 2.3 Larger object with smaller solid angle versus smaller image with larger solid angle.

$$d\Omega_1 \, dA_1 = d\Omega_2 \, dA_2. \tag{2.2}$$

Thus étendue (the product of dA and dΩ) is a conserved quantity, as illustrated in Figure 2.2. When one is dealing with regions with different indices of refraction n_1 and n_2, the correct expression is actually

$$n_1^2 \, d\Omega_1 \, dA_1 = n_2^2 \, d\Omega_2 \, dA_2. \tag{2.3}$$

Étendue in this sense is a conserved quantity in a perfect optical system, giving the well-known result that, for an imaging system, a smaller image size corresponds to a larger solid angle, as illustrated in Figure 2.3. In an imperfect optical system, étendue will increase.

Since étendue is conserved, if energy is conserved (no loss due to absorption), then *brightness* is also conserved. Strictly speaking these conservation laws require three qualifications: (1) that we are comparing regions of the same index of refraction ($n_1 = n_2$), (2) that there are no losses (absorptive, reflective), and (3) that we restrict ourselves to a narrow "elementary beam" (that is, this breaks down for imperfect optics, where aberrations are present).

2.3 Moments of the specific intensity

Applying the method of moments to the specific intensity yields three quantities of interest. The zeroth order moment is the mean intensity J_ν, defined by

$$J_\nu = \frac{1}{4\pi} \int I_\nu(\theta) \, d\Omega, \tag{2.4}$$

which has units of W m^{-2} Hz^{-1}. As we will see, the mean intensity is related to the energy density of radiation. The first order moment is the monochromatic *flux* (spectral flux density or spectral irradiance), denoted F_ν or S_ν. Consider a differential surface area dA at arbitrary orientation, as shown in Figure 2.4. The energy flow from the direction θ (within small $d\Omega$) is

$$dF_\nu = I_\nu(\theta) \cos\theta \, d\Omega. \tag{2.5}$$

In other words, the projected area is the actual area reduced by a factor of $\cos\theta$. Taking contributions from all directions,

$$F_\nu = \int I_\nu(\theta) \cos\theta \, d\Omega. \tag{2.6}$$

The astrophysical unit of flux is the jansky, $1\,\mathrm{Jy} = 1$ f.u. $= 10^{-26}$ W m^{-2} Hz^{-1}. An isotropic field has no net flux ($I_\nu(\theta) = $ constant, $F_\nu = 0$). Also of interest sometimes is the second order moment, the momentum flux p_ν (pressure),

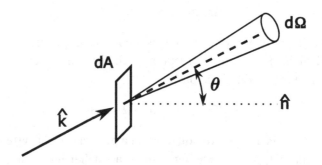

Figure 2.4 Flux through an aperture dA is reduced by the cosine of the angle between the unit wavevector and the unit normal.

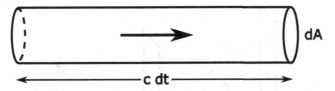

Figure 2.5 Radiation energy density is related to mean intensity.

$$p_\nu = \frac{1}{c} \int I_\nu(\theta) \cos^2\theta \; d\Omega. \qquad (2.7)$$

2.4 Energy density

The energy density in a radiation field is related to the zeroth moment of the specific intensity, the mean intensity. As illustrated in Figure 2.5, the energy passing through dA (per unit time, per unit solid angle) in a direction θ is that present in a cylinder of length c dt.

$$dE = u_\nu(\theta) \; dA \; c \; dt \; d\Omega \; d\nu. \qquad (2.8)$$

By definition

$$dE = I_\nu(\theta) \; dA \; dt \; d\Omega \; d\nu. \qquad (2.9)$$

Therefore the energy density per unit solid angle is

$$u_\nu(\theta) = I_\nu(\theta)/c \qquad (2.10)$$

and the total energy density is

$$u_\nu = \int u_\nu(\theta) \; d\Omega \qquad (2.11)$$

$$= \frac{1}{c} \int I_\nu(\theta) \; d\Omega \qquad (2.12)$$

$$= \frac{4\pi}{c} J_\nu. \qquad (2.13)$$

If and only if the radiation field is isotropic, we can write $u_\nu = 4\pi \, I_\nu/c$.

2.5 Flux from a surface of uniform brightness

From close up, the surface of a spherical object such as a star looks flat. So, what is the radiation field from such an object? Consider an infinite flat surface, as in Figure 2.6.

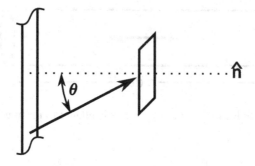

Figure 2.6 Integrated flux radiated by an infinite surface of uniform brightness.

$$F_\nu = \int I_\nu \cos\theta \, d\Omega \tag{2.14}$$

$$= \int \int I_\nu \cos\theta \sin\theta \, d\theta \, d\phi \tag{2.15}$$

$$= 2\pi \, I_\nu \int_0^{\pi/2} \sin\theta \, d\sin\theta \tag{2.16}$$

$$= 2\pi \, I_\nu \int_0^1 x \, dx \tag{2.17}$$

$$= \pi \, I_\nu. \tag{2.18}$$

This can be understood as an integral over $\Omega = 2\pi$ where $\langle \cos\theta \rangle = 1/2$.

What will happen as one moves away from the surface of the star and the curvature of the surface becomes apparent? At large distances we should expect the result

$$F_\nu = \Omega_S \, I_\nu, \tag{2.19}$$

where Ω_S is the solid angle subtended by the source.

2.6 Blackbody radiation

Blackbody radiation describes the properties of radiation inside a cavity or radiation emitted by a perfect absorber of temperature T. Consider a cubical cavity with sides of length L with periodic boundary conditions. The boundary conditions require that along each axis an integral number of wavelengths fit inside the length L. In other words, $\lambda_x = L/n_x$, where n_x is some integer. So the components of the wavevector $\vec{k}$ in Cartesian coordinates are

$$k_x = \frac{2\pi \, n_x}{L}, \quad k_y = \frac{2\pi \, n_y}{L}, \quad k_z = \frac{2\pi \, n_z}{L}. \tag{2.20}$$

The total wavenumber is then given by

$$k^2 = \left(\frac{2\pi}{L}\right)^2 (n_x^2 + n_y^2 + n_z^2), \tag{2.21}$$

where n_x, n_y, and n_z are all integers. These are the only allowed values of $\vec{k}$ which satisfy the boundary conditions. We refer to each such case as a mode.

Think of each mode as occupying a unit cell in a 3-dimensional space described by n_x, n_y, and n_z. If we now rescale along each axis by $2\pi/L$, in a space described by k_x, k_y, and k_z (k-space), each mode occupies a volume $(2\pi/L)^3$. In terms of frequency, and accounting for the existence of two independent polarizations, the number of allowed modes between frequencies ν and $\nu + d\nu$ is given by

$$dN_\nu = 2\left(\frac{L}{2\pi}\right)^3 d^3k. \tag{2.22}$$

The differential volume d^3k describes a thin spherical shell in k-space whose inner boundary corresponds to the frequency ν and whose outer boundary corresponds to $\nu + d\nu$. This can be rewritten as

$$dN_\nu = 2\left(\frac{L}{2\pi}\right)^3 k^2 \, dk \, d\Omega \tag{2.23}$$

$$= \frac{2L^3}{c^3} \nu^2 \, d\nu \, d\Omega, \tag{2.24}$$

since $k = 2\pi\nu/c$. Integrated over directions we get $\int d\Omega = 4\pi$ and letting the volume of the cube be $V = L^3$,

$$\frac{dN_\nu}{V} = \frac{8\pi}{c^3} \nu^2 \, d\nu. \tag{2.25}$$

Now we have to think quantum mechanically. Each of these modes represents a harmonic oscillator which can take on discrete amounts of energy which are multiples of $h\nu$. The number of these energy packets (photons) in a given mode, the occupation number of the mode, is $1/(e^{h\nu/kT} - 1)$. So the average energy per mode is $h\nu/(e^{h\nu/kT} - 1)$. Combining this with the density of modes gives an overall energy density of

$$u_\nu = \frac{dE_\nu}{V} = \frac{8\pi}{c^3} \frac{h\nu^3 \, d\nu}{e^{h\nu/kT} - 1}. \tag{2.26}$$

Since the radiation inside the cavity is isotropic, the specific intensity in frequency units is

$$I_\nu = \frac{c}{4\pi} u_\nu = \frac{2h\nu^3}{c^2} \frac{1}{e^{h\nu/kT} - 1}, \tag{2.27}$$

which is known as the Planck function and often written $B_\nu(T)$. In wavelength units this is

$$I_\lambda = \frac{2hc^2}{\lambda^5} \frac{1}{e^{hc/\lambda kT} - 1}.$$ (2.28)

At the *surface* of a blackbody (either an infinite surface or very close)

$$F_\nu = \frac{2\pi}{c^2} \frac{h\nu^3}{e^{h\nu/kT} - 1},$$ (2.29)

which is the radiated flux density. Integrating to get the flux

$$F = \int_0^\infty F_\nu \, d\nu = \frac{2\pi}{c^2} h \left(\frac{kT}{h}\right)^4 \int_0^\infty \frac{x^3 dx}{e^x - 1} = \frac{2\pi}{c^2} h \left(\frac{kT}{h}\right)^4 \frac{\pi^4}{15} = \sigma T^4,$$ (2.30)

a familiar result, the Stefan–Boltzmann law. The Stefan–Boltzmann constant (be careful of differing conventions on this definition) is

$$\sigma = \frac{2\pi^5 k^4}{15c^2 h^3}.$$ (2.31)

2.7 Atmospheric extinction (calibration)

Atmospheric extinction is present at all wavelengths. In the visible it is due primarily to Rayleigh scattering and aerosols. In the radio and infrared it is primarily due to water vapor. In each case the optical depth is proportional to the *air mass*, which for a plane-parallel atmosphere (small zenith angle) gives

$$\tau(\nu) \approx \tau_0(\nu) \sec z.$$ (2.32)

Essentially, at the zenith one is looking through a certain amount of atmosphere, which we call one air mass. And at non-zero zenith angle one is looking through a depth of atmosphere which is increased by sec z, as shown in Figure 2.7. The intensity reaching the ground is exponentially dependent on the optical depth,

$$I(z) = I_0 \, e^{-\tau} = I_0 \, e^{-\tau_0 \sec z}.$$ (2.33)

This provides a method for calibrating the atmospheric extinction. *If* one assumes that the extinction is stable, one can measure the received intensity at a variety of zenith angles. Then, by plotting log I versus sec z (air mass), the slope of the plot gives τ_0 and the intercept, which is essentially an extrapolation to zero air mass, gives the incident intensity I_0. This is known as Bouguer's method and is illustrated in Figure 2.7. In practice, atmospheric extinction is both time and site dependent, so one must measure frequently.

At microwave frequencies one does something similar using atmospheric *emission* (if the atmosphere absorbs, it also radiates). Assume the radiated emission to

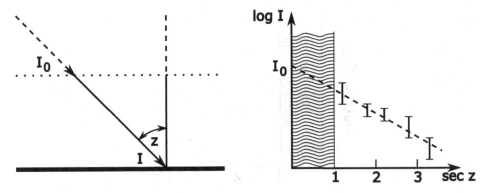

Figure 2.7 (Left) Path length through Earth's atmosphere, and hence opacity, increases as the secant of the zenith angle. (Right) Bouguer's method of deriving intensity corrected for atmospheric extinction by extrapolating from the physically accessible region (sec z > 1) to sec z = 0. The slope of the plot gives the zenith opacity.

be blackbody radiation. At these frequencies the Planck function can be simplified since $h\nu \ll kT_B$.

$$I_\nu = B_\nu(T_B) \approx 2\,\frac{\nu^2}{c^2}\,kT_B. \tag{2.34}$$

In this way (at low frequencies) one can measure I_ν in temperature units, and we refer to the intensity in terms of its *brightness temperature*. Assume the radiating air molecules are at some uniform temperature T_{atm}. By alternately observing the sky emission and an ambient temperature source at temperature T_{amb}, one finds that the measured continuum power

$$P_{sky}(z) \propto T_{atm}(1 - e^{-\tau}), \tag{2.35}$$

$$P_{amb} \propto T_{amb}. \tag{2.36}$$

Now if $T_{amb} \approx T_{atm}$, then

$$P_{amb} - P_{sky}(z) \propto T_{amb} - T_{atm}(1 - e^{-\tau}) \approx T_{atm}\,e^{-\tau}. \tag{2.37}$$

So by observing $P_{amb} - P_{sky}(z)$ at a variety of zenith angles and plotting $\log(P_{amb} - P_{sky}(z))$ versus sec z, one can generate a graph like that shown in Figure 2.8. This is sometimes called the *sky-dip* or *tipper* method. The slope of the plot again gives τ_0, in this case *without* observing any astronomical source. Water vapor, the dominant source of opacity, can vary rapidly. So again, this method must be repeated frequently.

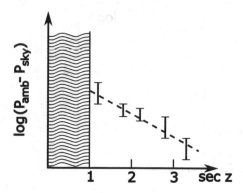

Figure 2.8 Sky-dip method for calculating zenith opacity from the slope of a semi-log plot.

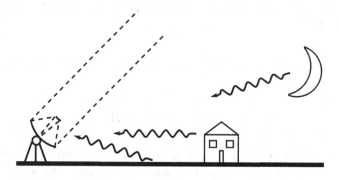

Figure 2.9 Radiation entering sidelobes of a radio telescope.

2.8 Absolute calibration

Uncalibrated data are worthless. Too many beginning astronomers, unfortunately, spend inadequate time and attention on calibration. The issue of calibration is very band and instrument specific. So there is little that we can say in general other than that absolute calibration is difficult in all wavelength ranges. A few examples will need to suffice for now.

Radio telescopes have strong sidelobes, due to diffraction. Some of these sidelobes pick up emission from nearby objects on the ground. Others receive power from the sky, but very far from the direction of interest. Some possible situations are shown in Figure 2.9. We want to know the sensitivity for on-axis sources (in the "main lobe"). But the sensitivity (antenna) pattern varies with distance from the telescope, until the *far field* is reached at a distance $R \gtrsim D^2/\lambda$, which can be many kilometers. Therefore proper radio calibration requires the use of astronomical sources! Planets can be useful calibration sources, but what intensities are expected? Planetary surfaces (e.g. Mars) can be quasi-blackbody emitters, but one needs to know the surface emissivity, which is frequency dependent. Surface temperature also varies (the subsolar point is hotter). For gaseous planets, the height of the absorbing (emitting) layer varies with frequency, therefore so does T_B.

Near the other end of the electromagnetic spectrum, consider calibration of an x-ray telescope. As we will see, Wolter telescopes will have some geometrical collecting area, but the *effective* collecting area will vary rapidly with energy due to K-shell absorption edges of the high-Z materials used as mirror coatings. Solid state detectors will have energy dependent efficiencies. There will be varying charged-particle backgrounds from cosmic rays and Earth's radiation belts to contend with. The detector may also be sensitive to ultraviolet radiation leakage. And detectors can deteriorate from any pre-flight calibration after prolonged exposure to radiation over multi-year mission lifetimes. Calibrations of x-ray systems to some degree may be checked by observations of white dwarfs with known characteristics.

2.9 Photometric magnitudes

The Johnson and Morgan (1953) system for visible photometry designated the spectral bands U(ltraviolet), B(lue), and V(isual). Later additions included R(ed), I(infrared), and longer wavelength infrared bands J, H, K, L, and M (Johnson, 1966; Bessell & Brett, 1988). Magnitudes are defined by

$$m_{U,B,V,...} = -2.5 \log\left(\frac{F_\lambda}{F_{\lambda_0}}\right) = -2.5 \log\left(\frac{F_\nu}{F_{\nu_0}}\right). \tag{2.38}$$

For each of the different wavelength bands a bandpass function is specified (often simplified to a central wavelength and a width) along with different zero points (specified by different values of F_{λ_0} and F_{ν_0}). We ignore here any flux variations across the bandpass. There have been various implementations of the "Johnson" system with different central wavelengths, bandwidths, and zero points. The true bandpasses are also affected by air-mass dependent atmospheric absorption, particularly the short wavelength end of the U band and most of the infrared bands. Some sample filter shapes are shown in Figure 2.10. Specifications for the Johnson–Cousins–Glass system are described in Table 2.1. The various zero points are chosen according to the Vega system, such that the colors U−B, B−V, . . . are zero for main sequence A0 stars (T = 10 800 K). That is, so that stars of this temperature will have the same magnitude in all spectral bands. Hotter stars will be brighter in the short wavelength bands and cooler stars will be fainter in the short wavelength bands. There is also, among many others, the narrower band Strömgren–Crawford system (Strömgren, 1966) with band designations u(ltraviolet), v(iolet), b(lue), y(ellow), and β. The β band covers the H_β line and has both a narrow and a wide

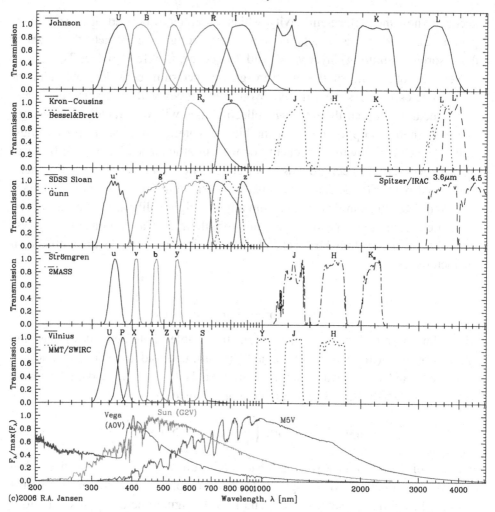

Figure 2.10 Filter response of some of the standard photometric systems (Jansen, 2006).

bandwidth version. The letter designations do *not* match those of the Johnson system; they even occur in different order! Other systems have been designed so that equal magnitudes correspond to equal flux densities F_ν (e.g. Oke, 1964).

The Sloan Digital Sky Survey (SDSS), a large compilation of stellar and galactic photometry, uses a different, narrower set of photometric bands, described in Table 2.2 and shown in Figure 2.10. The filters cover 300 nm to 1 µm with minimal overlap, enabling the determination of so-called photometric redshifts, based on the composite spectral energy distribution of a galaxy, rather than the redshift of a paticular spectral line from an individual star. The bands are named u(ltraviolet), g(reen), r(ed), i(nfrared), and z (for redshift).

Table 2.1. *Johnson–Cousins–Glass photometric system*

Band[a]	λ_0	F_{λ_0} (W m^{-2} μm^{-1})	F_{ν_0} (Jy)
U	366 nm	4.18×10^{-8}	1790
B	438	6.32×10^{-8}	4063
V	545	3.63×10^{-8}	3636
R	641	2.18×10^{-8}	3064
I	798	1.13×10^{-8}	2416
J	1.22 μm	3.15×10^{-9}	1589
H	1.63	1.14×10^{-9}	1021
K	2.19	3.96×10^{-10}	640
L	3.45	7.08×10^{-11}	285
M	4.75	2.04×10^{-11}	158

[a] Data from Bessell *et al.* (1998) except for band M, which is from Bessell & Brett (1988).

Table 2.2. *SDSS bands*

Band	$\lambda_0{}^a$ (nm)	$\Delta\lambda_0^a$ (nm)	$F_{\nu_0}{}^b$ (Jy)
u	352	48	3631
g	480	100	3631
r	625	96	3631
i	767	106	3631
z	911	123	3631

[a] Without telluric extinction.
[b] SDSS magnitudes are on a so-called asinh system, which does not correspond exactly with Equation 2.38, although the differences are small for all but the faintest objects.

Exercises

2.1 How many square degrees are there in a steradian? You may feel this is obviously, but please do an integral, for example to calculate the number of square degrees in 4π sr.

2.2 Calculate the étendue (m^2 degrees^2) of

a. the Palomar 200 inch telescope prime focus Large Format Camera
   (www.astro.caltech.edu/palomar/200inch/instruments.html);

b. the 48 inch Palomar Schmidt telescope (the Samuel Oschin telescope)
   (www.astro.caltech.edu/palomar/sot.html);

c. the 2.5 m Sloan Digital Sky Survey camera (www.sdss.org/);

d. the 10 meter Keck I telescope (assume an 8 arcmin × 8 arcmin field of
   view);

e. the 8.4 meter LSST, allowing for a rather large central blockage of 5.1
   meters, and assuming a 1.75 degree radius field of view.

2.3   A sphere of radius 0.1 meters radiates as a blackbody at a temperature of
      300 K. It is viewed by a detector system 1000 m away. The entrance aperture
      of the detector is 1 cm in radius.[1]

a. Calculate the specific intensity I_ν at wavelengths of 1 μm and 10 μm.

b. Calculate the flux density F_ν at the entrance aperture for each wavelength.

c. Assuming 50% detection efficiency and bandwidths of 1% ($\delta\lambda = 0.01\,\lambda$),
   calculate the power received by the detector at each wavelength.

d. Calculate the corresponding number of photons per second for each
   wavelength.

2.4   Consider an idealized bandpass filter centered at a wavelength λ_0 with a
      bandwidth $\Delta\lambda$. It has perfect transmission in the range

$$\lambda_0 - \frac{\Delta\lambda}{2} < \lambda < \lambda_0 + \frac{\Delta\lambda}{2} \tag{2.39}$$

and zero transmission outside this range. Through this filter you view a
blackbody source in the Rayleigh–Jeans limit. A naive estimate of the sig-
nal passed by the filter would be a power proportional to $I_{\lambda_0}\Delta\lambda$. This would
be wrong since the specific intensity varies across the passband. Show that
the fractional error introduced is of order

$$\frac{5}{6}\left(\frac{\Delta\lambda}{\lambda_0}\right)^2. \tag{2.40}$$

How big is this error if $\Delta\lambda/\lambda_0 = 0.2$?[2]

2.5   Calculate the flux density F_ν at a distance r from the center of a sphere of
      uniform brightness B_ν. The radius of the sphere is R ($r \geq R$). Comment on
      two aspects of your result: the functional dependence on r, and the value of
      F_ν at the surface of the sphere ($r = R$).

2.6   Consider an H II region which is optically thin from radio out to visible
      frequencies. Suppose further that its thermal bremsstrahlung spectrum has

[1] Adapted from Rieke (2002).
[2] Adapted from Rieke (2002).

a constant specific intensity, I_ν, over this frequency range. If it has a flux density of 10 mJy at radio frequencies, calculate its visible magnitude, m_V. Note that

$$m_V = -2.5 \log \left[\frac{F_\lambda}{F_{\lambda_0}} \right], \qquad (2.41)$$

where

$$F_{\lambda_0} = 3.63 \times 10^{-8} \text{ W m}^{-2} \text{ μm}^{-1} \qquad (2.42)$$

for the V photometric band ($\lambda_0 = 0.545$ μm).[3]

---

[3] Adapted from Lèna *et al.* (1998).

3

Positional astronomy

3.1 Fundamental reference system

The official system used for positional astronomy was introduced in 1976 by the International Astronomical Union (IAU). The changes made at that time included full consistency with the SI system of units (Le Système International d'Unités) and new experimental values for the fundamental constants (e.g. $GM_\odot$). It became a fully relativistic system, and a new standard (reference) epoch J2000.0 was introduced. This system was first implemented in *The Astronomical Almanac* for 1984 (and detailed in the "Supplement" in that volume, pp. S5–S38). In 1991 the treatment of space-time coordinates was further revised. An exhaustive description of the entire system is given in the *Explanatory Supplement to the Astronomical Almanac* (Seidelmann, 2006). Outdated and deprecated concepts include the epoch B1950.0, Besselian day numbers, E-terms of aberration, GMT, and ephemeris time (ET).

Further refinement was required after the astrometry mission Hipparcos provided significantly improved measurements of stellar positions. In this chapter we will focus first on those aspects of positional astronomy required for general uses such as "Where do I point my telescope?" Later we will introduce some aspects of precision astrometry. The most demanding applications require a relativistic treatment which goes well beyond what we are able to cover here.

3.2 Time systems

3.2.1 Atomic time

The fundamental system of time is international atomic time, TAI (Temps Atomique International). It is based on a worldwide weighted average of numerous atomic clocks, most of which are cesium clocks. Cesium clocks utilize a hyperfine transition ($F = 4 \rightarrow 3$) of the $^2S_{1/2}$ ground state of ^{133}Cs, as shown in Figure 3.1.

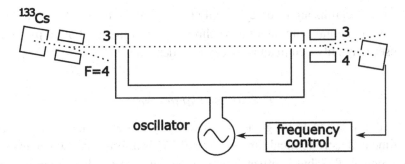

Figure 3.1 In a classical cesium clock a beam of ^{133}Cs atoms is emitted from an oven. The atoms pass through a magnet which selects those in the F = 3 hyperfine state. As the beam passes through a resonant cavity, some undergo a transition to F = 4. There are two microwave interaction regions, the separation between which (and the speed of the atoms) determines the resolution (Ramsey, 1949). A second magnet selects F = 4 atoms, which are then counted. This signal is used to control the frequency of the microwave oscillator.

The second is defined as 9 192 631 770 periods of this transition. The time system defined by this is considerably more stable than Earth's rotation. Individual cesium clocks have frequency stability of order 1.5–3 $\times 10^{-14}$ per day.[1] An ensemble of cesium clocks is even more stable. We will not discuss here the required synchronization of these clocks beyond noting that clocks at different altitudes (at different depths in Earth's gravitational potential well) will necessarily experience time passing at different rates. TAI is now referenced to the passage of time at Earth's geoid (mean sea level).

To be useful, atomic time must be made available (broadcast) worldwide. Secondary standards are then able to keep time locally, with occasional adjustment. Quartz oscillators are most commonly used to provide inexpensive, reasonably stable secondary standards. The time standard in most general use is Coordinated Universal Time, UTC (Temps Universel Coordonné). UTC transpires at the same rate as TAI, but is discontinuous due to the introduction of *leap seconds* (see below). UTC is broadcast in the USA in a variety of forms including both shortwave (WWV: 2.5, 5.0, 10.0, and 15.0 MHz) and longwave (WWVB: 60 kHz) radio. Other frequency bands are used elsewhere in the world. Varying propagation delays limit the accuracy of these methods to about 1 ms. The "atomic clocks" that consumers can buy are simply clocks equipped with radio receivers for synchronization. Accuracy down to the level of about 10 ns is available via the Global Positioning System (GPS). A convenient form of time transfer is the Internet, using NTP (Network Time Protocol), which can provide an accuracy of 1–50 ms. The site

---

[1] Recent improvements in laser-cooled atomic fountain clocks provide even better stability.

www.time.gov claims an accuracy of only 0.4 s. Finally, the CDMA cellular tele-
phone system maintains an internal synchronization at the level of 10 μs, although
an end-user may only obtain an accuracy of order 1 ms.

3.2.2 Astronomical time scales

Since Earth's rotation is irregular, it is necessary to define the linkage between
atomic time and astronomical time systems. UTC is a time scale which evolves at
the same rate as TAI, but is offset from it by an integral number of empirical leap
seconds,

$$\text{UTC} = \text{TAI} - \Delta\text{AT}. \tag{3.1}$$

Presently $\Delta\text{AT} = 34\,\text{s}$ (it was last updated on December 31, 2008). The
times at which such leap seconds have been introduced may be viewed at
ftp://maia.usno.navy.mil/ser7/tai-utc.dat. Although UTC includes the introduction
of leap seconds, GPS time remains locked to TAI, at a fixed offset of 14 s.

   The actual rotation of the Earth is described by Universal Time (UT1), which
obviously will not behave as simply as UTC.

$$\text{UT1} = \text{UTC} + \Delta\text{UT1}, \tag{3.2}$$

where ΔUT1 and therefore UT1 are determined, retroactively, from observations.
Information on ΔUT1 is available from ftp://maia.usno.navy.mil/ser7/ser7.dat.
Predictions of ΔUT1 accurate to ~ 0.1 s are available in some broadcasts.

   For some purposes it is necessary to consider general relativistic effects using
space-time coordinates for the solar system barycenter or for the geocenter.
We begin with terrestrial time (TT), which has a simple relationship to TAI,

$$\text{TT} = \text{TAI} + 32.184\,\text{s}. \tag{3.3}$$

Geocentric coordinate time (TCG), which is for a reference frame not in Earth's
gravitational potential, differs from TT by a constant rate

$$\text{TCG} \approx \text{TT} + 7 \times 10^{-10} \times \text{time}, \tag{3.4}$$

where "time" is the elapsed time in seconds since 1997 Jan 01 0^{h}. Barycentric coordinate time (TCB), similarly, is for a reference frame moving with the barycenter of
the solar system but not affected by the Sun's gravitational potential. Its calculation
requires a 4-vector transformation,

$$\text{TCB} = \text{TCG} + \text{relativistic corrections}. \tag{3.5}$$

Use of the dynamical times TDT and TDB, introduced in 1976, has been
deprecated since 1991.

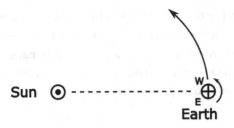

Figure 3.2 The concept of sidereal time as illustrated by a view from the north celestial pole. Directions east and west are shown for an observer on the daytime side of Earth. As Earth orbits the Sun, each day the Earth must rotate through more than 360° to bring the Sun back to the meridian.

3.2.3 Sidereal time

A time scale based on the apparent motions of the stars across the sky needs to differ in rate from a solar time scale, as illustrated in Figure 3.2. In one year the Earth rotates on its axis ~366.25 times. But since during that year Earth also goes through one revolution around the Sun, there are only ~365.25 times cycles of day and night. The Sun crosses the meridian only ~365.25 times in one year. Let the term "mean sidereal day" refer to one mean rotation of Earth. In that sense the sidereal day is both simpler and more fundamental than the solar day, which additionally depends on Earth's revolution around the Sun. A sidereal day is shorter, by about 4 minutes, than a solar day,

$$\frac{\text{sidereal day}}{\text{solar day}} \approx \frac{\text{year}/366.25}{\text{year}/365.25} \approx 0.99727. \tag{3.6}$$

A mean sidereal day is approximately $23^h56^m04^s$ of solar time. Or viewed another way, the Sun moves from west to east through the field of stars at an average rate of ~4 minutes per day. Formal definitions will follow, but for now think of Greenwich sidereal time, both mean (GMST) and apparent (GAST) as quantities which are tabulated in *The Astronomical Almanac* for various values of UT (e.g. GAST or GMST on 2009 Jan 1, 0^h UT). The difference, known as the "equation of the equinoxes" (a terrible name) is also tabulated,

$$\text{GAST} = \text{GMST} + \text{equation of equinoxes}. \tag{3.7}$$

The difference between these two definitions of sidereal time relates to the nutation of the Earth.

3.2.4 Solar time

Earth moves around the Sun at a non-uniform rate due to the ellipticity of Earth's orbit. So the Sun appears to move through the sky at different rates at different times during the year. For this reason, it has been convenient to develop a fictitious *mean solar position*, which differs from the position of the *true* Sun. This difference is described by the *equation of time* (another terrible name), which can also be obtained from *The Astronomical Almanac*. It is represented in graphical from by the *analema* (www.analema.com).

3.3 Spherical astronomy

3.3.1 Spherical coordinates (in general)

Astronomers use a number of different spherical coordinate systems. Define a spherical coordinate system as shown in Figure 3.3, where θ is a polar angle down from the z-axis and ϕ is an azimuthal angle measured in the xy-plane starting from the x-axis and increasing towards the y-axis. The relationship between the Cartesian and spherical systems is given by

$$z = r \cos \theta, \tag{3.8}$$
$$x = r \sin \theta \cos \phi, \tag{3.9}$$
$$y = r \sin \theta \sin \phi. \tag{3.10}$$

In general, a spherical coordinate system can be defined by the choice of the polar axis ($\theta = 0$) and the zero point of azimuth ($\phi = 0$). In some cases astronomers like to use the angle $90° - \theta$ instead of θ and sometimes $-\phi$ instead of ϕ.

It is often easier to manipulate (x, y, z) than (θ, ϕ). Yet we like to use spherical coordinates since for many purposes (e.g. pointing a telescope), we are interested only in the projection onto the celestial sphere, so we can ignore r. In that case, we can retain the advantages of the Cartesian system by using *directional cosines*. To do so we set $r = 1$ (use the unit sphere) and express directions in terms of the cosines of three angles, essentially the projections onto the Cartesian axes. These

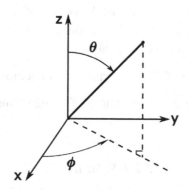

Figure 3.3 Conversion between Cartesian coordinates and spherical coordinates with the pole along the z-axis and azimuth measured from the x-axis.

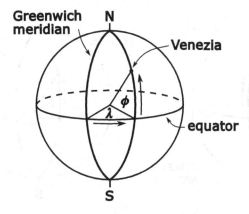

Figure 3.4 Latitude and longitude on Earth.

three directional cosines cannot be independent since the sum of their squares must equal unity. With r = 1, only two degrees of freedom remain.

3.3.2 *Latitude and longitude*

On Earth, by convention latitude (ϕ) is measured north or south (+ or −) from the equator, and longitude (λ) is measured east from the Greenwich meridian, as shown in Figure 3.4. Locations west of Greenwich have negative longitudes. Both latitude and longitude may be expressed in degrees, but astronomers often find it useful to express longitudes and related quantities in hours, where 1 hour is 15° (24 hours is 360°). Note that this is not an inertial coordinate system; it rotates with Earth.

3.3.3 *Equatorial coordinates*

Equatorial coordinates are similar to the latitude/longitude system, but on the celestial sphere. Declination (δ) is measured north or south (+ or −) from the celestial equator, as shown in Figure 3.5. Right ascension (α) is measured east from the vernal equinox (Υ, to be defined below). This roughly approximates an inertial system.

Since the Earth rotates, a rotating version of this coordinate system is also useful. The hour angle (h) is the angular distance measured (along the equator) west (!) from the local meridian to the hour angle circle of the source. Local apparent sidereal time is *defined* as the hour angle of the true vernal equinox ($\alpha = 0$).

$$h = \text{LAST} - \alpha_{\text{apparent}}, \tag{3.11}$$

$$\text{LAST} = \text{GAST} + \lambda, \tag{3.12}$$

$$\text{LMST} = \text{GMST} + \lambda. \tag{3.13}$$

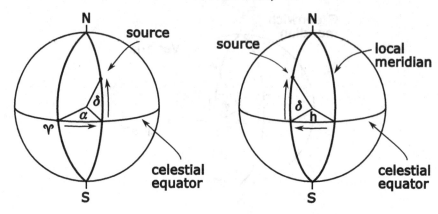

Figure 3.5 Fixed (left) and rotating (right) equatorial coordinate systems.

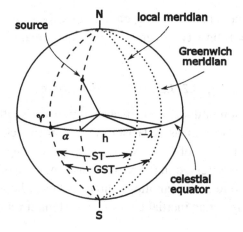

Figure 3.6 Equatorial coordinates showing meridians (dotted; red in electronic version) which rotate with the Earth and lines of constant right ascension (dashed; blue) which remain fixed.

Equivalently, sidereal time is the right ascension of stars which are transiting ($h = 0$). For a given location both sidereal time and hour angle are time-like quantities; they increase as time passes. These relationships are shown in Figure 3.6.

3.3.4 Horizon coordinate system (alt/az)

The horizon and the direction of the zenith depend on one's location on Earth's surface, as shown in Figure 3.7. Altitude (a) is an angle measured up from the horizon, towards the zenith (also called elevation). Zenith distance (z) is measured down from the zenith:

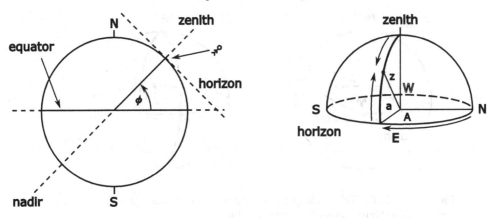

Figure 3.7 (Left) Location of an observer on Earth's surface at latitude ϕ. (Right) That observer's horizon coordinate system.

$$z = 90° - a. \tag{3.14}$$

The azimuth (A) is measured from north clockwise (towards east) in the plane of the horizon (the definition of A is not fully standardized).

3.3.5 Conversion formulae (alt/az ↔ ha/dec)

It is fairly easy to convert between the rotating equatorial (hour angle/declination) system and the horizon (altitude/azimuth) coordinate system. The only additional piece of information needed is the observer's latitude ϕ.

$$\cos a \sin A = - \cos \delta \sin h, \tag{3.15}$$
$$\cos a \cos A = \sin \delta \cos \phi - \cos \delta \cos h \sin \phi, \tag{3.16}$$
$$\sin a = \sin \delta \sin \phi + \cos \delta \cos h \cos \phi, \tag{3.17}$$
$$\cos \delta \cos h = \sin a \cos \phi - \cos a \cos A \sin \phi, \tag{3.18}$$
$$\sin \delta = \sin a \sin \phi + \cos a \cos A \cos \phi. \tag{3.19}$$

The right ascension α can be calculated from the hour angle h and the local sidereal time. A way to obtain these conversion formulae is discussed below. Similar formulae may be obtained for conversion between any two spherical coordinate systems.

3.3.6 Ecliptic coordinates

The ecliptic is the apparent path of the Sun (the plane of Earth's orbit). It is inclined 23.5° to the celestial equator. The vernal equinox (♈) is the intersection of the

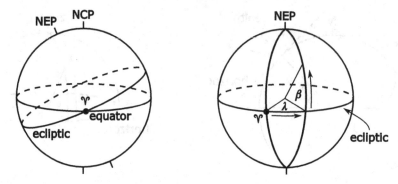

Figure 3.8 (Left) Relationship between equatorial and ecliptic coordinate systems. (Right) Definitions of ecliptic coordinates.

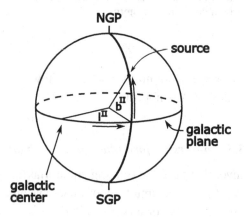

Figure 3.9 Definition of galactic coordinate system.

equator and the ecliptic (the point at which the Sun crosses the equator, going from south to north). Ecliptic latitude (β) is measured + and − from the ecliptic. Ecliptic longitude (λ) is measured east along the ecliptic from ♈, as shown in Figure 3.8. The effects of zodiacal dust will be most pronounced at small ecliptic latitudes.

3.3.7 Galactic coordinates

Galactic latitude (b^{II}) is measured + and − from the galactic plane and galactic longitude (l^{II}) is measured east along the plane from the nominal center of the Galaxy, as shown in Figure 3.9. Of course the Galaxy does not have a well-defined plane. The definitions of what constitutes the "plane" and the "center" are both adopted by convention. The convention has changed from an earlier system (b^{I}, l^{I}) and could conceivably change again. The galactic plane is inclined 62.6° to the celestial equator. The relationship between galactic and equatorial coordinates is shown in Figure 3.10. Interstellar extinction is high at small galactic latitudes. Observing near the galactic poles minimizes interstellar extinction, making such regions particularly attractive for extragalactic astronomy.

3.3.8 Spherical trigonometry

It is possible to conduct *spherical trigonometry* by carefully defining the concepts of angle and length for figures on the surface of a sphere. The "length" of a spherical arc (great circle) $\widehat{AB}$ is the angle $\angle AOB$, where O is the origin of the sphere. The "angle" between two spherical arcs (great circles) is the dihedral angle A between the planes AOC and AOB. These concepts are shown in Figure 3.11. A spherical triangle is formed by three great circle arcs $\widehat{AB}$, $\widehat{BC}$, and $\widehat{CA}$. Let "a" denote the length of the arc opposite to A (i.e. the arc length BC), etc.

Spherical trigonometry does *not* follow the laws of plane trigonometry (in the spherical triangle shown, $A + B + C > 180°$). The mathematics for spherical trigonometry is rather cumbersome, but may be derived from two equations:

$$\cos a = \cos b \cos c + \sin b \sin c \cos A, \tag{3.20}$$

$$\frac{\sin A}{\sin a} = \frac{\sin B}{\sin b} = \frac{\sin C}{\sin c}. \tag{3.21}$$

For this reason, and as discussed in the following section, there is little need now to use spherical trigonometry directly.

3.3.9 Rotation matrices

With computers it is often more convenient to express positions in rectangular coordinates and perform matrix multiplications. In the xy-plane, a rotation through an angle ϕ is given by

$$\begin{vmatrix} x' \\ y' \end{vmatrix} = \begin{vmatrix} \cos \phi & \sin \phi \\ -\sin \phi & \cos \phi \end{vmatrix} \begin{vmatrix} x \\ y \end{vmatrix}, \tag{3.22}$$

as shown in Figure 3.12. In three dimensions,

$$R_z(\phi) = \begin{vmatrix} \cos \phi & \sin \phi & 0 \\ -\sin \phi & \cos \phi & 0 \\ 0 & 0 & 1 \end{vmatrix} \tag{3.23}$$

represents a right-hand rotation by an angle ϕ around the z-axis. A general 3-dimensional rotation is constructed by a series of rotations about the cardinal axes. This generates a product of rotation matrices, which may be multiplied to produce a single 3×3 matrix. The conversion formulae between equatorial and horizon coordinate systems are generated by such a rotation.

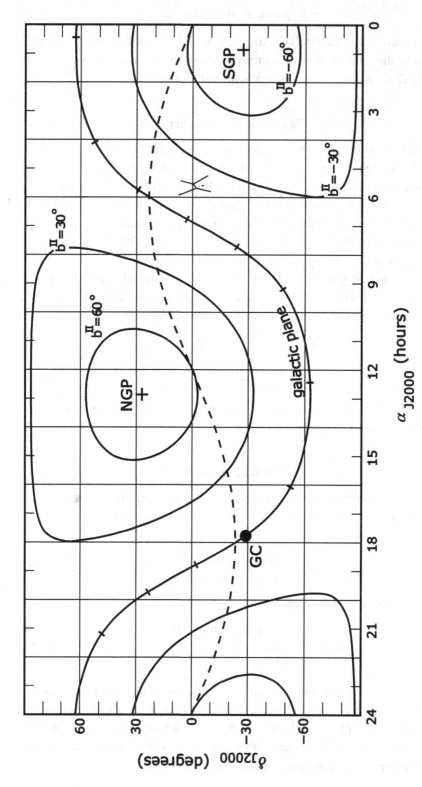

Figure 3.10 Relationship between galactic and equatorial coordinates. The galactic plane is shown (red in electronic version) with tic marks every 30 degrees of galactic longitude, and the north and south galactic poles are marked NGP and SGP. The ecliptic plane is shown dashed (blue).

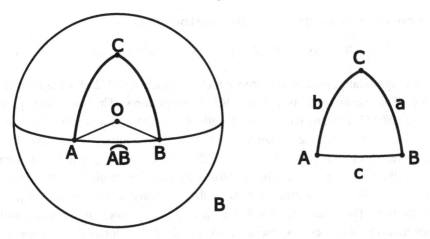

Figure 3.11 Spherical trigonometry illustrated by points A, B, and C on the surface of a sphere centered on the origin, O.

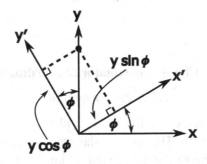

Figure 3.12 Rotation around the z-axis by an angle ϕ, illustrating the projected components in the primed system of a point at coordinates (0, y) in the unprimed system.

3.4 Epoch

Epoch refers to an instant in time, e.g. 2009 March 31 (0^h UT). Notice that by convention an epoch is given with the most significant part (years) first, followed in descending order by months, days, hours, minutes, and seconds. Another way to specify an epoch is by the Julian date, a system reckoned from the year 4713 BC (no connection with Julius Caesar or the "Julian" calendar). A Julian day begins at *noon* UT,

$$\text{JD} = 2454921.5 \text{ for 2009 March 31 } (0^h \text{ UT}). \tag{3.24}$$

The current "standard epoch" is J2000.0 (Julian),

$$\text{J2000.0} = \text{JD } 2451545.0 = \text{2000 Jan 1.5.} \tag{3.25}$$

The previous "standard epoch" was B1950.0 (Besselian),

$$B1950.0 = JD\ 2433282.423 = 1950\ Jan\ 0.923. \qquad (3.26)$$

Another significant epoch is the proper motion epoch J1991.25 for the Hipparcos catalog. The Julian year is exactly 365.25 days long. The Besselian year is 365.242 198 781 days, which was the length of a tropical year at B1900.0. A tropical year is the time for the "mean Sun" to move from equinox to equinox. This is inconvenient for two reasons: the length of the year is irrational, and it is no longer accurate (it does not correspond to the current length of the tropical year). An epoch is used to specify an astronomical event or the positions of astronomical objects at a specific time. For example, we will use the concept of epoch in discussing stellar proper motions. An epoch is also used to specify the orientation of a coordinate system. The Earth's axis precesses and nutates, affecting both the equator and the equinox. Positions for the epoch J2000.0 refer to the coordinate system defined by the *mean* equator and equinox for J2000.0. This will be our standard equator and equinox.

3.5 Changes in equatorial coordinates

3.5.1 Proper motion

The solar velocity with respect to the local standard of rest (LSR), an average of nearby stars, is ~15 km s^{-1}. The local stellar velocity dispersion is ~50 km s^{-1}. The differential galactic rotation is $\lesssim 25$ km s^{-1} kpc^{-1}. The combination of these effects produces "proper motion." At a distance of 100 pc

$$\mu(\text{typ}) \approx 0.1'' \text{ yr}^{-1}. \qquad (3.27)$$

Barnard's star, a nearby star (1.8 pc) famous for its high proper motion, has

$$\mu(\text{max}) = 10.31'' \text{ yr}^{-1}. \qquad (3.28)$$

Values for proper motion are often included in star catalogs, and may be used as follows:

$$\alpha = \alpha_0 + (t - t_0) \frac{\mu_\alpha}{100}, \qquad (3.29)$$

$$\delta = \delta_0 + (t - t_0) \frac{\mu_\delta}{100}, \qquad (3.30)$$

where t is in years and μ is in arcsec/century. Strictly speaking, proper motion is not a linear effect. Once a proper motion has become as large as a few degrees, higher order terms must be considered.

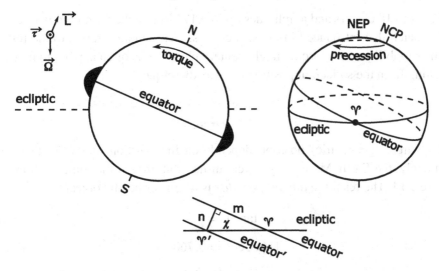

Figure 3.13 The Sun, Moon, and planets, lying on or near the ecliptic, apply a torque $\vec{\tau}$ on Earth due to Earth's equatorial bulge. Due to Earth's angular momentum $\vec{L}$, this results in a precession $\vec{\Omega}$ (left), which pulls the north celestial pole (NCP) out of the plane of the figure and causes it to circle the north ecliptic pole (NEP) (right). Near the vernal equinox the equator drops, and the vernal equinox shifts along the ecliptic by an annual amount χ, equivalent to a shift of m along the equator and n perpendicular to the equator (bottom).

3.5.2 Precession

The differential gravitational pull of the Sun, Moon, and planets on Earth's equatorial bulge produces a torque $\vec{\tau}$ on the Earth, as shown in Figure 3.13. This torque *attempts* to bring the equator in line with the ecliptic. The Earth has angular momentum $\vec{L}$ and therefore precesses at a rate $\vec{\Omega}$, where the quantities are related by

$$\vec{\tau} = \frac{d\vec{L}}{dt} = \vec{\Omega} \times \vec{L}. \tag{3.31}$$

General precession is a steady state motion with a 26 000 year period. It is not quite the same as the precession taught in physics classes, because the torque on the Earth is not steady. Precession produces a steady shift of the equator and the equinox. For 2009.5, the total rate of precession along the ecliptic is $\chi = 50.2902''$ yr^{-1}. At the position of the vernal equinox, this may be decomposed into a precession in declination of n $= 20.0412''$ yr^{-1} and one in right ascension of m $= 46.1243''$ yr^{-1}. Keep in mind that this is a shift in the *coordinate system*.

3.5.3 Nutation

The torques produced by the Moon, Sun, and planets are variable. The resulting quasi-periodic motions are referred to as *nutation*, with periods ranging from

4.7 days to 18.6 years and amplitudes up to $\pm 17''$. Again the terminology does not quite match that introduced in physics classes. Here nutation refers to the results of the variability of the torque. In elementary physics it is often applied to motions resulting from the sudden imposition of a steady torque.

3.5.4 Parallax

An apparent (geocentric) position depends on the position of the Earth (more accurately, the Earth–Moon barycenter) in its orbit around the Sun, as shown in Figure 3.14. The resulting *annual parallax* is larger for nearby objects:

$$\tan \Pi = \frac{a}{D},\tag{3.32}$$

$$\Pi(\alpha \text{Cen}) = 0''.760.\tag{3.33}$$

A parsec is the distance to an object which exhibits 1 arcsec of annual parallax, so the ratio of the length of the parsec to the AU is the same as the number of arcsec in a radian:

$$1 \text{ pc} = \frac{360 \times 3600}{2\pi} \approx 2.06 \times 10^5 \times 1 \text{ AU}.\tag{3.34}$$

For objects within the solar system, parallax can be quite large. In such cases one also needs to consider *diurnal parallax* by taking into account the observer's position on Earth's surface and Earth's 24 hour rotation.

3.5.5 Aberration of starlight

A moving observer sees a shift in apparent position due to special relativity. This effect is present for any type of observation (e.g. a radio interferometer), but may

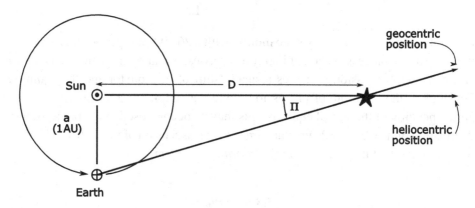

Figure 3.14 Parallax Π for a star at distance D.

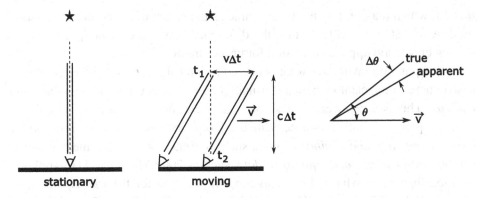

Figure 3.15 Aberration of starlight illustrated by the difference between a stationary and a moving observer with a zenith tube (left and center), and the general case for a line of sight not orthogonal to the direction of motion (right).

be most easily conceptualized using a zenith tube, as illustrated in Figure 3.15. A stationary observer points the tube straight up to see a star directly overhead. A moving observer must tip the tube forward, since the light takes a finite time to travel the length of the tube, during which time the bottom of the tube has moved. In the general case,

$$\tan \Delta\theta = \frac{v \sin\theta}{c + v \cos\theta}, \qquad (3.35)$$

$$\Delta\theta \approx \frac{v}{c} \sin\theta. \qquad (3.36)$$

Earth's mean orbital velocity of 29.8 km s^{-1} gives a maximum annual aberration of $\Delta\theta \approx \pm 20''$. On the equator, the velocity due to Earth's spin is 0.46 km s^{-1}, giving a maximum diurnal aberration of $\Delta\theta \approx \pm 0''.32$.

3.5.6 Reduction of celestial coordinates (overview)

In planning or analyzing the coordinates of an observation, one generally starts with UTC or with local civil time for a particular time zone, which one can readily correct to UTC. If one is in the USA, keep in mind that daylight savings time is kept in many locations from the second Sunday in March to the first Sunday in November. Knowing UTC, one can then look up either GMST or GAST. Mean sidereal time is the hour angle of the mean equinox. Apparent sidereal time is the hour angle of the true equinox. Local sidereal time, either mean or apparent (LMST or LAST), can then be calculated using the longitude of the observation point.

In planning an observation of a particular source, one might take its coordinates from some catalog. This catalog will have some epoch for the object's proper motion, which may or may not be the same as the epoch of the coordinate system used. A first step would be to take the difference between current epoch and the catalog epoch and apply a correction for proper motion.

At this point one will have what is known as a mean position for the coordinate system used in the catalog, which is usually referred to as the *catalog equator and equinox*. This is a heliocentric position. One then applies precession, in either a single step or two steps. If two steps are used, one first obtains a *mean position* for the *mean equator and equinox of year*, such as 2009.5, and then a mean position for the *mean equator and equinox of date*, such as 2009 Mar 31 0^h UT. Applying nutation then gives what is known as the *true position* for the *true equator and equinox of date*. These are still heliocentric positions. The effects of aberration and parallax account for Earth's position in the solar system. Applying these terms one obtains what is know as the *true position*, which is a geocentric position. If one makes further correction for the observatory's displacement from the geocenter, one obtains what are called *topocentric coordinates*.

3.5.7 Gravitational deflection of light

According to general relativity, light rays passing in the vicinity of a large mass such as the Sun will have their paths bent, due to the curvature of space-time. At the limb of the Sun, for Einstein's version of general relativity,

$$\Delta\theta\,(\text{limb}) = 1\rlap{.}''749 \tag{3.37}$$

as shown in Figure 3.16. At an angular distance of 10° from the Sun

$$\Delta\theta\,(10°) = 0\rlap{.}''047, \tag{3.38}$$

and at an angular distance of 90° from the Sun

$$\Delta\theta\,(90°) = 0\rlap{.}''004. \tag{3.39}$$

This is rather important (a big effect) in radio astronomy, since it is relatively easy to obtain high positional precision in interferometric measurements and it is practical to observe objects in directions rather close to the Sun.

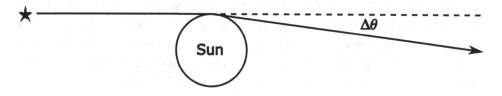

Figure 3.16 Gravitational deflection of light at the limb of the Sun.

3.5.8 Refraction

Refraction refers to the change in direction of propagation due to Earth's atmosphere. For a plane-parallel atmosphere with many thin layers with indices of refraction n_i, as shown in Figure 3.17, Snell's law gives

$$n_0 \sin \theta_0 = n_1 \sin \theta_1 = n_2 \sin \theta_2 = \ldots = \sin \theta, \qquad (3.40)$$

where θ is the angle of incidence (zenith angle) of the ray above the atmosphere and θ_i are the angles of propagation in the various layers. If θ_0 is the observed angle, at the ground, and we express θ as θ_0 plus a small correction $\Delta\theta$,

$$n_0 \sin \theta_0 = \sin \theta \qquad (3.41)$$
$$\approx \sin(\theta_0 + \Delta\theta) \qquad (3.42)$$
$$\approx \sin \theta_0 + \Delta\theta \cos \theta_0, \qquad (3.43)$$
$$\Delta\theta = (n_0 - 1) \tan \theta_0. \qquad (3.44)$$

Taking into account the curvature of the atmosphere adds a third order term,

$$\Delta\theta = A \tan \theta_0 + B \tan^3 \theta_0. \qquad (3.45)$$

The magnitude of refraction $(n_0 - 1)$ depends on frequency, atmospheric pressure, temperature, and water vapor content. Since the index of refraction of air is higher at the blue end of the visible spectrum, blue light is refracted more than red light:

$$n_0 - 1(\text{optical}) \approx 0.000\,293 \approx 60''. \qquad (3.46)$$

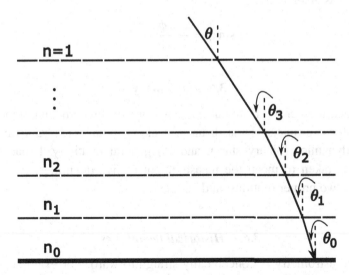

Figure 3.17 Refraction in a plane-parallel atmosphere with layers of indices $\{n_i\}$.

In the radio portion of the spectrum, the index is very dependent on water vapor content:

$$n_0 - 1 (<100 \text{ GHz}) \approx 65\text{--}75''. \tag{3.47}$$

Refraction affects altitude only (not azimuth). Decomposing its effect into celestial coordinates,

$$\Delta\alpha = \pm R \sec \delta \sin q, \tag{3.48}$$

where the + sign refers to $12^h < h < 24^h$ and the − sign refers to $0^h < h < 12^h$.

$$\Delta\delta = \pm R \cos q \tag{3.49}$$

with the + sign for $\phi < 0$ and the − sign for $\phi > 0$, and where $\sin q = \cos \phi \sin h / \sin z$. More details on appropriate values of the refraction constant are available in Allen (2001) for optical wavelengths and Thompson *et al.* (2001) at radio wavelengths.

3.5.9 Parallactic angle

Under some circumstances, the orientation of an instrument will rotate with respect to the equatorial coordinate system as a telescope tracks an object across the sky. An example is a radio telescope with an altitude/azimuth mount. Since most radio receivers are polarization sensitive, this is particularly important when observing polarized radiation. The amount of rotation is referred to as the parallactic angle, q. Its value was given above,

$$\sin q = \frac{\cos \phi \sin h}{\sin z}. \tag{3.50}$$

3.6 Astrometry

Accurate measurements of stellar positions are of fundamental importance to studies of, for example, (1) the distance scale (parallax), (2) galactic structure, (3) stellar dynamics (binary stars), and (4) general relativity. Broadly defined, *astrometry* is taken to mean the measurement of five quantities: two positional coordinates, two proper motions, and parallax.

3.6.1 Historical techniques

Narrow field astrometry is conceptually straightforward. For example, one takes a photographic plate (nowadays a CCD image) and measures relative positions. There are plenty of complications (telescope aberrations, spatial stability of

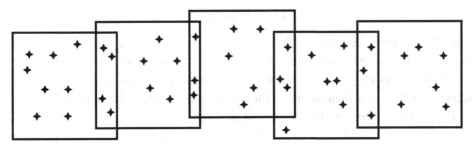

Figure 3.18 Wide field astrometric errors build up as the square root of the number of narrow field images.

photographic emulsion, etc.), but it is possible to obtain positional accuracies of order 0.01″ over small fields (~10′). This is true even in the presence of image blurring due to seeing, since one can measure the centroid of a stellar image to a small fraction of the seeing disk.

A more substantial problem is to obtain accurate relative positions of stars in very different parts of the sky (wide field astrometry). You cannot do this accurately using a series of overlapping narrow fields due to the accumulation of errors. Random errors increase as $\sqrt{N}$ and systematic errors as N, where N is the number of fields observed, as shown in Figure 3.18.

A meridian circle (transit telescope) uses the timing of a stellar transit to determine right ascension. A mechanical dial readout is used to determine declination. The accuracy is approximately 0.1″, but it is susceptible to systematic errors such as axis misalignment, refraction, etc.

A photographic zenith tube (PZT) uses reflection of light off of the surface of a bath of mercury to determine the local vertical. As with the meridian circle, the time of transit determines the right ascension. However, this technique is restricted to stars with declination approximately equal to the latitude. A full astrometric system requires PZTs at a variety of latitudes.

The astrolabe is similar to the PZT in its use of a mercury bath to determine the local vertical. However, optics are arranged so that the observer sees stars which are at exactly 60° elevation. With an astrolabe a large fraction of the sky is observable from one site in one night.

Radio interferometry (VLBI) has revolutionized astrometry by bringing precisions of order 0.001 arcsec to wide field astrometry. There is the additional advantage that radio interferometers can see distant quasars, which form a good inertial reference frame (a frame tied to nearby stars is not necessarily inertial). The problem is that most normal stars are poor emitters of radio waves and so cannot be observed directly.

3.6.2 Hipparcos

The ESA satellite Hipparcos (Perryman *et al.*, 1997) was designed to measure positions, proper motions, and parallaxes of 118 000 stars to an accuracy of 0.002 arcsec. As a byproduct it produced accurate space-based photometry of these stars and astrometric data on an additional 2.5×10^6 stars at the level of 0.06″.

The optics of the satellite combined the light from two fields of view separated on the sky by 58°. The focal plane of the telescope contained an occulting (modulating) grid followed by an image-dissector tube detector. Positional information was therefore encoded in the phase of the modulation of the stellar flux. As the satellite spun and the spin axis slowly changed, eventually all pairs of fields separated by 58° would be compared. Extensive computer analysis was used to recover the relative stellar positions. The system sensitivity was sufficient to make the astrometric survey complete for stars down to $m_V \approx 9$. Some stars as faint as 12.4^m were measured. Characteristics of the data are enumerated in Table 3.1. This precision was sufficient to demonstrate that previous reference systems such as the FK5 were not inertial and had significant zonal errors (inconsistency of different declination zones). The precision of Hipparcos required the definition of a new reference system, the International Celestial Reference System (ICRS). Through linked observations of extragalactic sources, the ICRS is nearly an inertial system.

Positional uncertainties increase with the time elapsed since the mean epoch of the observations. At the present epoch such uncertainties are dominated by the uncertainty in the proper motions. A new satellite mission is needed to reduce the error in proper motion. The GAIA (Global Astrometric Interferometer for Astrophysics) mission of the ESA is scheduled to be launched in 2011 into an orbit around the L2 Lagrangian point of the Earth–Sun system. Its focal plane will be covered by CCDs with of order 10^9 pixels, significantly more than the 120 megapixels of the SDSS. GAIA is expected to produce astrometric data and multicolor photometry of roughly 10^9 stars with an accuracy of order 2–20 microarcsec (μas) for stars brighter than 15^m. GAIA will also contain a spectrometer to measure radial velocities.

Table 3.1. *Hipparcos precision*

mean epoch	J1991.25
median error in position	0.0007″
median error in parallax	0.00097″
median proper motion error	0.0008″ per year
deviation from inertial	±0.00025″ per year

Exercises

3.1 Derive the formulae for transformation between alt/az and declination/hour angle coordinate systems using the method of rotation matrices. First, set up a right-hand coordinate system $\hat{i}, \hat{j}, \hat{k}$ with $\hat{k}$ towards the zenith and $\hat{i}$ and $\hat{j}$ in the horizontal plane, as shown in Figure 3.19.

$$\begin{vmatrix} x \\ y \\ z \end{vmatrix} = \begin{vmatrix} -\cos a \sin A \\ -\cos a \cos A \\ \sin a \end{vmatrix}. \tag{3.51}$$

a. Set up a similar coordinate system $\hat{i}', \hat{j}', \hat{k}'$ such that $\hat{i} = \hat{i}'$ and $\hat{k}'$ is towards the north celestial pole. Express a 3-dimensional vector in the primed coordinate system in terms of δ and h.

$$\begin{vmatrix} x' \\ y' \\ z' \end{vmatrix} = ? \tag{3.52}$$

b. Describe the 3×3 rotation matrix which represents a rotation about the $\hat{i} = \hat{i}'$ axis, transforming (x, y, z) into (x', y', z').
c. Carry out the matrix multiplication to derive three trigonometric relationships between a, A, h, δ, and ϕ.
d. Compare with the set of five relationships given in Section 3.3.5. Where might the other two formulae come from?

3.2 The *Positions and Proper Motions (PPM) Star Catalogue* for the Epoch J2000.0 (Röser & Bastian, 1991) lists a star (designated #153068) as having a position

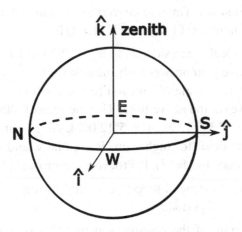

Figure 3.19 Altitude/azimuth coordinate system.

$$\alpha_0(J2000.0) = 7^h39^m18^s.113, \tag{3.53}$$
$$\delta_0(J2000.0) = +5°13'30''.06 \tag{3.54}$$

and a proper motion of

$$\mu_\alpha = -0.0475 \text{ s yr}^{-1}, \tag{3.55}$$
$$\mu_\delta = -1.023'' \text{ yr}^{-1}. \tag{3.56}$$

a. Calculate the correction for proper motion between J2000.0 and J2009.5.
b. Using the formulae at the top of page B54 of *The Astronomical Almanac* (2009), precess the J2000.0 coordinates to mean positions for the epoch J2009.5. Begin by calculating the time difference between the two epochs in Julian centuries,

$$T = (J2009.5 - J2000.0)/100 \tag{3.57}$$
$$= (JD2455014.875 - JD2451545.0)/36525. \tag{3.58}$$

Then calculate the precession constants M and N as directed in the middle of the page (including terms up to T^3 is sufficient). Then go to the formulae near the top of the page. Do not be confused by the references to α_m and δ_m (the coordinates for the mean epoch). "Mean" in this case refers to the mean of J2000.0 and J2009.5 (J2004.75). Since J2000.0 and J2009.5 are close together you can omit the calculation of α_m and δ_m and use α_0 and δ_0 instead. Finally, include your proper motion correction from part a.

3.3 The city of Urbana, Illinois, is located at a latitude of 40°06'20''.2 north and a longitude of $5^h52^m53^s.9$ (88°13'28''.5) west. Urbana is in the Central time zone, six time zones west of Greenwich. So it is necessary to add 6 hours to CST (Central Standard Time) to convert to coordinated universal time (UTC = CST + 6^h). Ignore ΔUT1 (assume UT1 = UTC).

a. What is the local mean sidereal time (LMST) in Urbana on 2009 Mar 7 21^h CST? Give your answer to the nearest 0.1 s.
b. What is the hour angle of Procyon (the star whose coordinates you precessed, above) at the above time? Use the mean J2009.5 coordinates you derived: $\alpha = 7^h39^m47^s.93$, $\delta = 5°12'00''.4$. Give your answer to the nearest minute of time (for higher precision, one would need to precess the coordinates back to Mar 7). Is Procyon east or west of the meridian?

3.4 Derive formulae for general precession, valid for time scales much smaller than the precessional period.

a. Make a diagram of the celestial sphere and a right-handed rectangular coordinate system, with the north celestial pole along $\hat{k}$ and the vernal equinox along $\hat{\imath}$. Show the equator and the ecliptic. Sketch the motion of the north celestial pole through its entire 25 800 year precessional period.

b. Calculate the angle through which the pole moves in one year. In what direction does it move? The rate of motion of the pole is generally referred to as the annual precession in declination.

c. What happens to the vernal equinox? Calculate the rate of motion of the vernal equinox. This is generally referred to as the annual precession in right ascension.

d. Using successive multiplications by rotation matrices to describe these two shifts in the coordinate systems, derive the relationship between precessed and unprecessed coordinates. Compare with the formulae at the top of page B54 of the *Almanac*. (It may seem more natural to you to use three rotations: transformation to ecliptic coordinates, rotation around the ecliptic pole, and then transformation back to equatorial coordinates. Feel free to do so if you like. The rotation matrix you derive would then be exact, although it should reduce to the same result for small precession angles.)

4

Fourier transforms

Fourier transforms constitute an important class of data analysis tools, which also underlie much of what we will be doing in optics and statistics. So we will take some time in introducing them and the various theorems associated with them. We will look carefully at Rayleigh's theorem, the properties of convolutions, the Wiener–Khinchin (autocorrelation) theorem, and Shannon's sampling theorem. We will also discuss Fourier transforms in more than one dimension and related integral transforms such as the Hankel transform. The classic reference work in this field is Bracewell (2000). We will adhere to Bracewell's conventions regarding the placement of factors of 2π (with the factor of 2π in the Fourier kernel), which we consider to give the most straightforward versions of the theorems and the Fourier transform pairs.

4.1 Fourier series

Consider an arbitrary function $f(x)$ defined on the interval $(-0.5, 0.5)$. It can be represented by the series expansion

$$f(x) = \frac{a_0}{2} + \sum_{n=1}^{\infty} (a_n \cos 2\pi nx + b_n \sin 2\pi nx),\tag{4.1}$$

where the coefficients a_n and b_n may be found by multiplying the function by the appropriate sine or cosine and integrating:

$$a_n = 2 \int_{-1/2}^{1/2} f(x) \cos 2\pi nx \, dx,\tag{4.2}$$

$$b_n = 2 \int_{-1/2}^{1/2} f(x) \sin 2\pi nx \, dx.\tag{4.3}$$

The set of basis functions $\{\cos 2\pi nx, \sin 2\pi nx\}$ are orthogonal:

$$\int_{-1/2}^{1/2} \cos 2\pi nx \, \sin 2\pi mx \, dx = 0, \tag{4.4}$$

$$\int_{-1/2}^{1/2} \cos 2\pi nx \, \cos 2\pi mx \, dx = \frac{1}{2} \delta_{nm}, \tag{4.5}$$

$$\int_{-1/2}^{1/2} \sin 2\pi nx \, \sin 2\pi mx \, dx = \frac{1}{2} \delta_{nm} \quad (n \neq 0), \tag{4.6}$$

where the Kronecker delta function is defined as

$$\delta_{nm} = \begin{cases} 1 & n = m \\ 0 & n \neq m. \end{cases} \tag{4.7}$$

This may be generalized to other intervals such as $(-L/2, L/2)$:

$$f(x) = \frac{a_0}{2} + \sum_{n=1}^{\infty} \left(a_n \cos\frac{2\pi nx}{L} + b_n \sin\frac{2\pi nx}{L} \right), \tag{4.8}$$

$$a_n = \frac{2}{L} \int_{-L/2}^{L/2} f(x) \cos\frac{2\pi nx}{L} \, dx, \tag{4.9}$$

$$b_n = \frac{2}{L} \int_{-L/2}^{L/2} f(x) \sin\frac{2\pi nx}{L} \, dx. \tag{4.10}$$

And this may also be expressed in complex form,

$$f(x) = \sum_{n=-\infty}^{\infty} \tilde{a}_n \, e^{i2\pi nx/L}, \tag{4.11}$$

$$\tilde{a}_n = \frac{1}{L} \int_{-L/2}^{L/2} f(x) \, e^{-i2\pi nx/L} \, dx, \tag{4.12}$$

which introduces the concept of negative frequencies. A cosine function will be viewed as composed of both positive and negative frequency components. We use the tilde to indicate a complex quantity, which would be the case for real functions $f(x)$. Later the distinction becomes irrelevant. Many functions we deal with are potentially complex, so we will omit the tilde.

Fourier series possess many simple properties. For example, if $f(x)$ is an even function such that $f(x) = f(-x)$, then only cosine terms are present ($b_n = 0$). If $f(x)$ is an odd function such that $f(-x) = -f(x)$, then only sine terms are present ($a_n = 0$). Consider a square wave of period $L = 2\pi$, as in Figure 4.1, defined by

$$f(x) = \begin{cases} 1 & 0 < x < \pi \\ -1 & -\pi < x < 0. \end{cases} \tag{4.13}$$

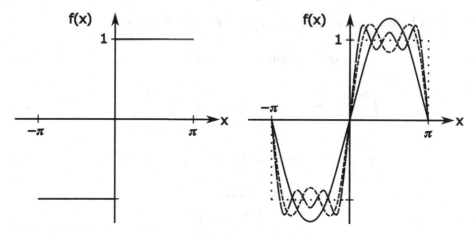

Figure 4.1 Approximating a square wave as a finite Fourier series with $n = 1$ (solid black), $n = 3$ (short dashes; red in electronic version), and $n = 5$ (long dashes; blue).

The Fourier coefficients of $f(x)$ are

$$a_n = 0, \tag{4.14}$$

$$b_n = \frac{1}{\pi} \int_{-\pi}^{\pi} f(x) \sin nx \, dx \tag{4.15}$$

$$= \frac{2}{\pi} \int_{0}^{\pi} \sin nx \, dx \tag{4.16}$$

$$= \begin{cases} 0 & \text{even } n \\ \frac{4}{\pi n} & \text{odd } n. \end{cases} \tag{4.17}$$

So the original function is a sum of odd sine terms,

$$f(x) = \frac{4}{\pi} \left(\sin x + \frac{\sin 3x}{3} + \frac{\sin 5x}{5} + \cdots \right). \tag{4.18}$$

4.2 Fourier integrals

By increasing the length of the interval L to infinity we can write

$$f(x) = \int_{-\infty}^{\infty} F(s) \, e^{i2\pi sx} \, ds, \tag{4.19}$$

$$F(s) = \int_{-\infty}^{\infty} f(x) \, e^{-i2\pi sx} \, dx. \tag{4.20}$$

This form with 2π in the kernel is the notation we will use, and we will consider the $-i$ transform to be the forward transform (sometimes also $\mathcal{F}$ or $\rightharpoonup$ or $\tilde{f}$) and

Table 4.1. *Symmetry properties of Fourier transform pairs*

f(x)	F(s)
Real and even	Real and even
Real and odd	Imaginary and odd
Imaginary and even	Imaginary and even
Complex and even	Complex and even
Complex and odd	Complex and odd
Real and asymmetrical	Complex and Hermitian
Imaginary and asymmetrical	Complex and anti-Hermitian
Real even plus imaginary odd	Real
Real odd plus imaginary even	Imaginary
Even	Even
Odd	Odd

the $+i$ transform to be the inverse transform (sometimes $\mathcal{F}^{-1}$ or $\leftarrow$). We will also often use lower case letters to represent functions and upper case to indicate their forward Fourier transforms (as above). Fourier transform pairs possess symmetry properties, which are presented in Table 4.1, from Bracewell (2000). Fourier transforms also obey simple rules for scalar multiplication, $\mathcal{F} af(x) = a\,\mathcal{F}f(x)$, and additivity, $\mathcal{F}(f(x) + g(x)) = \mathcal{F}f(x) + \mathcal{F}g(x)$. More complicated rules involving Fourier transforms are discussed in the following sections.

4.2.1 Relationship to the Dirac delta (impulse) function

Consider a Fourier transform followed by an inverse transform,

$$f(x) = \int_{-\infty}^{\infty} ds\, e^{i2\pi sx} \int_{-\infty}^{\infty} f(x')\, e^{-i2\pi sx'}\, dx' \tag{4.21}$$

$$= \int_{-\infty}^{\infty} dx'\, f(x') \left[\int_{-\infty}^{\infty} e^{i2\pi s(x-x')}\, ds \right]. \tag{4.22}$$

The quantity in brackets must vanish except at $x = x'$, in order for this to be true for *any* function f. A conventional definition of the Dirac delta function is $\delta(x) = 0$ for $x \neq 0$ and yet the delta function integrates to unity,

$$\int_{-\infty}^{\infty} \delta(x)\, dx = 1. \tag{4.23}$$

The integral representation of the Dirac delta function is

$$\delta(x) = \int_{-\infty}^{\infty} e^{i2\pi sx}\, ds. \tag{4.24}$$

4.2.2 Parseval's theorem (Rayleigh's theorem)

This theorem defines the concept of a power spectrum. In the time domain,

$$\int_{-\infty}^{\infty} |f(t)|^2 \, dt = \int_{-\infty}^{\infty} dt \, f(t) \, f^*(t) \tag{4.25}$$

$$= \int_{-\infty}^{\infty} dt \int_{-\infty}^{\infty} F(\nu) \, e^{i2\pi\nu t} \, d\nu \int_{-\infty}^{\infty} F^*(\nu') \, e^{-i2\pi\nu' t} \, d\nu' \tag{4.26}$$

$$= \int_{-\infty}^{\infty} d\nu \, F(\nu) \int_{-\infty}^{\infty} d\nu' \, F^*(\nu') \int_{-\infty}^{\infty} e^{i2\pi(\nu-\nu')t} \, dt \tag{4.27}$$

$$= \int_{-\infty}^{\infty} d\nu \, F(\nu) \int_{-\infty}^{\infty} d\nu' \, F^*(\nu') \, \delta(\nu - \nu') \tag{4.28}$$

$$= \int_{-\infty}^{\infty} d\nu \, F(\nu) \, F^*(\nu). \tag{4.29}$$

So ultimately,

$$\int_{-\infty}^{\infty} |f(t)|^2 \, dt = \int_{-\infty}^{\infty} |F(\nu)|^2 \, d\nu. \tag{4.30}$$

Let's revisit the problem of an exponentially decaying oscillator. Consider a function $f(t)$ which is zero for $t < 0$ and equals $e^{-\Gamma t/2} \sin 2\pi \nu_0 t$ for $t \geq 0$, as shown in Figure 4.2. Its Fourier transform is

$$F(\nu) = \int_{-\infty}^{\infty} f(t) \, e^{-i2\pi\nu t} \, dt \tag{4.31}$$

$$= \int_{0}^{\infty} e^{-\Gamma t/2} \, e^{-i2\pi\nu t} \sin 2\pi \nu_0 t \, dt \tag{4.32}$$

$$= \int_{0}^{\infty} e^{-\Gamma t/2} \, e^{-i2\pi\nu t} \frac{1}{2i} \left(e^{i2\pi\nu_0 t} - e^{-i2\pi\nu_0 t} \right) dt \tag{4.33}$$

$$= \frac{1}{2} \left(\frac{1}{2\pi\nu + 2\pi\nu_0 - i\Gamma/2} - \frac{1}{2\pi\nu - 2\pi\nu_0 - i\Gamma/2} \right). \tag{4.34}$$

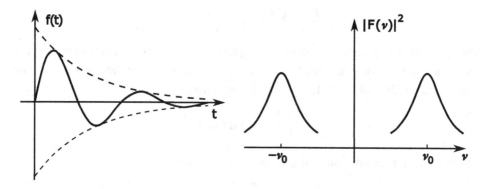

Figure 4.2 An exponentially decaying wave and its power spectrum.

Considering for now only positive frequencies, near ν_0, and with small amounts of damping ($\Gamma \ll 2\pi\nu_0$),

$$F(\nu) \approx -\frac{1}{2}\frac{1}{2\pi\nu - 2\pi\nu_0 - i\Gamma/2}, \tag{4.35}$$

$$|F(\nu)|^2 \approx \frac{1}{4}\frac{1}{(2\pi\nu - 2\pi\nu_0)^2 + (\Gamma/2)^2}, \tag{4.36}$$

which is a Lorentzian profile. The negative frequencies provide the same relative spectral content. Normalizing so that $\int_{-\infty}^{\infty}\phi(\nu)d\nu = 1$,

$$\phi(\nu) = \frac{1}{\pi}\frac{\Gamma/4\pi}{(\nu - \nu_0)^2 + (\Gamma/4\pi)^2}. \tag{4.37}$$

For an electric field f(t), the power is proportional to $|f(t)|^2$, and through Parseval's theorem, to $|F(\omega)|^2$. This leads to the concept of frequency content or spectral power density.

4.2.3 Properties of Fourier transforms

Some crucial properties of Fourier transforms are presented in Table 4.2 (Bracewell, 2000). The addition property was mentioned earlier. Similarity says that a narrow function in the time domain corresponds to a broad function in the spectral domain, and vice versa. Notice that an additional scaling in amplitude is required with the similarity property. Translation in either the time or frequency domain corresponds to a phase winding (a modulation) in the opposite domain. And differentiation contains a boost of any high frequency components and a reduction of low frequency components.

Table 4.2. *Properties of Fourier transforms*

Addition	$f(x) + g(x)$	$\rightleftharpoons$	$F(s) + G(s)$
Similarity	$f(ax)$	$\rightleftharpoons$	$\frac{1}{|a|}F\left(\frac{s}{a}\right)$
	$\frac{1}{|b|}f\left(\frac{x}{b}\right)$	$\rightleftharpoons$	$F(bs)$
Translation	$f(x-a)$	$\rightleftharpoons$	$e^{-i2\pi as}F(s)$
	$f(x)\,e^{i2\pi ax}$	$\rightleftharpoons$	$F(s-a)$
Derivative	$\frac{df(x)}{dx}$	$\rightleftharpoons$	$i2\pi\,s\,F(s)$

4.2.4 Convolution

The *convolution* of two functions $f_1(x)$ and $f_2(x)$ is defined as

$$g(x) = \int_{-\infty}^{\infty} f_1(u)\, f_2(x - u)\, du \tag{4.38}$$

and written as

$$g(x) = f_1 * f_2. \tag{4.39}$$

Convolutions have the properties of commutativity, associativity, and distributivity:

$$f * g = g * f, \tag{4.40}$$

$$f * (g * h) = (f * g) * h, \tag{4.41}$$

$$f * (g + h) = f * g + f * h. \tag{4.42}$$

One of the most important properties of convolutions is that the Fourier transform of a convolution of two functions equals the product of their individual Fourier transforms. And the Fourier transform of a product of functions is the convolution of their Fourier transforms.

$$f_1(x) * f_2(x) \rightleftharpoons F_1(s)\, F_2(s), \tag{4.43}$$

$$f_1(x)\, f_2(x) \rightleftharpoons F_1(s) * F_2(s). \tag{4.44}$$

4.2.5 Autocorrelation (Wiener–Khinchin theorem)

The *cross correlation* of two functions $f_1(x)$ and $f_2(x)$ is defined as

$$g(x) = \int_{-\infty}^{\infty} f_1^*(u)\, f_2(x + u)\, du = \int_{-\infty}^{\infty} f_1^*(u - x)\, f_2(u)\, du. \tag{4.45}$$

Although equivalent, we prefer the first of these definitions, as it emphasizes that the difference between convolution and cross correlation is essentially a sign change (and a complex conjugation, for complex functions). We will indicate cross correlations symbolically as

$$g(x) = f_1 \star f_2. \tag{4.46}$$

Cross correlations are not commutative. The behavior of cross correlations under Fourier transforms is shown below, as well as the special case of autocorrelations, which will be particularly important. Written this way, we can say that the Fourier transform of an autocorrelation is equal to the power spectrum, a result known as the Wiener–Khinchin theorem.

$$f_1(x) \star f_2(x) \rightleftharpoons F_1^*(s)\, F_2(s), \tag{4.47}$$

$$f(x) \star f(x) \rightleftharpoons |F(s)|^2. \tag{4.48}$$

An important property of autocorrelation functions is that they peak at $s = 0$, known as zero-lag.

4.2.6 Common functions and Fourier transform pairs

We introduce in Table 4.3 (Bracewell, 2000) a menagerie of common functions that will prove useful, along with their names and common symbols. In Table 4.4 we list some common Fourier transform pairs. Note that these pairs are only applicable for the conventions we have adopted (our definitions of forward and reverse transforms).

Table 4.3. *Symbols for common functions*

boxcar, top hat, or rectangle	$\Pi(x)$	$= \begin{cases} 1 & \|x\| < \frac{1}{2} \\ 0 & \|x\| > \frac{1}{2} \end{cases}$
triangle	$\Lambda(x)$	$= \begin{cases} 1 - \|x\| & \|x\| < 1 \\ 0 & \|x\| > 1 \end{cases}$
Heaviside step function	$H(x)$	$= \begin{cases} 1 & x > 0 \\ 0 & x < 0 \end{cases}$
even impulse pair	$II(x)$	$= \frac{1}{2}\delta(x + \frac{1}{2}) + \frac{1}{2}\delta(x - \frac{1}{2})$
odd impulse pair	$I_I(x)$	$= \frac{1}{2}\delta(x + \frac{1}{2}) - \frac{1}{2}\delta(x - \frac{1}{2})$
comb (or shah)	$III(x)$	$= \sum_{n=-\infty}^{\infty} \delta(x - n)$
sinc	$\text{sinc}(x)$	$= \dfrac{\sin \pi x}{\pi x}$

Table 4.4. *Fourier transform pairs*

$e^{-\pi x^2}$	$\rightleftharpoons$	$e^{-\pi s^2}$
1	$\rightleftharpoons$	$\delta(s)$
$\Pi(x)$	$\rightleftharpoons$	$\text{sinc}(s)$
$\Lambda(x)$	$\rightleftharpoons$	$\text{sinc}^2(s)$
$\cos \pi x$	$\rightleftharpoons$	$II(s)$
$\sin \pi x$	$\rightleftharpoons$	$i\,I_I(s)$
$III(x)$	$\rightleftharpoons$	$III(s)$
$H(x)$	$\rightleftharpoons$	$\frac{1}{2}\delta(s) - \dfrac{i}{2\pi s}$

4.2.7 Aliasing and Shannon's sampling theorem

One of the most common methods of taking data is to measure (sample) some continuous function f(t) at regular intervals Δt. However, this results in unavoidable confusion as to what frequency components are present. For example, if $\Delta t = 1$ ms and we measure $+1, -1, +1, -1, \ldots$, is the frequency 500 Hz or 1500 Hz (see Figure 4.3)?

Shannon's theorem says that if $F(\nu)$ is known *a priori* to be limited to a finite bandwidth such that $F(\nu) = 0$ for all frequencies $|\nu| \geq \nu_{max}$, then the samples *fully specify* f(t) as long as the sampling is done at least as fast as what is known as the *Nyquist rate*. The Nyquist rate, which is a property of the band-limited signal, is defined as $2\nu_{max}$, so one needs to sample at a rate

$$\nu_s > 2\nu_{max}, \tag{4.49}$$

requiring the spacing between the samples to be

$$\Delta t < \frac{1}{2\nu_{max}}. \tag{4.50}$$

Somewhat confusingly, the frequency $\nu_s/2$ is called the *Nyquist frequency* even though it is half of the frequency at which one samples.

The following discussion is along the lines of that given by Thompson *et al.* (2001). The sampled function g(t) is a product of the original f(t) and the shah function,

$$g(t) = f(t) \, III \left(\frac{t}{\Delta t} \right), \tag{4.51}$$

where the scaling to an interval of width Δt gives

$$\frac{1}{\Delta t} \, III \left(\frac{t}{\Delta t} \right) = \sum_{n=-\infty}^{\infty} \delta(t - n\Delta t). \tag{4.52}$$

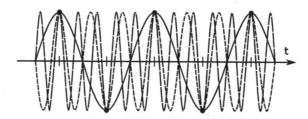

Figure 4.3 Aliasing: samples at 1 ms intervals (dots; red in electronic version) cannot distinguish between signals at 500 Hz (black), 1500 Hz (short dashes; blue), or 2500 Hz (long dashes; green).

Taking the Fourier transform of g(t) and applying the convolution theorem,

$$G(v) = F(v) * III(v\Delta t) \tag{4.53}$$

where

$$III(v\Delta t) = \frac{1}{\Delta t} \sum_{m=-\infty}^{\infty} \delta(v - \frac{m}{\Delta t}). \tag{4.54}$$

This convolution with the shah function causes a *replication* of the spectrum $F(v)$ at intervals $\Delta v = (\Delta t)^{-1}$, as shown in Figure 4.4. If $(\Delta t)^{-1} < 2v_{max}$, the replicas overlap and information is lost. This corruption of portions of the spectrum by overlapping replicas is known as *aliasing*. If $(\Delta t)^{-1} > 2v_{max}$, there is no loss of information. To recover f(t), one must further multiply with a boxcar function to remove the replicas,

$$G(v) \Pi (v\Delta t) = F(v). \tag{4.55}$$

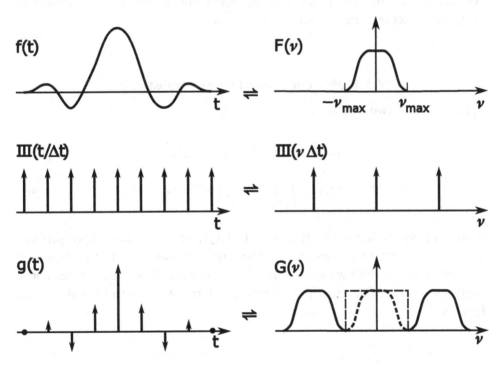

Figure 4.4 In the time domain a function f(t) is sampled at intervals Δt by multiplying it with the sampling function $III(t/\Delta t)$. In the frequency domain this corresponds to a convolution of the Fourier transform with a replication function $III(v\Delta t)$. If the signal is band limited and the samples are frequent enough, it is possible to recover the original Fourier transform (short dashes; red in electronic version) by multiplying by a boxcar (long dashes; blue) and from that recover the original signal. Adapted from Thompson *et al.* (2001).

Take the inverse Fourier transform of this to recover f(t) as a convolution of the sampled data with the sinc function,

$$f(t) = g(t) * \frac{1}{\Delta t} \text{sinc}\left(\frac{t}{\Delta t}\right), \qquad (4.56)$$

where

$$\text{sinc}\left(\frac{t}{\Delta t}\right) = \frac{\sin \pi t/\Delta t}{\pi t/\Delta t}. \qquad (4.57)$$

The sinc function essentially provides a formula for interpolating between the samples. If the conditions of the sampling theorem are satisfied, this interpolation is not an estimate or a guess, it is exact. In the real world, the conditions of the theorem can only be satisfied approximately, so the interpolation will only be approximately correct (although often very close to being exact). Finally, remember the definition of the sinc function, $\text{sinc}\, x = \sin \pi x/(\pi x) \neq \sin x/x$!

4.3 Higher-dimensional Fourier transforms

For functions in two dimensions,

$$f(\vec{x}) = \int \int_{-\infty}^{\infty} F(\vec{s})\, e^{i2\pi \vec{s}\cdot\vec{x}}\, d^2\vec{s}, \qquad (4.58)$$

$$F(\vec{s}) = \int \int_{-\infty}^{\infty} f(\vec{x})\, e^{-i2\pi \vec{s}\cdot\vec{x}}\, d^2\vec{x}. \qquad (4.59)$$

If the function is factorable, $f(\vec{x}) = f_x(x)f_y(y)$, then one can simply perform a pair of 1-dimensional transforms. Otherwise, one can refer to tabulations of 2-dimensional Fourier transform pairs. The Fourier transform of a 2-dimensional Gaussian is still a Gaussian. The Fourier transform of a constant is still a delta function.

4.3.1 Hankel (Fourier–Bessel) transforms

If there is rotational symmetry about the origin in two dimensions, then we have a function of only the radial coordinate $r = \sqrt{x^2 + y^2}$ and $f(\vec{x}) = f(r)$. Its Fourier transform is also a function of only the radial coordinate $q = \sqrt{s_x^2 + s_y^2}$ and $F(\vec{s}) = F(q)$.

$$F(\vec{s}) = F(q) = \int\int_{-\infty}^{\infty} f(r) \, e^{-i2\pi \vec{s} \cdot \vec{x}} \, dx \, dy \tag{4.60}$$

$$= \int_0^{\infty} \int_0^{2\pi} f(r) \, e^{-i2\pi qr \cos(\theta - \phi)} \, r \, dr \, d\theta \tag{4.61}$$

$$= \int_0^{\infty} f(r) \left[\int_0^{2\pi} e^{-i2\pi qr \cos\theta} \, d\theta \right] r \, dr \tag{4.62}$$

$$= 2\pi \int_0^{\infty} f(r) \, J_0(2\pi qr) \, r \, dr. \tag{4.63}$$

Note the Bessel function kernel in this transform. The inverse transform is given by

$$f(r) = 2\pi \int_0^{\infty} F(q) \, J_0(2\pi qr) \, q \, dq. \tag{4.64}$$

For us, the most important example of a Hankel transform will be the Airy pattern. For rotational symmetry in n dimensions,

$$F(q) = \frac{2\pi}{q^{n/2-1}} \int_0^{\infty} f(r) \, J_{n/2-1}(2\pi qr) \, r^{n/2} \, dr, \tag{4.65}$$

where q and r are understood to be n-dimensional radial coordinates. From this we can see that the 3-dimensional spherically symmetric Fourier transform is

$$F(q) = \frac{2\pi}{q^{1/2}} \int_0^{\infty} f(r) \, J_{1/2}(2\pi qr) \, r^{3/2} \, dr, \tag{4.66}$$

which, using Bessel function identities, can also be written as

$$F(q) = 4\pi \int_0^{\infty} f(r) \, \text{sinc}(2qr) \, r^2 \, dr. \tag{4.67}$$

Exercises

4.1 We derived the Fourier series representation for the square wave

$$f(x) = \begin{cases} +1 & 0 < x < \pi \\ -1 & -\pi < x < 0, \end{cases} \tag{4.68}$$

$$f(x) = \frac{4}{\pi} \left(\sin x + \frac{\sin 3x}{3} + \frac{\sin 5x}{5} + \cdots \right). \tag{4.69}$$

For any fixed point x in the interior of the interval $0 < x < \pi$, the nth partial sum, $f_n(x)$, will converge on $+1$ if one considers sufficiently large n,

$$f_n(x) = \frac{4}{\pi} \left(\sin x + \frac{\sin 3x}{3} + \frac{\sin 5x}{5} + \cdots + \frac{\sin nx}{n} \right). \tag{4.70}$$

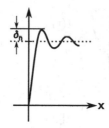

Figure 4.5 Illustration of Gibbs' phenomenon.

However, for *any* value of n, the partial sum $f_n(x)$ will overshoot the value +1 in the manner shown in Figure 4.5. This is known as Gibbs' phenomenon.

a. Find the value of x corresponding to the first maximum in $f_n(x)$.
b. Calculate a series expression for δ_n, the amount of overshoot.
c. Evaluate your expression for δ_n for n = 1, 3, and 5. Your results should converge towards the value 0.179.

4.2   The boxcar function $\Pi(x)$ (also known as the top-hat or rectangle function) is defined by

$$\Pi(x) = \begin{cases} 0 & x < -1/2 \\ 1 & -1/2 < x < 1/2 \\ 0 & x > 1/2 \end{cases}. \qquad (4.71)$$

The triangle function $\Lambda(x)$ is defined by

$$\Lambda(x) = \begin{cases} 0 & x < -1 \\ 1 - |x| & -1 < x < 1 \\ 0 & x > 1 \end{cases}. \qquad (4.72)$$

Use these definitions and the Fourier transform

$$\frac{\sin x}{x} \rightleftharpoons \pi \Pi(\pi s) \qquad (4.73)$$

to prove that

$$\left(\frac{\sin x}{x}\right)^2 \rightleftharpoons \pi \Lambda(\pi s). \qquad (4.74)$$

4.3   The Heaviside unit step function H(x) is defined to be

$$H(x) = \begin{cases} 1 & x > 0 \\ 0 & x < 0 \end{cases}. \qquad (4.75)$$

a. By direct evaluation of the convolution, show that

$$H(x) * \left[e^x H(x)\right] = (e^x - 1)H(x). \qquad (4.76)$$

b. Show that convolution with the Heaviside unit step function is equivalent to integration in the following sense:

$$H(x) * f(x) = \int_{-\infty}^{x} f(x')dx'. \tag{4.77}$$

4.4 Using the convolution theorem, give a fully simplified expression for

$$e^{-ax^2} * e^{-bx^2}, \tag{4.78}$$

where $*$ is the symbol for convolution.[1]

4.5 A sinusoidal signal at a frequency of 30 Hz is connected to the input of a broadband amplifier which introduces noise at all frequencies up to a cutoff frequency of 1000 Hz. The output is sampled at a rate of 200 Hz. State *all* noise frequencies which are aliased onto the 30 Hz signal frequency. Remember that a *real* signal (or noise component) of frequency ν may be represented as a sum of *complex* signals at frequencies $\pm\nu$. (Hint: remember to think of both positive and negative frequencies, and consider the "replication" action of sampling.)

4.6 Show that the two forms given for the 3-dimensional Hankel transform are equivalent.

---

[1] Adapted from Lèna *et al.* (1998).

5

Detection systems

5.1 Interaction of radiation and matter

In order to study electromagnetic radiation, it is necessary for the radiation to interact in some fashion with some physical "detector." If we think of the radiation in terms of photons, there are three types of interactions available to us: the photoelectric effect, Compton scattering, and pair production. Variations on these effects form the basis of photon detectors. Electromagnetic radiation can also be thought of in terms of waves, and in some situations the wave picture is more appropriate for understanding the detection process. Wave detectors could either measure the electromagnetic field directly or measure the power transfer from electromagnetic energy into thermal energy. There is no strict dividing line between these views since electromagnetic radiation always retains both particle and wave characteristics. But roughly speaking, the particle viewpoint is more useful at high frequencies, where the photons are energetic, and the wave viewpoint is more useful at lower frequencies. One characteristic which helps determine which viewpoint may be more useful is $\langle n \rangle$, the average photon occupation number of the modes of the radiation field.

5.2 Photoelectric effect

For an isolated atom, there is a threshold energy for removing a bound electron, the ionization potential, which varies depending on what shell the electron occupies. The shells are designated K, L, M..., depending on the principal quantum number ($n = 1, 2, 3, \ldots$). For hydrogen, the ionization potential from $n = 1$ corresponds to an ultraviolet photon (the Lyman limit at 91.2 nm or 13.6 eV). But for heavier elements the K-shell ionization shifts rapidly into the x-ray regime

$$E_K \propto Z^2. \tag{5.1}$$

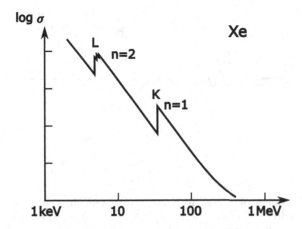

Figure 5.1 X-ray total attenuation cross section for Xe, which is dominated by photoelectric absorption at these energies. Compton scattering becomes dominant above about 400 keV. Based on the NIST XCOM database.

See Figure 5.1 for the ionization cross section of Xe. The cross section peaks just above threshold for each shell, but then drops rapidly ($\sim v^{-3}$) at higher energy due to the difficulty in transferring the excess photon momentum to the nucleus. For $n > 1$ there is subshell structure (2s, $2p_{1/2}$, $2p_{3/2}$, ...). The photoelectric effect is important in the design of x-ray proportional counters.

When other atoms are present, as in molecules and solids, the electronic energy levels will be very different, as will the photoelectric cross sections. For solids in vacuum, the thresholds can be ~ 1 eV and depend on the crystalline structure and the nature of the surface. The ionization potential in this case is usually called the work function. Photon absorption efficiencies approach 100% in the visible and ultraviolet, but the overall device efficiencies are limited by the electron escape probabilities. In a semiconductor a photon can be thought of as "ionizing" an atom, producing a "free" electron which remains in the conduction band of the lattice. Thresholds are of order 0.1–1 eV for intrinsic semiconductors and of order to 0.01–0.1 eV for extrinsic semiconductors. The latter photon energies correspond to infrared photons. Photochemistry is somewhat similar in that photons produce localized ionization or electronic excitation.

5.3 Compton scattering

In Compton scattering a photon scatters off of a free (or bound) electron, yielding a scattered photon with a new, lower frequency and a new direction, as shown in Figure 5.2. For an unbound electron initially at rest,

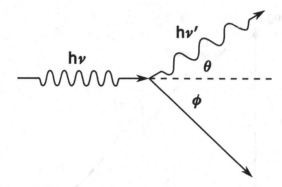

Figure 5.2 A photon of energy $h\nu$ scatters off an electron, after which it has a smaller energy $h\nu'$ and travels at an angle θ with respect to its initial direction. The electron travels at an angle ϕ with respect to the initial photon direction.

$$\nu' = \nu \left[1 + \frac{h\nu}{m_e c^2} (1 - \cos \theta) \right]^{-1}, \tag{5.2}$$

$$\lambda' = \lambda + \frac{h}{m_e c} (1 - \cos \theta), \tag{5.3}$$

where $h/(m_e c)$ has units of length and equals 0.0024 nm. Low energy photons lose little energy; high energy photons (gamma rays) lose much of their energy. The wavelength increases by of order 0.0024 nm, independent of wavelength. The Compton cross section for free electrons is given by the integrated Klein–Nishina formula,

$$\sigma_C = \sigma_T \frac{3}{4} \left\{ \frac{1+\alpha}{\alpha^2} \left[\frac{2(1+\alpha)}{1+2\alpha} - \frac{1}{\alpha} \ln(1 + 2\alpha) \right] \right.$$
$$\left. + \frac{1}{2\alpha} \ln(1 + 2\alpha) - \frac{1 + 3\alpha}{(1+2\alpha)^2} \right\}, \tag{5.4}$$

where $\alpha = h\nu/(m_e c^2)$ and the Thomson cross section is

$$\sigma_T = \frac{8\pi}{3} \left(\frac{e^2}{4\pi \epsilon_0 m_e c^2} \right)^2 = 6.65 \times 10^{-25} \text{ cm}^2. \tag{5.5}$$

For $h\nu \ll m_e c^2$ ($\alpha \ll 1$) the Compton cross section approaches the Thomson cross section, as shown in Figure 5.3, and for $h\nu \gg m_e c^2$

$$\sigma_C \approx \sigma_T \frac{3}{8} \frac{m_e c^2}{h\nu} \ln \frac{2h\nu}{m_e c^2}. \tag{5.6}$$

The Compton scattering cross section is largest at small energy and decreases monotonically with energy. At low energies there are many scattering events, but

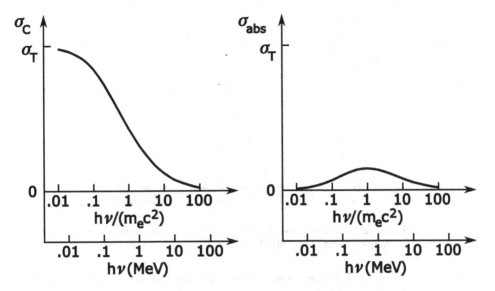

Figure 5.3 (Left) Compton scattering cross section for free electrons, representing the number of scatterings per electron s^{-1} divided by the number of incident photons cm^{-2} s^{-1}. (Right) Compton absorption cross section, representing the energy absorbed per electron s^{-1} divided by the incident energy cm^{-2} s^{-1}.

very little energy is lost. So the *energy* absorption cross section is small in Compton scattering at low energy (because little energy is transferred to the electron). The energy absorption cross section rises to a peak for photon energies around 1 MeV, and declines at higher energy (because there are few scattering events).

5.4 Pair production

Photons with energies in excess of $2m_ec^2$ are able to produce electron-positron pairs. An interaction with a nucleus is needed to balance momentum. The pair production cross section rises, starting at 1.022 MeV, and reaches an approximately constant value at high photon energy (in the gamma ray region of the spectrum) as shown in Figure 5.4. Cross sections scale with the square of the atomic number,

$$\sigma_P \approx \alpha \, Z^2 \, \sigma_T, \tag{5.7}$$

where here α is the fine structure constant and σ_T again is the Thomson cross section. For lead and tungsten at high energies the pair production cross section is of order 3×10^{-23} cm^2 per atom.

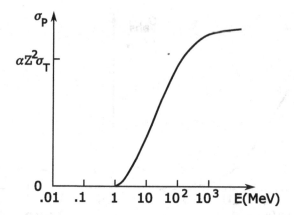

Figure 5.4 Pair production cross section in lead; shape is similar for other chemical elements. Based on the NIST XCOM database.

5.5 Electromagnetic wave interactions

In principle, at low enough frequencies (quasi-statically) one could measure the electric field strength directly. In practice, for frequencies from 10 MHz to 1 THz (10^7–10^{12} Hz), detection systems consist of antennas (which couple the free space electromagnetic wave into some type of waveguide or circuit), plus amplifiers (sometimes), plus some type of non-linear device. A non-linear device has a non-linear relationship between applied voltage and current. If the device response is proportional to the voltage squared, then it is linearly proportional to the wave intensity. Such a device may be used for detection (direct measurement of the power) or mixing (frequency translation).

Such a system will exhibit purely classical behavior as long as the photon energy ($h\nu$) is much smaller than $e\,\delta V$ (where δV is the voltage width of the non-linearity). For semiconductors $\delta V \approx 0.1$ V, so this is always a classical process for radio frequencies. Quantum effects are visible only with superconducting devices ($\delta V \ll 1$ mV) and then only at the highest frequencies ($\gtrsim 10^{11}$ Hz).

5.6 Optical and ultraviolet detectors

5.6.1 Photomultipliers

Photomultipliers today are of critical importance in high energy astrophysics, as we will see. But they retain one characteristic important for some types of optical work, namely a rapid response time ($\lesssim 1$ ns). It is relatively easy to understand their method of operation, so we will begin here.

A photomultiplier tube has a transparent vacuum window to admit optical and ultraviolet radiation, as illustrated in Figure 5.5. The radiation then strikes a

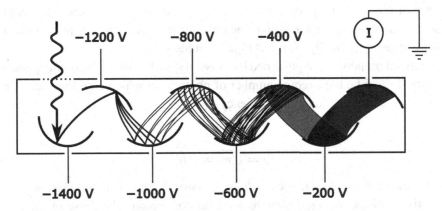

Figure 5.5 Schematic view of a photomultiplier with a photon (red line in electronic version) entering the tube through a thin vacuum window (dotted). The photocathode then releases a photoelectron (blue) which is accelerated towards a dynode where it produces secondary electrons (green). The cascade continues and produces a measurable current pulse at the anode. Actual electron trajectories depend on the placement and shape of the electrodes and on space charge effects. Optimal electron trajectories contribute to high efficiency and short response time, therefore magnetic shielding is typically required (Hamamatsu Photonics, 2006).

photocathode surface which has a potential energy barrier (work function) of a few eV. A photon of sufficient energy can overcome this barrier and liberate one electron from such a surface. An electron will not necessarily be emitted; the photon energy may be shared between several electrons and ultimately dissipated as heat. Typical commercial photocathodes have quantum efficiencies up to 25% at visible wavelengths, and a variety of spectral responses are available. Quantum efficiencies can be much higher at ultraviolet wavelengths.

A photomultiplier operates by multiplying the number of electrons through a cascade process. If the photocathode emits an electron, the electron is accelerated through a potential drop of order 100–200 volts. This *primary* photoelectron, now energetic, then strikes a second surface (dynode) from which it is able to liberate several *secondary* electrons. The average number of secondary electrons, g, may be about 4. This number will be subject to statistical fluctuations. But g should be large enough so that the probability of zero secondary electrons (which would terminate the cascade) is small. These secondary electrons are accelerated through a similar potential drop, and this process repeats through some number of stages, n. The total gain $G = g^n$, which might be of order $4^{10} \approx 10^6$. Because of statistical fluctuations, the final number of electrons in the cascade may range from 5×10^5 to 1.5×10^6. One generally sets some minimum threshold for the number of cascade electrons so that incidents in which an electron is spontaneously emitted by one of the dynodes may be ruled out. Other than in relation to this threshold, the final number of electrons is ignored in this *pulse counting mode*. If one only counts pulses, a photomultiplier has good linearity, limited primarily by pulse overlaps (pulse widths are typically 10 ns) at high counting rates.

The signal properties are governed by Poisson statistics of the incoming photons. In a time T, let $\bar{n}$ be the average number of photons, in which case the uncertainty in n is $\sigma_n = \sqrt{\bar{n}}$. For a quantum efficiency η,

$$\bar{n}' = \eta \, \bar{n}, \tag{5.8}$$

$$\sigma_{n'} = \sqrt{\eta \bar{n}} = \sqrt{\bar{n}'}. \tag{5.9}$$

The nature of Poisson statistics will be discussed later. But for now note that the statistics of the *detected* photons are also Poissonian, like those of the initial radiation field. We refer to this as *photon noise*.

Additional noise results from a *dark current* due to, among other things, thermionic emission (thermally emitted electrons) with a current density

$$j = \frac{4\pi e \, m_e k^2}{h^3} \, T^2 \, e^{-W/kT} \tag{5.10}$$

where W is the work function. This current is significant at room temperature ($\sim$nA at the first dynode), but may be reduced by cooling. The average value of the dark current (or the average number of dark pulses) may simply be subtracted. But fluctuations in the dark current can be a significant source of noise. A larger work function will decrease the dark current, but at a cost to the response at the red end of the spectrum. Since the dark current will be proportional to the area of the photocathode, the size should be kept as small as possible while remaining consistent with constraints imposed by the optical system.

5.6.2 *Other electron multiplication devices*

Another type of electron multiplication device is the *microchannel plate*, in which electrons travel through long narrow holes in a dielectric, with their numbers multiplying with every collision with the walls of the channel. The rate of collisions can be increased by having the holes curved or by having stacks with holes slanted in alternating directions (a so-called Z-stack). A high voltage is applied across the plate in order to accelerate the electrons between collisions. Typical gains are of order 10^6, similar to photomultipliers. But in addition, the microchannel plate preserves spatial information, making it useful as an imaging device. In one configuration, the microchannel plate is preceded by a thin semi-transparent photocathode on which an optical image is focussed. The photoelectrons are ejected from the back side of the photocathode (the side in vacuum, facing the microchannel plate). After the multichannel plate one could place a phosphorescent anode,

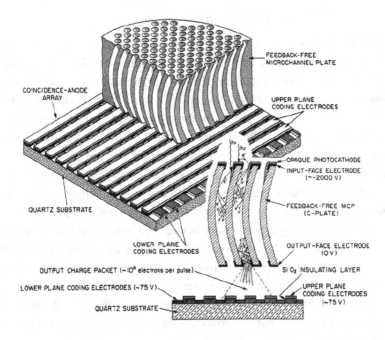

Figure 5.6 A Multi-Anode Microchannel Array (Timothy, 1983).

which converts each electron in the cascade into multiple photons at the wavelength characteristic of the phosphor. There are some difficulties with this approach since the photons are emitted into a large solid angle.

An alternative is to follow the microchannel plate with an anode consisting of crossed sets of multiple parallel "wires" (MAMA: Multi-Anode Microchannel Array), an example of which is shown in Figure 5.6. The centroid of the electrodes in the two anode planes which receive the most current determines the spatial location of the current pulse, and the corresponding spot where the photon hits the photocathode. A MAMA is an example of a photon-counting camera. An image is built up by recording the time of arrival and location of individual photons. The Hubble Space Telescope's STIS (Space Telescope Imaging Spectrograph) contains two MAMA detectors for ultraviolet wavelengths plus a CCD camera for visible wavelengths. STIS was repaired in 2009 during Servicing Mission 4 (SM4). The Hubble ACS (Advanced Camera for Surveys) has an ultraviolet detector called the SBC (Solar Blind Channel), which is also a MAMA. The ACS was also repaired during SM4. The new COS (Cosmic Origins Spectrometer), installed during SM4, uses a MAMA for its near-ultraviolet channel and a microchannel plate with double-delay line anodes for its far-ultraviolet channel.

5.6.3 Solid state detectors

A unit cell of crystalline silicon is a face-centered cubic structure, like diamond. Each silicon atom has four nearest neighbors in a tetrahedral configuration. It can be thought of as being bound to each of those neighbors by a single covalent bond containing two electrons, one contributed by each of the atoms. Although each atom really has four nearest neighbors in three dimensions, it is often useful to represent it in two dimensions with a picture in which the silicon atoms form a regular square grid. The limitations of this picture are that this is not a 2-dimensional slice out of a 3-dimensional structure, and it does not accurately represent the atomic arrangement beyond that of nearest neighbors. Nearest neighbors of any particular atom are not simultaneously nearest neighbors of another atom, although the 2-dimensional picture suggests that they would be. Highly purified silicon is known as an *intrinsic semiconductor* and is a poor conductor of electricity since the electrons are tied up in valence bonds. However, if one of those bonds is broken by absorption of a photon or by thermal excitation, then an electron is raised in energy into the conduction band, leaving behind a hole, as in Figure 5.7. Both the electron and hole are mobile charge carriers, although they may have very different mobilities.

An *extrinsic semiconductor* is formed if one of the silicon atoms is replaced by an impurity atom with a different number of valence electrons. For example, arsenic atoms have five valence electrons. So when an arsenic atom is placed in a silicon crystal, there is an extra electron not tied up in the valence bonding, which therefore is a free carrier. This is known as n-type doping, since it provides excess negative charge carriers. Boron has three valence electrons, so a boron impurity leaves a hole as a free carrier. This is p-type doping since the hole carries positive charge.

Figure 5.7 (Left) Intrinsic silicon with valence electrons (blue in electronic version) and a thermally activated conduction band electron (red) and hole (green circle). (Center) Extrinsic n-doped silicon with an arsenic atom donating a free electron (red). (Right) Extrinsic p-doped silicon with a boron atom contributing a hole (green circle).

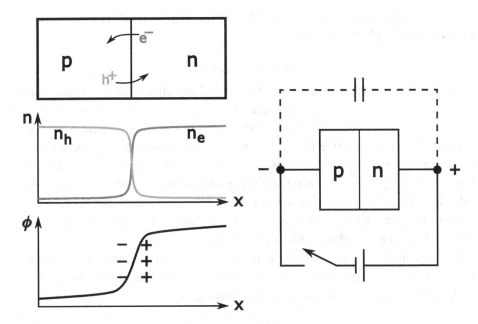

Figure 5.8 (Left) If p-type and n-type silicon are brought together, holes will diffuse across the junction from the p-type material and electrons will diffuse from the n-type material. An equilibrium will be reached in which holes are the majority carrier on the p side and electrons the majority carrier on the n side. The region around the junction will be relatively depleted of free carriers. A gradient in the potential ϕ will be established due to the fixed lattice charges left behind. (Right) The equivalent circuit of a pn junction contains a capacitance associated with the depletion region. Reverse bias establishes a stored charge on the capacitance.

Silicon is an elemental semiconductor from group IV of the periodic table. It is also possible to make compound semiconductors, for example, binary III-V compounds such as GaAs and InSb. It is also possible to make II-VI semiconductors as well as ternary and higher compounds (for example, HgCdTe).

Consider what would happen if a piece of p-type silicon and a piece of n-type silicon were joined together, as in Figure 5.8. At the boundary initially there would be a high concentration of free electrons on the n side but none on the p side. So some of these electrons would diffuse across the boundary into the p side. There would likewise be a high concentration of holes on the p side, and some of these would diffuse across the boundary to the n side. The donor and acceptor atoms would be left behind with static positive and negative charges, which would set up an electrostatic field. After enough carriers had diffused across, this field would prevent further diffusion. The region around the boundary would be depleted of the majority charge carriers (electrons on the n side and holes on the p side) and is appropriately called the *depletion region*. By applying reverse bias (positive voltage to the n side and negative voltage to the p side) one can increase the size of the depletion region.

Silicon diode detectors

By applying reverse bias to a pn junction, one is effectively storing charge on the equivalent of a parallel plate capacitor (the depletion region is an insulator and the p and n regions are conductors). There may also be stray capacitance present. Imagine then disconnecting the bias. If photons are absorbed, let's say, within the depletion region, electron-hole pairs are produced. The electrostatic field within the depletion region will sweep the electrons to the n side and the holes to the p side, decreasing the amount of stored charge. After some time one could reapply the bias, restoring the original charge, and the current flow would reveal how many photons had been absorbed in the depletion region.

The sensitivity of this technique is limited by thermodynamic fluctuations. In thermodynamic equilibrium at a temperature T, the uncertainty in the stored charge (the charge fluctuations at fixed voltage) is given by

$$(\Delta Q)^2 = kT\,C. \tag{5.11}$$

This is known as kTC noise. For a temperature of 150 K and a capacitance of 1 pF,

$$\Delta Q \approx 280\ \mathrm{e}^-, \tag{5.12}$$

which is relatively high if one wants to detect individual photons. In order to be limited by photon statistics rather than kTC noise, one would need of order 10^5 photons ($\sqrt{10^5} \approx 300$). There is also dark noise from thermally activated leakage currents (which depend exponentially on temperature).

Charge-coupled devices

Instead of a pn junction, consider starting with a substrate of p-type silicon, on the surface of which an insulating oxide (SiO$_2$) is grown, as in Figure 5.9. Then deposit small, thin (semi-transparent) metallic electrodes on top of the oxide. Each electrode defines an MOS (metal-oxide-semiconductor) capacitor. Positive bias applied to the electrodes creates depletion regions (depleted of holes, the majority carriers in p-type silicon) which serve as storage regions for electrons (the minority carriers). A charge-coupled device (CCD) consists of a 2-dimensional array of such pixels.

The detection characteristics of CCD arrays depend on the method of illumination and certain physical characteristics of the manufacturing. In all cases photons enter the semiconductor and are absorbed in or near the depletion region. If the absorption occurs inside the depletion region, the electron is drawn towards the positively charged electrode and trapped by the oxide. The hole is expelled from

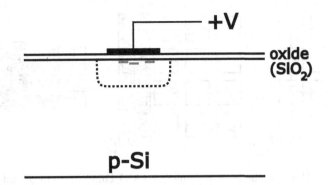

Figure 5.9 A metal-oxide-semiconductor (MOS) device. Positive voltage applied to the metal creates a depletion region (dashed line) by repelling the holes in p-type silicon. Electrons are drawn towards the metallization and trapped by the oxide layer.

the depletion region. If the absorption occurs outside the depletion region, it is necessary for the electron to diffuse to the boundary of the depletion region before the electron recombines (which would reduce the detection efficiency). Some CCDs are *front* illuminated: the radiation passes through the semi-transparent electrode. These are typically *thick CCDs* which have enhanced response in the red portion of the spectrum since the thick device is able to contain several optical absorption lengths, even at long wavelengths where the absorption length is typically longest. Quantum efficiencies are typically of order 70%. *Thinned CCDs* have been etched away from the underside, the back, and are typically *back* illuminated. They have poorer red response because the thickness of the remaining material is only of the same order as the absorption length in the red. But they have better blue response and higher peak quantum efficiencies, typically nearly 90%.

The method of signal readout gives CCDs their name. CCDs are essentially shift registers, which preserve the integrity of the trapped charge bundles with charge-transfer efficiencies of order $\eta_{CT} \approx 0.99999$ per shift as the packets are shifted across the device in a "bucket brigade" technique to a readout amplifier. The readout generally takes place after the exposure and can be in the form of, for example, sequential readout of the final column of the CCD followed by a single step of all rows over by one column to repopulate the final column, as shown in Figure 5.10. This is iterated until the entire device is read out. An alternative technique is used in the Sloan Digital Sky Survey in which the telescope is stationary and the stars drift across the focal plane and the CCD (drift scan) in synchronism with the rate of charge packet shift across the CCD.

Advantages of CCD detectors for astronomy include high quantum efficiency, good linearity, low readout noise ($\sim 3 \, e^-$ RMS), and large numbers of pixels

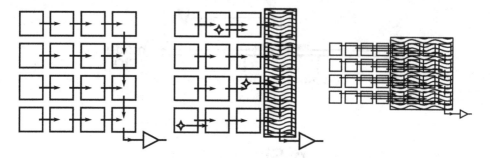

Figure 5.10 Three methods of reading out CCD chips: (Left) Post-integration readout beginning with rightmost column (blue in electronic version) followed by column shifts (red), repeated until entire chip has been read out. (Center) Charge packets are shifted along chip (red) at the same rate as the stellar field (green). (Right) Entire field of view is shifted (red) into readout registers behind an opaque screen. Field can be read out while another integration proceeds.

(the Sloan CCD camera has a total of 120 megapixels, enough to do simultaneous multicolor photometry over wide fields). Electron storage capacity can be of order 10^5–10^6 electrons per pixel, which gives a large, but limited dynamic range. Image defects can be caused by cosmic ray hits and by spillover from overfull packets due to bright sources. One does not typically use CCDs for high speed photometry since the readout takes time.

Careful data reduction techniques for CCDs include measurement of dark frames (which need to be subtracted) and "flat fields" (images under uniform illumination) by which the images are divided to obtain gain-corrected images.

The above discussion is somewhat oversimplified. There are varieties of semi-conductor manufacturing processes and varieties of readout techniques for CCDs. A user needs to have a detailed understanding of the characteristics of the actual instrument being used. CCDs are discussed in greater detail by Rieke (2002), Jansen (2006), and many others.

Many Hubble Space Telescope instruments have used CCDs. As mentioned above, these include the STIS (Space Telescope Imaging Spectrograph), the ACS (Advanced Camera for Surveys), and the ultraviolet–visible channel of WFC3 (Wide Field Camera 3). Both STIS and ACS were repaired during Servicing Mission 4, and WFC3 was installed at that time (along with COS).

5.7 Infrared astronomy

Our view of the universe in the infrared is very different than that at visible wavelengths. Interplanetary and interstellar extinction drop very rapidly as one moves from the visible to the infrared, meaning that the infrared is better suited for viewing along lines of sight in the galactic plane and for viewing deeply embedded objects. One example of the latter is IRC +10216, an evolved star heavily obscured by surrounding dust. It has a greater flux at 5 μm than any other source outside the solar system, $M(5.0\,\mu m) \approx -5^m$, yet it is difficult to observe at visible wavelengths except with a large telescope and a sensitive detector ($V(550\,nm) \approx +18^m$). Another example is the ability to image stars in orbit around the galactic center (Ghez *et al.*, 1998; Genzel & Karas, 2007). Also important is the ability to use the infrared to observe star forming regions such as those in Orion and Taurus. There are also some unique atomic and molecular lines visible in the infrared, such as those of shocked molecular hydrogen (H_2). In the mid-infrared around 10 μm and beyond, an observer is faced with the problem of strong competing thermal emission from everything ranging from the telescope to the zodiacal dust and warm interstellar dust. Ground-based infrared observers face atmospheric absorption bands of CO_2, O_3, and H_2O, which make the atmosphere opaque through large portions of the infrared.

Thermal emission from the telescope is an additional challenge. Each telescope mirror may have an emissivity $\epsilon \approx 1$–2%, making it important to minimize the number of telescope mirrors in order to minimize thermal background, as in Figure 5.11. This usually means a Cassegrain telescope, with only two mirrors. The secondary mirror should be *small* compared to the primary, since thermal radiation will enter the beam in direct proportion to the ratio of areas of the primary and secondary mirrors. Effectively, only the annular portion of the beam (as determined by the hole in the primary) will see the "cold" sky. The remainder of the beam will see some portion of the "warm" surrounding environment unless that portion of the beam is blocked with a cold aperture stop at an exit pupil within the cryogenic detector system. One also wants to have a *slightly*

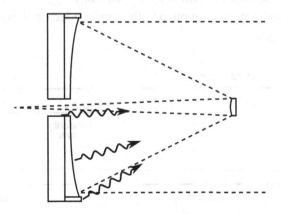

Figure 5.11 Three possible sources of thermal background entering the beam of an infrared Cassegrain telescope.

undersized secondary mirror (or equivalently, an oversized primary) to minimize emission entering the beam from supporting structure at the edges of the mirror. However, doing so will limit the field of view of the telescope. Thermal background is a particular problem since it will vary due to small temperature changes, variations in atmospheric emissivity, etc. Such effects can be minimized by use of a chopping secondary mirror and synchronous detection. These variations tend to have 1/f spectral character (more slow variations than rapid variations), implying that imposing as rapid a modulation as possible will be most effective. Most of these problems can be minimized by using cooled telescopes in space. In the recent past there was the NICMOS (Near Infrared Camera and Multi-Object Spectrometer) instrument on the Hubble Space Telescope. Current dedicated infrared space telescopes include Spitzer (formerly known as SIRTF) and Herschel.

5.7.1 Infrared photoconductors

In the infrared one often uses semiconductors as photoconductive devices. In an intrinsic photoconductor, an absorbed photon can excite an electron into the conduction band and leave a hole in the valence band, as in Figure 5.12. For infrared astronomy the most important intrinsic photoconductor is probably InSb, which has a bandgap of ~ 0.2 eV, well matched to the energies of infrared photons with wavelengths of 6 μm or less. In contrast, extrinsic photoconductors have impurity levels whose energies can be selected by appropriate choice of dopants. Photons of rather smaller energies (longer wavelengths) are able to excite free carriers in such systems. For infrared astronomy important examples of extrinsic photoconductors include Si:As (silicon doped with arsenic) and, for long wavelength applications, Ge:Ga. In some cases stress is applied to the semiconductor to extend the response to longer wavelengths.

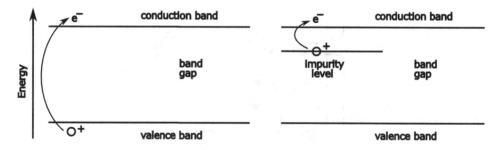

Figure 5.12 Energy level diagrams for intrinsic (left) and extrinsic (right) photoconductors.

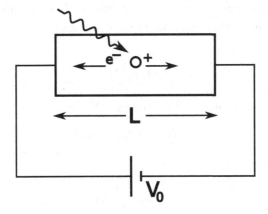

Figure 5.13 Photoconductive gain $G' > 1$ corresponds to carriers making more than one trip through the circuit before recombining.

Assume that one has a photoconductor with quantum efficiency $\eta \approx 50\%$. If the applied power is P, corresponding to a photon rate of $P/h\nu$ photons per second, a current will flow which is equal to

$$I = \eta \, \frac{P \, e \, G'}{h\nu},$$ (5.13)

as shown in Figure 5.13, where G' is a quantity known as the photoconductive gain (the number of trips the carriers make through the circuit before recombining). One can speak of the responsivity, S, as the ratio of current to applied power,

$$S = \eta \, \frac{e}{h\nu} \, G'.$$ (5.14)

If the average lifetime of a carrier is τ and carriers are accelerated to an average velocity v,

$$G' = \frac{v \, \tau}{L}$$ (5.15)

$$= \frac{\mu \, E \, \tau}{L}$$ (5.16)

$$= \frac{V_0 \, \mu \, \tau}{L^2}.$$ (5.17)

The *mobility* μ, defined as the ratio of velocity to applied electric field E, is a property of the semiconductor. The electric field is the applied voltage V_0 divided by the length L of the semiconductor. Such detectors are cooled to prevent thermal generation of carriers, to reduce Johnson noise, and to optimize the mobility, μ.

The noise-equivalent-power (NEP) of a photoconductor is that signal power which gives a signal to noise ratio S/N = 1 for a 1 Hz bandwidth (i.e. a 1 second integration). In an ideal system this would be limited by the statistical fluctuations in the background radiation (BLIP – background limited infrared photodetector). For a background power P_B, the number of background photons detected in time T is

$$N = P_B \frac{1}{h\nu} \eta T. \tag{5.18}$$

Poisson statistics says the uncertainty in N is $\sqrt{N}$,

$$\sigma_N = \sqrt{N} = (P_B \eta T / h\nu)^{1/2}. \tag{5.19}$$

But from the definition of NEP,

$$\sigma_N = NEP \frac{1}{h\nu} \eta T. \tag{5.20}$$

Equating these we get

$$NEP = \left(\frac{P_B \, h\nu}{\eta \, T} \right)^{1/2} \tag{5.21}$$

$$= \left(\frac{2 P_B \, h\nu}{\eta} f_c \right)^{1/2}, \tag{5.22}$$

where the bandwidth is $2f_c = 1/T$. In practice, an ideal photoconductor will have $\sqrt{2}$ more noise, since fluctuations of equal magnitude are produced by the statistics of the generation and recombination of free carriers, known as G-R noise.

$$NEP = 2 \left(\frac{P_B \, h\nu}{\eta} f_c \right)^{1/2} \quad \text{(watts)}. \tag{5.23}$$

Although the above has dimensions of power, it is common to refer to the NEP as

$$NEP = 2 \left(\frac{P_B \, h\nu}{\eta} \right)^{1/2} \quad (\text{W}/\sqrt{\text{Hz}}). \tag{5.24}$$

The philosophy of blocked impurity band (BIB) detectors is to physically separate the functions of photon absorption and photoconduction and to optimize these regions separately. There is a heavily doped infrared-absorbing layer (typically Si:As), but this layer necessarily has high conductivity. The "blocking" layer is of high purity and consequently low conductivity. This allows the device to operate with high impedance, minimizing thermally generated current and Johnson noise. These devices also exhibit charge multiplication through an electron cascade effect.

5.7.2 NICMOS

The Near Infrared Camera and Multi-Object Spectrometer (NICMOS) was installed on the HST during the 1997 service mission of the space shuttle Discovery (STS 82). It contained three separate near-infrared (0.8–2.5 μm) cameras with

256×256 HgCdTe detector arrays. The cameras had different magnification scales. Each was able to operate in an imaging mode (with various filter bandwidths). Two cameras were also capable of imaging polarimetry, and one was capable of low resolution "slitless" *grism* spectroscopy (see Chapter 10). After installation, focus problems were discovered, which were thought to be due to a physical distortion of the cryostat. This distortion also created a reduced lifetime for the cryogen, which was officially exhausted on January 3, 1999. Operation of NICMOS was restored during servicing mission SM3B March 1–12, 2002, by the crew of the space shuttle Columbia (STS 109), who installed a closed-cycle cooling system. That was the last successful mission of space shuttle Columbia. Beginning in September 2008 NICMOS experienced numerous "safing" anomalies which prevented cryogenic operation. At the time this is written NICMOS is not available for use, and it is not clear whether further attempts will be made to revive it. Some of its capabilities are covered by the WFC3 Near-InfraRed (NIR) channel, which covers 0.85–1.7 μm with a 1024×1024 HgCdTe array.

5.7.3 Bolometers

Bolometers are thermal detectors. They measure the physical temperature rise ΔT of an absorber subject to a radiation field of intensity I_ν. Assume that the bolometer surface is black, $\epsilon = 1$, so that the incident power, $P = I_\nu A \Omega \Delta\nu$, is completely absorbed. The bolometer, which has some heat capacity C, is connected to a heat sink by some thermal conductance G, as shown in Figure 5.14. The heat flow from the bolometer into the heat sink is described by the differential equation

$$C \frac{d}{dt}(\Delta T) = I_\nu A \Omega \Delta\nu - G \Delta T. \qquad (5.25)$$

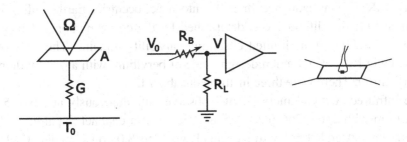

Figure 5.14 (Left) Thermal circuit for a bolometer receiving radiation through an étendue of $A\Omega$. The heat capacity of the bolometer is C. It is connected to a heat bath at temperature T_0 by a thermal conductance G. (Center) Circuit for biassing the bolometer R_B with a high impedance load resistor R_L. (Right) A composite bolometer with a substrate suspended by four wires, and attached to the bolometer itself, which has two wires.

For a radiation field applied at time t = 0, the solution of this equation is

$$\Delta T = \frac{I_\nu \, A \, \Omega \, \Delta \nu}{G} \left(1 - e^{-t/\tau}\right), \tag{5.26}$$

where the time constant $\tau = C/G$. For high sensitivity one would like a small heat conductance, G. However, this implies a long time constant. So one would also like a small heat capacity to keep τ reasonable ($\sim$10–100 ms). Note the conflicting requirements of large area and small heat capacity.

The temperature rise is sensed by the change in resistance of a semiconductor:

$$\Delta V \approx \frac{V_0 R_L \Delta R_B}{(R_B + R_L)^2} \approx V_0 \frac{R_B \alpha \Delta T}{R_L}, \tag{5.27}$$

where α is the temperature coefficient of the bolometer response ($\Delta R_B = R_B \, \alpha \, \Delta T$) and the load resistance is much larger than the bolometer resistance ($R_L \gg R_B$). Both R_B and α are large for semiconductors. Noise analysis is given in Rieke (2002).

A *composite bolometer* consists of a large area bismuth film on a crystalline substrate. Bismuth films are very black, and a large area can be provided to absorb radiation. But the film and substrate have a small heat capacity. A small germanium bolometer is provided to record the temperature change while adding minimal heat capacity. The thermal conductance is provided by wires used to suspend the device in the vacuum dewar plus the electrical leads to the germanium bolometer.

5.7.4 Spitzer

The Spitzer Space Telescope (formerly known as the Space InfraRed Telescope Facility, SIRTF) was launched in 2003 into a heliocentric, Earth-trailing orbit. Its primary mission lifetime was determined by the cryogen supply, which was exhausted in 2009. Some limited observing capability remains. The telescope is a lightweight Ritchey–Chrétien design made of beryllium with a 0.85 m diameter primary mirror. There were three instruments aboard.

The InfraRed Array Camera (IRAC) observes simultaneously at 3.6, 4.5, 5.8, and 8.0 μm with 256 × 256 pixel arrays. The 3.6 μm channel is paired with the 5.8 μm channel and the 4.5 μm channel with the 8.0 μm channel. Each pair observes a separate field of view and requires a single dichroic beamsplitter. The short wavelength detectors are InSb, whereas the longer wavelength detectors are Si:As. In both cases the detector chips are "bump-bonded" to a separate silicon readout layer. The Si:As detectors are the only parts of Spitzer that remain operational after the exhaustion of the cryogen.

The InfraRed Spectrograph (IRS) has capability for both low and high resolution spectroscopy from less than 10 μm to about 40 μm using both Si:As and Si:Sb detectors. The high resolution systems were echelle designs.

The Multiband Imaging Photometer (MIPS) contained a 128 × 128 pixel Si:As BIB array for 24 μm, a 32 × 32 pixel Ge:Ga array for 70 μm, and a 2 × 20 pixel stressed Ge:Ga array for 160 μm. The 70 μm channel was also capable of low resolution spectroscopy.

5.7.5 Herschel

The Herschel Space Observatory was launched in 2009 by the ESA along with the Planck spacecraft. Both spacecraft went into orbit around the Earth–Sun L_2 Lagrangian point. Herschel was originally named FIRST (Far Infrared and Sub-millimetre Telescope). It contains a 3.5 meter diameter cooled telescope and three scientific instruments.

The Heterodyne Instrument for the Far Infrared (HIFI) covers much of the wavelength range between 157 μm and 625 μm in a total of seven bands, with diffraction-limited angular resolution of $12'' - 41''$. The nature of radio-frequency technology, including heterodyne detection and acousto-optical and autocorrelator backend spectrometers, is discussed in Chapter 12.

The Photodetector Array Camera and Spectrometer (PACS) contains two bolometer arrays capable of simultaneous imaging of the same field in bands longward and shortward of 130 μm with arrays of 32 × 16 pixels and 64 × 32 pixels, respectively. It also contains an imaging spectrometer which may be used to observe a 5 × 5 pixel field with resolving power of order 1000–5000 covering both the 55–105 μm and the 105–210 μm bands simultaneously. The spectrometer employs both stressed and unstressed Ge:Ga photoconductors.

The Spectral and Photometric Imaging REceiver (SPIRE) contains both a three band imaging photometer and an imaging Fourier transform spectrometer. The basic nature of Fourier transform spectroscopy is discussed in Chapter 10. The SPIRE photometer and spectrometer both make use of hexagonally packed spiderweb bolometer arrays. Spiderweb bolometers are also a feature of the Planck HFI instrument, discussed below in Chapter 12. SPIRE's wavelength coverage of about 200–700 μm is at longer wavelengths than covered by PACS.

5.7.6 WFIRST

The WFIRST (Wide Field InfraRed Survey Telescope) was rated first among large space projects by the 2010 US National Research Council decadal review of astronomy and astrophysics. Its main scientific goals are the search for exoplanets and dark energy. The telescope will be of modest aperture (about 1.5 meters), but it will have a large field of view supporting multiple 2048 × 2048 HgCdTe arrays with a total of more than 100 megapixels.

Exercises

5.1 Prove that the integrated Klein–Nishina formula reduces to the Thomson cross section in the limit $\alpha \to 0$.

5.2 A photomultiplier is typically used in a pulse counting mode. Each anode current pulse has a finite temporal width t_0 (typically of order 10 ns). Assume that when a photon is detected it renders the tube dead (insensitive to further photon arrivals) for a time t_0, after which the tube automatically turns back on. Assume this is the only loss of efficiency and neglect noise.

   a. Calculate an expression for the efficiency as a function of photon arrival rate.
   b. Verify your expression gives a reasonable result for small photon flux.
   c. Verify your expression gives a reasonable result for large photon flux.
   d. Some types of counting circuits will work somewhat differently. Assume instead that there must be a *gap* of at least t_0 between photon arrival times in order for another photon to be counted. That is, assume that a rapid series of photon arrivals is able to keep the tube dead. From your knowledge of Poisson statistics, make an informed guess as to the efficiency as a function of photon arrival rate under this scenario.

5.3 An infrared bolometer consists of a detector of heat capacity C connected to a heat sink (thermal reservoir) at a temperature of 2.2 K. The thermal conductance of the connection is $G = 10^{-6}$ W K^{-1}. Ignore any Joule heating produced by electrical current flowing through the detector.

   a. What must the heat capacity be in order for the device to have a reasonable thermal time constant ($\tau = 0.01$ s)?
   b. The change in temperature produces a change in resistance of the sensing element and ultimately a change in the output voltage of the device. Let this rate of change be $dV/dT = 0.25$ V K^{-1}. What is the overall responsivity of the bolometer (how many volts per watt)?
   c. The above is true for slowly varying signals. How will this change if the infrared signal varies sinusoidally at a frequency of 10 Hz? (Hint: We discuss the effects of an electrical RC filter in Chapter 7. This is simply the equivalent *thermal* circuit.)

6

Orthodox statistics

In this chapter we focus on some aspects of orthodox (frequentist) statistics. Many of these topics in statistics are well covered by Lyons (1991) and Bevington & Robinson (2003) in greater detail than we do here. The chapter following this will deal with the nature of noise processes. Other topics in statistics are deferred to Chapter 13.

6.1 Probability distributions

Both discrete and continuous random variables may be described by probability distributions. Consider, first, a random variable x with a discrete set of possible values $\{x_i\}$. Any possible outcome x_i has an associated probability p_i. The laws of probability require that

$$\sum_i p_i = 1. \tag{6.1}$$

If the $\{x_i\}$ are real valued, one can define a *cumulative* or *integral* probability distribution, which we designate using $P(x)$, as the sum of the probabilities of all outcomes $\{x_i\}$ with $x_i < x$,

$$P(x) = \sum_i^{x_i < x} p_i. \tag{6.2}$$

The derivative of this function is the probability *density*, which we differentiate from the cumulative probability by using lower case,

$$p(x) = \frac{dP(x)}{dx}. \tag{6.3}$$

When we discuss probabilities, we will mostly be using this probability *density*. For a discrete random variable, $p(x)$ consists of a set of delta functions,

87

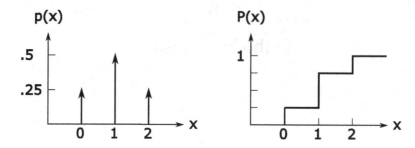

Figure 6.1 Probability density p(x) and cumulative probability P(x) for the numbers of "heads" achieved after two flips of a coin.

$$p(x) = \sum_i p_i \, \delta \, (x - x_i).$$ (6.4)

For example, if we flip two coins and for each "heads" record a score of 1, the possible total scores are 0, 1, and 2, with

$$p_0 = 0.25,$$ (6.5)
$$p_1 = 0.50,$$ (6.6)
$$p_2 = 0.25,$$ (6.7)

as shown in Figure 6.1. We can extend this concept to a continuous real variable x by defining p(x) as the derivative of P(x), as above. The quantity p(x) dx corresponds to the probability of having an event within some interval dx around x, hence the term "probability density." There are three special probability distributions which will be important to us.

6.1.1 Binomial distribution

Consider a series of coin flips, as shown in Figure 6.2, where n is the number of flips and k is the number of times the coin comes up heads. The binomial coefficient, sometimes referred to as "the number of combinations of n things taken k at a time," gives the number of possible outcomes with k heads, and is written as

$$\binom{n}{k} = \frac{n!}{k! \, (n-k)!}.$$ (6.8)

If the coin is fair (has a 50/50 chance of coming up heads), there is a total of 2^n possible outcomes which are equally likely. The probability of obtaining k heads is then

$$p_k = \binom{n}{k} \frac{1}{2^n}.$$ (6.9)

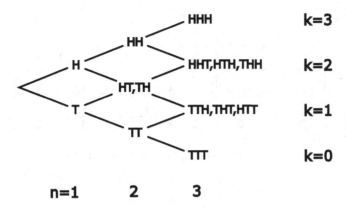

Figure 6.2 The binomial distribution, illustrated by three successive flips of a coin.

More generally, one might encounter a coin which is not weighted 50/50. If f is the probability of obtaining "heads" in each coin flip, then after n flips the probability of k heads is

$$p_k = \binom{n}{k} f^k (1 - f)^{n-k}. \tag{6.10}$$

The term coin, of course, can mean any random event with only two possible outcomes.

6.1.2 Poisson distribution

One may extend the concept of probability to multiple, discrete outcomes by taking a limiting case of the binomial distribution. The Poisson distribution corresponds to the binomial distribution in the limit $n \to \infty$, while $f \to 0$ in such a way that their product is a constant, $nf = a$. First, calculate

$$\lim_{n \to \infty} \binom{n}{k} = \frac{1}{k!} \lim_{n \to \infty} \frac{n!}{(n-k)!} = \frac{1}{k!} n^k, \tag{6.11}$$

where the last step is true since $n \gg k$. Similarly,

$$\lim_{n \to \infty} (1 - f)^{(n-k)} = \lim_{n \to \infty} (1 - f)^n = \lim_{f \to 0} (1 - f)^{a/f} = e^{-a}. \tag{6.12}$$

To understand the last equality above, do a binomial series expansion of $(1 - f)^{a/f}$ and compare it with the series expansion of e^{-a}. Putting it all together, the probability of obtaining a specific outcome k is

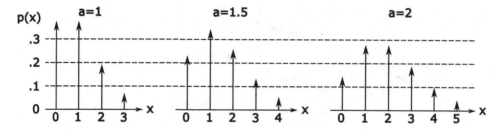

Figure 6.3 The Poisson distributions for $a = 1$, 1.5, and 2.

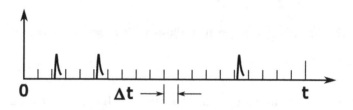

Figure 6.4 Subdivide the time interval from 0 to t into narrow bins of width Δt so that no two pulses occur in the same bin.

$$p_k = \lim_{n \to \infty} \binom{n}{k} f^k (1 - f)^{n-k} \tag{6.13}$$

$$= \frac{1}{k!} n^k f^k e^{-a} \tag{6.14}$$

$$= \frac{a^k e^{-a}}{k!}. \tag{6.15}$$

Examples are shown in Figure 6.3. In terms of the probability density,

$$p(x) = e^{-a} \sum_{k=0}^{\infty} \frac{a^k}{k!} \delta(x - k). \tag{6.16}$$

This situation is applicable to counting statistics. If radioactive decays occur at an average rate r, then after a time t, on average one expects to see $a = rt$ events. But one *may* see more or fewer events. What is the probability of seeing exactly k events? Subdivide the time interval t into n bins of width Δt, as shown in Figure 6.4. Take the limit $n = t/\Delta t \to \infty$ so that *at most* one event will occur in a time Δt with a probability $f = \Delta t \, r$ as $f \to 0$. The average number of events in the time interval from 0 to t is $a = nf = tr$. The probability of k events is $p_k = a^k e^{-a}/k!$. The *cumulative* probability distribution function for the Poisson case is related to the incomplete gamma function (Press *et al.*, 2007).

6.1.3 *Gaussian (normal) distribution*

The normal or Gaussian distribution is another limiting case of the binomial distribution. Take the limit $n \to \infty$ with f finite, so that $nf \to \infty$. The probability distribution will peak near $k = nf$, therefore also $k \to \infty$. Expand $n!$ using Stirling's formula

$$n! \approx \sqrt{2\pi n} \left(\frac{n}{e}\right)^n \left(1 + \frac{1}{12n} + \frac{1}{288n^2} + \cdots\right). \qquad (6.17)$$

Expand $k!$ and $(n-k)!$ similarly. Then

$$p_k = \binom{n}{k} f^k (1-f)^{n-k} \approx \frac{1}{\sqrt{2\pi n}} \left(\frac{k}{n}\right)^{-k-\frac{1}{2}} \left(\frac{n-k}{n}\right)^{-n+k-\frac{1}{2}} f^k (1-f)^{n-k} \qquad (6.18)$$

$$= \frac{1}{\sqrt{2\pi n}} \exp\left[-(k+\tfrac{1}{2})\ln\frac{k}{n} - (n-k+\tfrac{1}{2})\ln\frac{n-k}{n} + k\ln f + (n-k)\ln(1-f)\right]. \qquad (6.19)$$

Looking at small deviations ξ around nf, let $k = nf + \xi$ where $\xi \ll nf$. Then

$$p_k \approx \frac{1}{\sqrt{2\pi n}} \frac{1}{\sqrt{f(1-f)}} \exp\left[-\frac{1}{2}\frac{\xi^2}{nf(1-f)}\right]. \qquad (6.20)$$

If we make the notational substitution $\sigma^2 = nf(1-f)$ and pass from the discrete to the continuous case, we get

$$p(\xi) = \frac{1}{\sqrt{2\pi}\,\sigma} e^{-\xi^2/2\sigma^2}. \qquad (6.21)$$

Gaussians may have any mean value, so we shift the Gaussian to have a mean value of μ and get the conventional form of the Gaussian distribution,

$$p(x) = \frac{1}{\sqrt{2\pi}\,\sigma} e^{-(x-\mu)^2/2\sigma^2}. \qquad (6.22)$$

Our notational choices were well made, since σ may be seen to be the standard deviation of this distribution. The cumulative probability of the normal distribution may be written in terms of the error function,

$$P(x) = \int_{-\infty}^x p(k)\,dk \qquad (6.23)$$

$$= \frac{1}{2}\left[1 + \text{erf}\left(\frac{x-\mu}{\sqrt{2}\sigma}\right)\right], \qquad (6.24)$$

where we are using the convention that the error function is defined as

$$\text{erf}(x) = \frac{2}{\sqrt{\pi}} \int_0^x e^{-t^2}\,dt. \qquad (6.25)$$

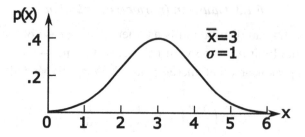

Figure 6.5 The Gaussian (normal) distribution for $\bar{x} = 3$, $\sigma = 1$.

The Gaussian distribution, or something close to it, is often encountered in experimental situations. An example is shown in Figure 6.5.

6.2 Moments of a probability distribution

Assume initially that we know the distribution $p(x)$. Applying the method of moments, the mean of the distribution is

$$\mu = \langle x \rangle = \int_{-\infty}^{\infty} x\, p(x)\, dx. \tag{6.26}$$

The variance is

$$\sigma^2 = \int_{-\infty}^{\infty} (x - \mu)^2\, p(x)\, dx = \langle x^2 \rangle - \langle x \rangle^2, \tag{6.27}$$

where σ is the standard deviation. The next higher central moment (i.e. with the mean subtracted) is related to the skew and the next is related to the kurtosis. Skew and kurtosis are often not robust indicators. The subject of robust estimation is discussed below in Section 6.8. The normal distribution is a two parameter distribution with a mean μ and a standard deviation σ, which are independent. The Poisson distribution is a one parameter distribution with a mean $\mu = a$, and a standard deviation $\sigma_x = \sqrt{a}$.

One can also apply the method of moments to a set of measurements. It is important to distinguish between the moments of the distribution, if known, and the moments of a data set.

6.3 Characteristic (moment-generating) function

In this section we will deviate from our conventions for the Fourier transform by omitting the factor of 2π from the kernel and by using the $+i$ transform as the forward transform. Taking the Fourier transform of the probability density,

$$\phi(k) = \int_{-\infty}^{\infty} p(x) \, e^{ikx} \, dx \tag{6.28}$$

$$= \int_{-\infty}^{\infty} p(x) \left[1 + ikx - \frac{1}{2!}k^2x^2 - \frac{i}{3!}k^3x^3 + \cdots \right] dx \tag{6.29}$$

$$= 1 + ik\langle x \rangle + \frac{(ik)^2}{2!}\langle x^2 \rangle + \frac{(ik)^3}{3!}\langle x^3 \rangle + \cdots. \tag{6.30}$$

In other words, this *characteristic* function $\phi(k)$ generates the moments of the distribution, which is why it is also known as the *moment-generating* function. For a normal distribution,

$$\phi(k) = \int_{-\infty}^{\infty} \frac{1}{\sqrt{2\pi}\,\sigma} \, e^{-(x-\mu)^2/2\sigma^2} \, e^{ikx} \, dx \tag{6.31}$$

$$= e^{ik\mu} \, e^{-k^2\sigma^2/2}. \tag{6.32}$$

Since the Fourier transform of a Gaussian is a Gaussian, the moment-generating function for the normal distribution is a Gaussian, multiplied by a complex phase factor due to the displacement of the mean from zero.

Let's look at some of the properties of this characteristic function. Let x have a probability density $p(x)$, and let y have probability density $q(y)$. Consider $z = f(x,y)$. The first two moments of z are

$$\langle z \rangle = \int \int_{-\infty}^{\infty} f(x, y) \, p(x) \, dx \, q(y) \, dy, \tag{6.33}$$

$$\langle z^2 \rangle = \int \int_{-\infty}^{\infty} [f(x, y)]^2 \, p(x) \, dx \, q(y) \, dy, \tag{6.34}$$

and the characteristic function of z is

$$\phi_z(k) = \int \int_{-\infty}^{\infty} e^{ik\,f(x,y)} \, p(x) \, dx \, q(y) \, dy. \tag{6.35}$$

Take as an example the function $f(x, y) = x + y$,

$$\phi_z(k) = \int \int_{-\infty}^{\infty} e^{ik(x+y)} \, p(x) \, dx \, q(y) \, dy \tag{6.36}$$

$$= \int_{-\infty}^{\infty} e^{ikx} \, p(x) \, dx \int_{-\infty}^{\infty} e^{iky} \, q(y) \, dy \tag{6.37}$$

$$= \phi_x(k) \, \phi_y(k). \tag{6.38}$$

The characteristic function of the sum of two independent random variables is the product of their individual characteristic functions.

6.4 Central limit theorem

The central limit theorem is largely attributable to Pierre-Simon Laplace. Consider a random variable x with a probability density $p(x)$, mean μ_x, variance σ_x^2, and unspecified higher moments. Subtracting the mean, we have the characteristic function

$$\phi_{x-\mu_x}(k) = \int e^{ik(x-\mu_x)} p(x) \, dx \tag{6.39}$$

$$= 1 - \frac{1}{2} k^2 \sigma_x^2 + O(k^3). \tag{6.40}$$

By considering $\phi_{x-\mu_x}$ we now have a function with zero mean (a power series expansion with no term proportional to k), a known second moment, and arbitrary higher moments. Now take n measurements of x and form the average,

$$a = \frac{1}{n} (x_1 + x_2 + \cdots + x_n). \tag{6.41}$$

What is the probability distribution of this average a, or of $a - \mu_x$? From the previous result for $f(x, y) = x + y$, generalize to

$$\Phi_{a-\mu_x}(k) = \left[\phi_{x-\mu_x}\left(\frac{k}{n}\right) \right]^n \tag{6.42}$$

$$= \left[1 - \frac{1}{2} \frac{k^2 \sigma_x^2}{n^2} + O\left(\frac{k^3}{n^3}\right) \right]^n, \tag{6.43}$$

and in the limit $n \to \infty$,

$$\Phi_{a-\mu_x}(k) = e^{-k^2 \sigma_x^2 / 2n}, \tag{6.44}$$

$$p(a) = \frac{\sqrt{n}}{\sqrt{2\pi} \sigma_x} e^{-n(a-\mu_x)^2 / 2\sigma_x^2}. \tag{6.45}$$

This is a normal distribution with mean $\mu_a = \mu_x$ and a standard deviation $\sigma_a = \sigma_x / \sqrt{n}$, *no matter what* the shape of the initial distribution $p(x)$. An example of a difficult initial distribution is shown in Figure 6.6. In other words, whenever you add or average a large number of measurements, the sum or average of the measurements approaches a normal (Gaussian) distribution. The width of the distribution narrows: the variance is reduced by a factor of n and the standard deviation is reduced by $\sqrt{n}$. This result, known as the *central limit theorem*, is valuable because we often choose to average or add results. And it explains why averaging is good; it gives a better estimate of the quantity we are trying to measure.

It is interesting to note that a Gaussian distribution is *least informative* in the information theoretic sense. Lacking additional information about the noise, a Gaussian is the best (least prejudiced) thing to assume.

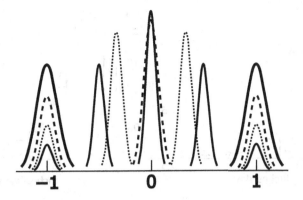

Figure 6.6 Qualitative convergence of a two-peaked distribution (thick black) towards a Gaussian after averaging two (dashed), three (dotted), and four (thin solid line; red in electronic version) draws.

But there are many situations in which the central limit theorem does not apply. For one thing, it works for random errors but not systematic errors. Systematic errors in general will not be representable by simple probability distributions. Consider, for example, the case of an incorrect scale factor, which no amount of averaging will correct. Even for random errors we need to be careful. Sometimes people misunderstand the central limit theorem and think that that it says something like "all random errors have Gaussian distributions." This is *false*. For one thing, the physical process producing the distribution may not be Gaussian. And mathematical operations other than addition do not make a distribution more Gaussian. The theorem does not apply to taking powers (x^n), multiplication of random variables (x times y), exponentiation (e^x), or any of a multitude of other operations. In fact, squaring takes Gaussian distributions and makes them non-Gaussian, as we will see in the next chapter. So understand carefully what the central limit theorem says. It has great value, but is not all encompassing.

6.5 Experimental data

An example of the central limit theorem as it applies to an actual data set is shown in Figure 6.7. Rainfall is the result of a complex set of physical phenomena involving, among other things, solar heating, cooling, humidity, wind, air turbulence, and the presence or absence of nucleation centers. No one can predict from first principles the probability distribution of whether or not it will rain, and if so, how much rain there will be. But let us assume that, on any given day, rainfall is governed by some underlying probability distribution. The observed statistics of daily

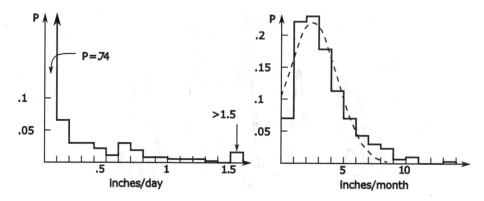

Figure 6.7 Rainfall statistics for Urbana, Illinois. (Left) For the year 1993, 74% of the days had less than 0.1 inches of rain. Six days had 1.5 inches or more of rain, which together added up to 15.32 inches of rain, 26% of the yearly rainfall. An outlier, a single day with 5.32 inches of rain, accounted for 10% of the yearly rainfall. (Right) Monthly rainfall statistics for 25 years (1984–2008). A qualitative fit using a truncated Gaussian is shown with a dashed line.

rainfall, in this example, follow a very lopsided distribution. But when averaged down to monthly rainfall statistics, the distribution is much smoother and more compact.

More generally, consider a random variable x described by some probability density $p(x)$ with *unknown* moments μ and σ. Unlike our earlier discussion, this is the more common case. If you knew the answer, why would you need to measure it? Measure n samples $x_1, x_2, \ldots, x_n$. We will explain later what *maximum likelihood* means. But for now, we state that the best (maximum likelihood) estimate of μ is

$$\bar{x} = \frac{1}{n} \sum_n x_i, \tag{6.46}$$

which we call the *sample mean*. The best *unbiassed estimate* (which we will also define later) of the variance σ^2 is

$$s^2 = \frac{1}{n-1} \sum_n (x_i - \bar{x})^2, \tag{6.47}$$

known as the *sample variance*. Although unbiassed, this is in fact not a maximum likelihood estimate. A maximum likelihood estimate would have a factor of $1/n$ in front. Although unbiassed and maximum likelihood both sound like desirable properties, one often cannot have both at the same time. Finally, note that if $p(x)$ is normal, then as we have just seen, $p(\bar{x})$ is also normal.

6.6 Chi-squared (χ^2) distribution

Assume a variable x is described by a probability distribution p(x) with *known* moments μ and σ. We make n measurements of x. Define a quantity known as chi-squared,

$$\chi^2 = \sum_{i=1}^{n} \left(\frac{x_i - \mu}{\sigma} \right)^2, \tag{6.48}$$

which is useful, among other things, for seeing how well a data set is described by a normal distribution. If we *assume* p(x) is a normal distribution, we can calculate the probability density of various values of χ^2 ($\chi^2 > 0$),

$$p_n \left(\chi^2 \right) = \frac{1}{2^{n/2} \Gamma(n/2)} \left(\chi^2 \right)^{\frac{n}{2}-1} e^{-\chi^2/2}. \tag{6.49}$$

The expectation value of χ^2 is the mean of the χ^2 distribution,

$$\langle \chi^2 \rangle = n, \tag{6.50}$$

which makes sense since each deviation from μ should be of order σ. Note that this is the *mean*, which is not equal to either the *mode* or the *median* of the distribution. One can define the *reduced* χ^2,

$$\chi_\nu^2 = \frac{\chi^2}{\nu}, \tag{6.51}$$

where if m parameters are determined from the data, then we refer to $\nu = n - m$ as the number of *degrees of freedom*. For example, m = 2 if the mean and the standard deviation are derived from the data instead of being assumed *a priori*. Then

$$\langle \chi^2 \rangle = \nu, \tag{6.52}$$

$$\langle \chi_\nu^2 \rangle = 1. \tag{6.53}$$

The chi-squared distribution is broad, with a variance

$$\sigma^2(\chi^2) = 2\nu. \tag{6.54}$$

The tail of the cumulative probability distribution of χ^2 is given by

$$P \left(\chi_\nu^2 > \chi_0^2 \right) = \int_{\chi_0^2}^{\infty} p_\nu \left(\chi^2 \right) d\chi^2. \tag{6.55}$$

A brief tabulation of this function is given in Table 6.1, with more extensive tabulations given in standard texts such as Bevington & Robinson (2003). As can be seen, the range of likely values of χ_ν^2 decreases with increasing ν. The cumulative probability $P \left(\chi_\nu^2 > 1 \right)$ depends on ν but is in the range 30–50%. The cumulative

Table 6.1. *Values of χ_ν^2 for which the probability of χ_ν^2 exceeding that value is equal to the probability at the top of each column. These values depend on ν, the number of degrees of freedom.*

ν	$P = 0.99$	0.95	0.90	0.50	0.10	0.05	0.01	0.001
1	0.0002	0.004	0.016	0.455	2.706	3.841	6.635	10.827
2	0.010	0.052	0.105	0.693	2.303	2.996	4.605	6.908
3	0.038	0.117	0.195	0.789	2.084	2.605	3.780	5.423
4	0.074	0.178	0.266	0.839	1.945	2.372	3.319	4.617
5	0.111	0.229	0.322	0.870	1.847	2.214	3.017	4.102
6	0.145	0.273	0.367	0.891	1.774	2.099	2.802	3.743
7	0.177	0.310	0.405	0.907	1.717	2.010	2.639	3.475
8	0.206	0.342	0.436	0.918	1.670	1.938	2.511	3.266
9	0.232	0.369	0.463	0.927	1.632	1.880	2.407	3.097
10	0.256	0.394	0.487	0.934	1.599	1.831	2.321	2.959
20	0.413	0.543	0.622	0.967	1.421	1.571	1.878	2.266
30	0.498	0.616	0.687	0.978	1.342	1.459	1.696	1.990
40	0.554	0.663	0.726	0.983	1.295	1.394	1.592	1.835
50	0.594	0.695	0.754	0.987	1.263	1.350	1.523	1.733
100	0.701	0.779	0.824	0.993	1.185	1.243	1.358	1.494

probability $P\left(\chi_\nu^2 > 2\right) \lesssim 1\%$ for $\nu > 15$. If the value of χ^2 is too improbable then something is wrong, either with our data, our estimates of the errors, or the assumption that the error distribution is normal.

The adoption of the reduced chi-squared does not remove all dependence on the number of degrees of freedom ν. So other texts such as Lyons (1991) and Abramowitz & Stegun (1970) avoid the use of χ_ν^2 entirely and leave everything in terms of χ^2. Although this has some cost in terms of the required dynamic range of the table, it is appealing for its simplicity and is the form we prefer. This version of the cumulative probability distribution of chi-squared is given in Table 6.2.

Chi-squared is often used not just for determining the mean and standard deviation of a set of numbers, but for fitting a function to a set of data. Consider the independent variable $\{x_i\}$ to be error free and the dependent variable $\{y_i\}$ to have uncertainties $\{\sigma_i\}$. In this case $f(x)$ is a parameterized functional fit to the data, and the definition of chi-squared is

$$\chi^2 = \sum_{i=1}^{n} \left(\frac{f(x_i) - y_i}{\sigma_i} \right)^2. \tag{6.56}$$

The functional parameters are calculated by minimizing χ^2 (Lyons, 1991). Unreasonable values of χ^2 may indicate an inappropriate choice of fitting function.

Table 6.2. *Values of χ^2 for which the probability of χ^2 exceeding that value is equal to the probability at the top of each column. These values depend on ν, the number of degrees of freedom.*

ν	$P = 0.99$	0.95	0.90	0.50	0.10	0.05	0.01	0.001
1	0.0002	0.004	0.016	0.455	2.706	3.841	6.635	10.828
2	0.020	0.103	0.211	1.386	4.605	5.991	9.210	13.816
3	0.115	0.352	0.584	2.366	6.251	7.815	11.345	16.266
4	0.297	0.711	1.064	3.357	7.779	9.488	13.277	18.467
5	0.554	1.145	1.610	4.351	9.236	11.071	15.086	20.515
6	0.872	1.635	2.204	5.348	10.645	12.592	16.812	22.458
7	1.239	2.167	2.833	6.346	12.017	14.067	18.475	24.322
8	1.646	2.733	3.490	7.344	13.362	15.507	20.090	26.125
9	2.088	3.325	4.168	8.343	14.684	16.919	21.666	27.877
10	2.558	3.940	4.865	9.342	15.987	18.307	23.209	29.588
20	8.260	10.851	12.443	19.337	28.412	31.410	37.566	45.315
30	14.954	18.493	20.599	29.336	40.256	43.773	50.892	59.703
40	22.164	26.509	29.051	39.336	51.805	55.759	63.691	73.402
50	29.707	34.764	37.689	49.335	63.167	67.505	76.154	86.661
100	70.065	77.930	82.358	99.334	118.498	124.342	135.807	149.449

6.7 Student's t-distribution

Suppose that we were told that $\{x_i\}$ were drawn from a Gaussian distribution and we were presented with the data set $\{2, 3, 4\}$. Using formulae from Section 6.5 we would estimate the mean of the distribution by the sample mean, $\bar{x} = 3$, and estimate the variance σ^2 by the sample variance, $s^2 = 1$. Suppose we were now asked "What is the probability that the underlying mean of the distribution $\mu > 4$?" It would be tempting to evaluate the tail of a Gaussian distribution such as

$$\int_4^\infty \frac{1}{\sqrt{2\pi}\,s} e^{-(x-3)^2/2s^2} \, dx = \frac{1}{2}\left[1 - \text{erf}\left(\frac{x-3}{\sqrt{2}\,s}\right)\right] = 0.159. \qquad (6.57)$$

Being told that this was incorrect, it might then occur to us that we should use a narrower Gaussian since the uncertainty in the mean is less than s by $\sqrt{n}$. We would get a probability of 0.042, which would still be incorrect. The basic problem is that s is only an *estimate* of σ. This situation calls for the use of Student's t-distribution rather than the Gaussian distribution. Define

$$t = \frac{\bar{x} - \mu}{s/\sqrt{n}}. \qquad (6.58)$$

Table 6.3. *Fractional value in a one-sided tail of Student's*
t-distribution

ν	t = 0.5	1.0	1.5	2.0	2.5	3.0
1	0.352	0.250	0.187	0.148	0.121	0.102
2	0.333	0.211	0.136	0.092	0.065	0.048
3	0.326	0.196	0.115	0.070	0.044	0.029
4	0.322	0.187	0.104	0.058	0.033	0.020
5	0.319	0.182	0.097	0.051	0.027	0.015
10	0.314	0.170	0.082	0.037	0.016	0.007
20	0.311	0.165	0.075	0.030	0.011	0.004
Gaussian	0.309	0.159	0.067	0.023	0.006	0.001

This quantity t is distributed according to Student's t-distribution

$$p(t, \nu) = \frac{\Gamma\left(\frac{\nu+1}{2}\right)}{\sqrt{\nu\pi}\,\Gamma\left(\frac{\nu}{2}\right)} \left(1 + \frac{t^2}{\nu}\right)^{-(\nu+1)/2}, \tag{6.59}$$

where $\nu = n-1$ is the number of degrees of freedom. A coarse tabulation of Student's t is given in Table 6.3. So to answer the sample question posed above, the probability is 0.113. Student's t-distribution approaches a Gaussian distribution as $\nu \to \infty$. One important feature of the t-distribution is that for small numbers of degrees of freedom it falls off much less rapidly than a Gaussian, as illustrated in Figure 6.8. There are additional uses of the t-distribution, discussed in standard statistics textbooks.

6.8 Robust estimation

We mentioned that the sample mean $\bar{x}$ was the best estimator of the central moment μ of a distribution. It is best in a maximum likelihood sense (see Chapter 13) if one assumes that the probability distribution is Gaussian. This can be derived by finding the value of $\bar{x}$ for which χ^2 is minimized. However, $\bar{x}$ is clearly *not* best if one requires immunity to the effects of large fluctuations, such as shown in Figure 6.9. For example, the median of a data set is more robust than the mean. It is less subject to change when a large fluctuation occurs.

The field of robust statistics deals with estimators that are more immune to fluctuations. It is particularly useful when there are errors which are not Gaussian distributed, often known as *outliers*. What is desired is a penalty function which realistically reflects the probability of outlying events. In the least-squares case the penalty function was quadratic. Each data point contributed a penalty in χ^2 proportional to the square of the deviation of that point from the mean. We would

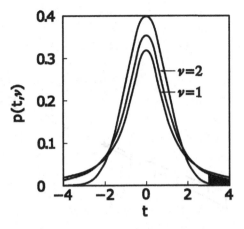

Figure 6.8 Probability density function for Student's t-distribution with $\nu = 1$ (red in electronic version), 2 (blue), and for the Gaussian distribution (black). Also illustrated by the shaded area is the meaning of a one-sided tail to a distribution.

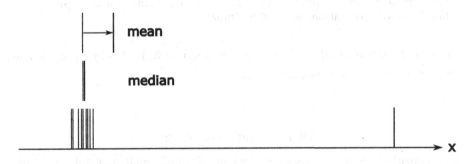

Figure 6.9 An outlier can greatly affect the average of a data set. The median is a more robust estimator than the mean.

like a penalty function which grows more gradually, if at all, beyond a certain point. Ideally the penalty function should be $-\log(\rho(x))$, where $\rho(x)$ is the true probability density. Possible empirical choices include

$$\sum_i \left| \frac{x_i - \bar{x}}{\sigma_i} \right|, \tag{6.60}$$

$$\sum_i \log \left(1 + \frac{1}{2} \left(\frac{x_i - \bar{x}}{\sigma_i} \right)^2 \right), \tag{6.61}$$

or almost any convex function. However, the minimization problem is considerably more complex in these cases and must be handled numerically. An example reflecting a typical astronomical situation is shown in Figure 6.10. The "outlying"

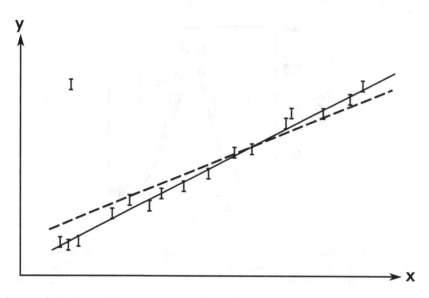

Figure 6.10 An outlier can greatly affect a least-squares fit, such as the straight line fit shown here with and without the outlier.

point strongly constrains the least-squares fit, and that fit is clearly wrong. Robust estimation gives a more reasonable result.

6.9 Propagation of errors

Let us consider how errors propagate when we apply mathematical operations. We will not consider the detailed shapes of the distributions, but only their moments: their means and especially their variances. Consider random variables u, $v, \ldots$, which have some probability distributions with means $\bar{u}, \bar{v}, \ldots$ and variances $\sigma_u^2, \sigma_v^2, \ldots$ Now consider a quantity x which is a function of these variables,

$$x = f(u, v, \ldots). \tag{6.62}$$

Roughly speaking, for small errors,

$$\bar{x} = f(\bar{u}, \bar{v}, \ldots). \tag{6.63}$$

Variations in the individual values $u_i, v_i, \ldots$ give rise to deviations of x_i from its mean value by

$$x_i - \bar{x} \approx (u_i - \bar{u})\frac{\partial x}{\partial u} + (v_i - \bar{v})\frac{\partial x}{\partial v} + \cdots. \tag{6.64}$$

The variance in x is given by

$$\sigma_x^2 = \lim_{n\to\infty} \frac{1}{n} \sum_{i=1}^{n} (x_i - \bar{x})^2 \qquad (6.65)$$

$$= \lim_{n\to\infty} \frac{1}{n} \sum_{i=1}^{n} (u_i - \bar{u})^2 \left(\frac{\partial x}{\partial u}\right)^2 + (v_i - \bar{v})^2 \left(\frac{\partial x}{\partial v}\right)^2 + 2(u_i - \bar{u})(v_i - \bar{v}) \frac{\partial x}{\partial u} \frac{\partial x}{\partial v} + \cdots. \qquad (6.66)$$

Defining the variances σ_u^2 and σ_v^2 in the usual way,

$$\sigma_u^2 = \lim_{n\to\infty} \frac{1}{n} \sum_{i=1}^{n} (u_i - \bar{u})^2, \qquad (6.67)$$

$$\sigma_v^2 = \lim_{n\to\infty} \frac{1}{n} \sum_{i=1}^{n} (v_i - \bar{v})^2. \qquad (6.68)$$

We also define a quantity known as the covariance σ_{uv}^2,

$$\sigma_{uv}^2 = \lim_{n\to\infty} \frac{1}{n} \sum_{i=1}^{n} [(u_i - \bar{u})(v_i - \bar{v})]. \qquad (6.69)$$

Putting these together we get

$$\sigma_x^2 \approx \sigma_u^2 \left(\frac{\partial x}{\partial u}\right)^2 + \sigma_v^2 \left(\frac{\partial x}{\partial v}\right)^2 + 2\sigma_{uv}^2 \left(\frac{\partial x}{\partial u}\right)\left(\frac{\partial x}{\partial v}\right). \qquad (6.70)$$

If, in addition, u and v are uncorrelated, which means that $\sigma_{uv}^2 = 0$, then

$$\sigma_x^2 = \sigma_u^2 \left(\frac{\partial x}{\partial u}\right)^2 + \sigma_v^2 \left(\frac{\partial x}{\partial v}\right)^2. \qquad (6.71)$$

If u and v are inherently correlated, or are calculated from other variables in a way which makes them correlated, then one must either measure or calculate the covariance and use the full formula for σ_x^2.

Here are a few specific examples. If we form a linear combination of the variables u and v, then we need to sum the variances, appropriately weighted by the constants a and b,

$$x = au + bv, \qquad \sigma_x^2 = a^2 \sigma_u^2 + b^2 \sigma_v^2. \qquad (6.72)$$

If we multiply or divide random variables, then we sum the *fractional* variances,

$$x = auv, \qquad \frac{\sigma_x^2}{x^2} = \frac{\sigma_u^2}{u^2} + \frac{\sigma_v^2}{v^2}, \qquad (6.73)$$

$$x = a\frac{u}{v}, \qquad \frac{\sigma_x^2}{x^2} = \frac{\sigma_u^2}{u^2} + \frac{\sigma_v^2}{v^2}. \qquad (6.74)$$

If we take a random variable and raise it to some power, then the fractional error is increased by that power,

$$x = au^b, \qquad \frac{\sigma_x}{x} = b\frac{\sigma_u}{u}. \tag{6.75}$$

And if we exponentiate,

$$x = ae^{bu}, \qquad \frac{\sigma_x}{x} = b\sigma_u. \tag{6.76}$$

Exercises

6.1 Derive Equation 6.20 from Equation 6.19. Be careful to keep all terms of order ξ^2.

6.2 Assume uncertainties in measured parallax follow a Gaussian distribution. Use the small angle approximation for parallax, $\Pi(\text{arcsec}) = 1/D(\text{pc})$.

    a. Star A has a measured parallax of 10 ± 1 mas (milliarcsecond). What is its distance and the uncertainties in the distance?

    b. Star B has a measured parallax of 1 ± 1 mas. What is its distance and the uncertainties in the distance?

    c. Star C has a measured parallax of -1 ± 1 mas. What is its distance and the uncertainty in the distance? Note that while a negative parallax is unphysical, a *measured* parallax (including errors) may be negative.

6.3 Calculate the expectation value of χ^2 by integrating Equation 6.49.

6.4 You are suspicious that the errors in your data are not Gaussian. A Gaussian probability distribution is defined by

$$p(x) = \frac{1}{\sqrt{2\pi}\sigma} e^{-(x-\mu)^2/2\sigma^2}. \tag{6.77}$$

    a. From the characteristic function $\phi(k)$, show that when $\mu = 0$ there are only even moments to the distribution.

    b. Still for the case $\mu = 0$, show how the fourth moment $\langle x^4 \rangle$ is related to the second moment $\langle x^2 \rangle$.

This is one test you might apply to check for Gaussianity.

7

Stochastic processes and noise

7.1 Stochastic process

Consider a real variable x whose value is time dependent and also random, $x(\xi,t)$, where ξ is some particular realization of $x(t)$. A *stochastic process* refers to the set of possible outcomes $x(\xi,t)$. Do not assume that for any particular value of ξ the process necessarily looks "random" or "noise like." It is possible that $\xi = 1$ corresponds to a sine wave, $\xi = 2$ corresponds to a square wave, etc. The randomness corresponds to the selection of a particular outcome, not to its time evolution. We cannot in general assign probabilities p_ξ to each particular outcome because there may be an infinite number of such outcomes, each with infinitesimal probability. We also cannot assign a probability density in the sense $p(\xi)\,d\xi$ because in general the ξ are not ordered, or even orderable.

Instead, for a fixed moment in time t, we can define the statistical properties of x by the usual probability density, $p(x,t)$. This enables us to speak of expectation values of x, such as the ensemble mean (the mean over the ensemble $\{\xi\}$),

$$\eta(t) = E\{x(t)\} = \int_{-\infty}^{\infty} x(\xi, t)\, p(x, t)\, dx, \tag{7.1}$$

and the ensemble variance,

$$\sigma^2(t) = E\{[x(t)]^2\} - \{E[x(t)]\}^2 \tag{7.2}$$

$$= \int_{-\infty}^{\infty} [x(\xi, t)]^2\, p(x, t)\, dx - [\eta(t)]^2. \tag{7.3}$$

In addition, the fact that x is time dependent makes it useful to consider how its values at two different times are related. We define the autocorrelation of x as the expectation value

$$R(t_1, t_2) = E\{x(t_1)x(t_2)\} \tag{7.4}$$

$$= \int \int_{-\infty}^{\infty} x_1 \, x_2 \, p(x_1, x_2, t_1, t_2) \, dx_1 \, dx_2, \tag{7.5}$$

and the autocovariance

$$C(t_1, t_2) = E\{[x(t_1) - \eta(t_1)] \, [x(t_2) - \eta(t_2)]\} \tag{7.6}$$

$$= \int \int_{-\infty}^{\infty} (x_1 - \eta_1)(x_2 - \eta_2) \, p(x_1, x_2, t_1, t_2) \, dx_1 \, dx_2 \tag{7.7}$$

$$= R(t_1, t_2) - \eta(t_1)\eta(t_2). \tag{7.8}$$

Note that

$$\sigma^2(t) = C(t, t) = R(t, t) - \eta(t)^2. \tag{7.9}$$

7.1.1 Stationary process

A *stationary process* is one in which $p(x, t)$ is independent of time, in which case its moments, such as $\eta(t)$, are also independent of time. This is roughly equivalent to saying that the autocorrelation $R(t_1, t_2)$ depends only on the time difference $\tau = t_2 - t_1$,

$$R(t_1, t_2) = R(\tau). \tag{7.10}$$

The autocorrelation of a real stationary process is an even function, $R(-\tau) = R(\tau)$, and $R(0)$ corresponds to the average power of the process. If a process is stationary, one can speak of its time average

$$\langle x \rangle = \lim_{T \to \infty} \frac{1}{T} \int_{-T/2}^{T/2} x(t) \, dt. \tag{7.11}$$

We will limit ourselves to considering only stationary processes.[1] Under some conditions the ergodic hypothesis applies, that is, time and ensemble averages are equivalent, $\langle x \rangle = \eta$.

7.2 Spectral density of a Poisson random process

In a Poisson random process, such as shot noise, one has discrete events occurring at random times. Assume that we have a stationary Poisson process. The autocorrelation function depends only on the time difference,

$$R(t_1, t_2) = R(\tau). \tag{7.12}$$

---

[1] Except for the brief description already given in Chapter 1.

For $\tau > 0$, events separated by τ are statistically independent, and therefore their autocorrelation cannot depend on τ and must equal some constant c_1:

$$R(\tau) = c_1. \tag{7.13}$$

For $\tau = 0$, each event correlates perfectly with itself, which can be described by a delta function times some constant c_2, so that

$$R(\tau) = c_1 + c_2\,\delta(\tau). \tag{7.14}$$

Both c_1 and c_2 depend only on the average counting rate r, in much the same way as $\bar{x}$ and σ_x (the mean number of counts seen after a time T and the standard deviation in that number) depended only on r. The only question is how c_1 and c_2 depend on r. Visualize this by considering time bins of width Δt where an event produces a signal of height $1/\Delta t$ (a rectangular pulse of unit area), as in Figure 7.1. The autocorrelation function can then be reduced to a summation. For an event rate r,

$$R = \int \int x_1\,x_2\,p(x_1, x_2, t_1, t_2)\,dx_1 dx_2. \tag{7.15}$$

For $\tau > 0$,

$$R = \sum\sum (\Delta t)^{-1}(\Delta t)^{-1}(r\,\Delta t)^2 = r^2. \tag{7.16}$$

And for $\tau = 0$,

$$R = \sum\sum (\Delta t)^{-1}(\Delta t)^{-1}(r\,\Delta t) = r(\Delta t)^{-1}. \tag{7.17}$$

In the limit $\Delta t \to 0$, this reduces to

$$R(\tau) = r^2 + r\,\delta(\tau). \tag{7.18}$$

For additional detail see Helstrom (1991) and Papoulis (1991).

By the Wiener–Khinchin theorem, the power spectral density is the Fourier transform of $R(\tau)$. Since the Fourier transform of a constant is a delta function and vice versa,

$$S(v) = r^2\,\delta(v) + r. \tag{7.19}$$

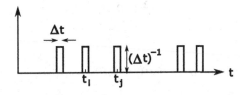

Figure 7.1 Autocorrelation of a Poisson random process, to be taken in the limit $\Delta t \to 0$.

The spectrum of a Poisson random process (shot noise) is white (flat) except at $\nu = 0$. The result is more complex if the events are not instantaneous but spread out in time (such as current pulses from a photomultiplier tube).

7.3 Spectral density of a Gaussian random process

A noise process $x(t)$ is called a Gaussian random process (GRP) if the probability density $p(x(t))$ and *all* joint (multivariate) distributions $p(x_1(t_1), x_2(t_2))$, $p(x_1(t_1), x_2(t_2), x_3(t_3))$, ... are Gaussian. A physical realization of such a process may be a very rapid succession of overlapping Poisson impulses, each of very small amplitude (Helstrom, 1991; Papoulis, 1991). By the central limit theorem, the sum of these impulses will follow a Gaussian distribution. Assume such a process is stationary and that $\eta(t) = 0$. It turns out that the *entire* process is then specified by the autocorrelation function (Mathews & Walker, 1970),

$$R(t_1, t_2) = E\{x(t_1)\,x(t_2)\}, \tag{7.20}$$

and for a stationary process,

$$R(\tau) = E\{x(t)\,x(t+\tau)\}. \tag{7.21}$$

The power spectral density (PSD) is the Fourier transform of this autocorrelation function (ACF),

$$S(\nu) = \mathcal{F}\,R(\tau). \tag{7.22}$$

Since Gaussian random processes can have a variety of ACFs, they can also have a variety of PSDs. Based on the model above, in which a Gaussian process is formed from rapid Poisson impulses, one might expect a white PSD. However, we can filter a Gaussian process so as to change its PSD while retaining its Gaussian statistics. In summary, a Gaussian *distribution* is specified by its mean and its variance. A Gaussian random *process* is specified by its mean and its autocorrelation function, R. If $\eta = 0$,

$$\sigma_x^2 = R(0). \tag{7.23}$$

If $\eta \neq 0$,

$$\sigma_x^2 = C(0) = R(0) - \eta^2. \tag{7.24}$$

A Gaussian random process may have any PSD. In contrast, remember that a Poisson random process has a white PSD and is specified by its counting rate r.

7.4 The transformation $y = x^2$

Assume the random variable x is real valued and normally distributed with zero mean and variance σ_x^2,

$$p_x(x) = \frac{1}{\sigma_x\sqrt{2\pi}}\, e^{-x^2/2\sigma_x^2}. \tag{7.25}$$

Now take the function $y = x^2$. There obviously is zero probability of observing $y < 0$. For $y > 0$, consider Figure 7.2, indicating the relationship between the width of an interval dx and the width of the corresponding interval dy,

$$\frac{dy}{dx} = 2x = 2\sqrt{y}. \tag{7.26}$$

Allowing for positive and negative values of x,

$$p_y(y)\, dy = 2\, p_x(x)\, dx. \tag{7.27}$$

The probability density for y ($y > 0$) is then given by

$$p_y(y) = \frac{1}{\sqrt{y}}\, p_x(x) = \frac{1}{\sigma_x\sqrt{2\pi y}}\, e^{-y/2\sigma_x^2}, \tag{7.28}$$

which is distinctly *non*-Gaussian, with moments

$$\langle y \rangle = \sigma_x^2, \tag{7.29}$$

$$\sigma_y^2 = 2\sigma_x^4. \tag{7.30}$$

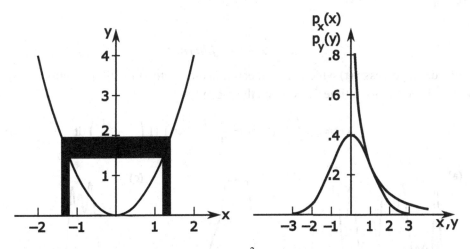

Figure 7.2 (Left) For the function $y = x^2$, intervals $x \pm dx$ (red in electronic version) are mapped into an interval $y \pm dy$ (blue). (Right) A Gaussian probability distribution $p_x(x)$ with $\sigma_x = 1$ (red) and the corresponding probability distribution $p_y(y)$ (blue).

If we are dealing with a Gaussian *process*, these results can be combined with results from earlier in this chapter to give

$$R_y(0) = 3\,R_x^2(0). \tag{7.31}$$

With a little more work it can be shown that

$$R_y(\tau) = R_x^2(0) + 2\,R_x^2(\tau). \tag{7.32}$$

7.5 Filtering

A filtering of some process $x(t)$ is defined by its convolution with some filter function $h(t)$, where to insure causality, $h(t) = 0$ for $t < 0$.

$$y(t) = x(t) * h(t) \tag{7.33}$$

$$= \int_{-\infty}^{\infty} x(t - \theta)\,h(\theta)\,d\theta. \tag{7.34}$$

By the convolution theorem,

$$\tilde{y}(s) = \tilde{x}(s)\,\tilde{h}(s). \tag{7.35}$$

In terms of power, $H(s) = |\tilde{h}(s)|^2$, the absolute square of the transfer function $\tilde{h}(s)$, describes how the power spectrum of the signal is modified, as shown in Figure 7.3. But the power transfer function $H(s)$ is not a complete description since it does not include any phase information. For a complete description of the filtering process we require $\tilde{h}(s)$ or $h(t)$.

7.5.1 Low pass filtering

Consider a process $x(t)$ which has a mean η and an autocorrelation function $R_x(t)$. After observing for a time T, estimate the mean by

$$x_T = \frac{1}{T} \int_0^T x(t)\,dt = \frac{1}{T} \int_{-\infty}^{\infty} x(t)\,\Pi\left(\frac{1}{2} - \frac{t}{T}\right) dt \tag{7.36}$$

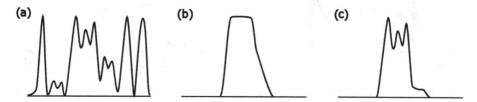

(a) **(b)** **(c)**

Figure 7.3 (a) Unfiltered power spectrum $|\tilde{x}(s)|^2$. (b) Filter passband $|\tilde{h}(s)|^2$. (c) Filtered power spectrum $|\tilde{y}(s)|^2$.

$$= \frac{1}{T} x(t) * \Pi \left(\frac{t}{T} \right) \bigg|_{t=T/2}. \tag{7.37}$$

The displacement of the boxcar by $T/2$ is trivial and we will ignore it. The uncertainty (variance) in x_T is given by

$$\sigma_{x_T}^2 = \frac{1}{T} \int_{-\infty}^{\infty} \left[R_x(\tau) - \eta^2 \right] \Lambda \left(\frac{\tau}{T} \right) d\tau. \tag{7.38}$$

The derivation of this result will be left for the exercises at the end of this chapter. It is clear that the variance drops as $1/T$, as expected, but to evaluate the integral further we would need more information about the nature of the process.

So now let x have $\eta = 0$ and a white spectrum ($S_x(f) = $ constant), and let x be filtered by an ideal low pass filter, as in Figure 7.4,

$$\tilde{y}(f) = \tilde{x}(f) \, \tilde{h}(f), \tag{7.39}$$

$$\tilde{h}(f) = \Pi \left(\frac{f}{2f_c} \right). \tag{7.40}$$

If we estimate the mean, as before,

$$y_T = \frac{1}{T} \int_0^T y(t) \, dt. \tag{7.41}$$

The variance in y_T is given by

$$\sigma_{y_T}^2 = \frac{1}{T} \int_{-\infty}^{\infty} S_y(f) \, |\mathrm{sinc}(Tf)|^2 \; T \, df. \tag{7.42}$$

The proof of this will also be left for the exercises at the end of the chapter. So for $T \gg 1/f_c$,

$$\sigma_{y_T}^2 \approx \frac{1}{T} S_y(0) = \frac{1}{T} S_x(0). \tag{7.43}$$

And because $\sigma_y^2 = R_y(0) = \int S_y(f) \, df = 2f_c S_x(0)$,

$$\sigma_{y_T}^2 = \frac{\sigma_y^2}{2f_c T}. \tag{7.44}$$

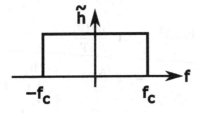

Figure 7.4 Ideal low pass filter.

7.6 Estimation in the presence of Gaussian noise

Consider a fixed signal x_S and random noise $x_N(t)$,

$$x(t) = x_S + x_N(t). \tag{7.45}$$

Assume that $x_N(t)$ has zero mean and an autocorrelation function $R(\tau)$. If $R(\tau)$ is strongly peaked at $\tau = 0$, then $S(\nu)$ can be approximately constant out to some frequency f_c. After some time T,

$$\bar{x} = \frac{1}{T} \int_0^T x(t)\, dt \to x_S, \tag{7.46}$$

and it will have a variance

$$\sigma_{\bar{x}}^2 = \frac{R(0)}{2 f_c T}. \tag{7.47}$$

Note that the noise decreases as $1/\sqrt{T}$. Since the noise is assumed to be Gaussian and we know $\bar{x}$ and σ_x^2, we know the entire distribution (all the moments).

7.7 Photon noise

We have seen that blackbody radiation can be described by an energy density

$$u_\nu = \frac{8\pi}{c^3} \frac{h\nu^3}{e^{h\nu/kT} - 1} \tag{7.48}$$

and a specific intensity

$$I_\nu = \frac{2h\nu^3}{c^2} \frac{1}{e^{h\nu/kT} - 1}. \tag{7.49}$$

If we restrict ourselves to an étendue of $A\Omega = \lambda^2$ and a single polarization, equivalent to considering a single mode, we get a mean power of

$$\bar{P}(\nu) = h\nu \frac{1}{e^{h\nu/kT} - 1}, \tag{7.50}$$

which is just the energy per photon times the photon occupation number. From thermodynamic considerations it can be shown that the fluctuations in the power are

$$\langle [P(\nu) - \bar{P}(\nu)]^2 \rangle = kT^2 \frac{d\bar{P}(\nu)}{dT} \tag{7.51}$$

$$= kT^2 \bar{P}(\nu)\, e^{h\nu/kT} \frac{h\nu}{kT^2} \frac{1}{e^{h\nu/kT} - 1} \tag{7.52}$$

$$= \bar{P}(\nu)\, h\nu \left[1 + \frac{1}{e^{h\nu/kT} - 1} \right]. \tag{7.53}$$

In the field of engineering one would usually write this as

$$\langle \left[P(v) - \bar{P}(v) \right]^2 \rangle = \bar{P}(v)\, h v \left[\frac{1}{2} + \frac{1}{2} \coth \frac{h v}{2kT} \right]. \tag{7.54}$$

In the limit $h v \gg kT$ we have photon noise. The photons are independent and obey Poisson statistics, with fluctuations proportional to the square root of the number of photons,

$$\langle \Delta P^2 \rangle = \bar{P} h v \tag{7.55}$$

$$\langle \Delta n^2 \rangle = \langle n \rangle. \tag{7.56}$$

In the limit $h v \ll kT$ we have thermal noise. There are many photons per mode, their arrival times are correlated (they are bosons,[2] so they are bunched), and the fluctuations are larger:

$$\langle \Delta P^2 \rangle = \bar{P} kT = \bar{P}^2 \tag{7.57}$$

$$\langle \Delta n^2 \rangle = \langle n \rangle^2. \tag{7.58}$$

An expression which correctly links these two limits is

$$\langle \Delta n^2 \rangle = \langle n \rangle \left(1 + \langle n \rangle \right). \tag{7.59}$$

7.8 Thermal noise

The blackbody power spectral density (power per mode) of thermal noise is approximately white at low frequencies, $h v \ll kT$,

$$P(v) \approx kT. \tag{7.60}$$

If this is rolled off with a low pass filter such as

$$H(v) = \Pi \left(\frac{v}{2 v_c} \right), \tag{7.61}$$

as in Figure 7.5, the resulting variance is

$$\sigma^2 = kT\, 2 v_c. \tag{7.62}$$

In real life, the coefficient is dependent on the exact shape of the filter cutoff. For an RC filter, where $v_c = 1/(2\pi RC)$,

$$H(v) = \frac{1}{1 + (v/v_c)^2}, \tag{7.63}$$

$$\sigma^2 = kT\, \pi\, v_c, \tag{7.64}$$

---

[2] For Fermi–Dirac statistics, this would be replaced by $1/(e^{E/kT} + 1)$.

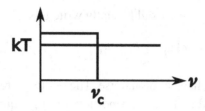

Figure 7.5 Thermal noise passed by a low pass filter.

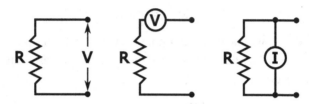

Figure 7.6 Johnson noise of a resistor R at temperature T and the Thévenin and Norton equivalent circuits.

with an autocorrelation function

$$R(\tau) = kT\,\pi\,\nu_c\,e^{-2\pi\,\nu_c|\tau|}. \tag{7.65}$$

A special case is Johnson noise (the open circuit voltage noise of a resistor), shown in Figure 7.6. The real resistor at temperature T may be represented by the Thévenin equivalent circuit containing an ideal (noiseless) resistor in series with a noise voltage source of amplitude V. Consider connecting this equivalent circuit, via a transmission line of impedance R and electrical length L, to a load resistor of resistance R, also at temperature T (Nyquist, 1928). The voltage source sees an impedance 2R, resulting in a current of V/(2R). The impedance matched circuit can be thought of as containing two modes, in each of which there is a traveling wave carrying power kT $\Delta\nu$ in opposite directions. (If L > c/$\Delta\nu$ there are more modes and more energy on the transmission line, but it is delivered at the same rate.) Equating the power dissipated in the original resistor with the power delivered, we get

$$\langle V^2\rangle = 4kT\,R\,\Delta\nu. \tag{7.66}$$

Using instead the Norton equivalent circuit containing a noise current source we would get

$$\langle I^2\rangle = 4kT\,\Delta\nu/R. \tag{7.67}$$

Exercises

7.1 Consider a time-dependent variable x(t) which consists of some fixed signal plus some stationary random noise,

$$x(t) = s + n(t). \tag{7.68}$$

Assume that $\langle n(t) \rangle = 0$. Define the running mean $y_T(t)$ as the average of x(t) over the preceding T seconds,

$$y_T(t) = \frac{1}{T} \int_{t-T}^{t} x(\theta) \, d\theta. \tag{7.69}$$

a. Formulate this running mean as a convolution and show that this constitutes a linear filtering of x(t) in the sense of Equation 7.33.
b. Calculate the transfer function $\tilde{h}(f)$ and its bandwidth Δf,

$$\Delta f = \int_{-\infty}^{\infty} |\tilde{h}(f)|^2 df. \tag{7.70}$$

This *running* mean is subtly different than the time averaging we did in Section 7.5.1.[3]

7.2 White noise is filtered by a narrow bandpass filter centered at frequency ν_0 with width $\Delta\nu$ ($\Delta\nu \ll \nu_0$). Its power transfer function is given by

$$H(\nu) = \frac{1}{2} \left[e^{-(\nu-\nu_0)^2/2(\Delta\nu)^2} + e^{-(\nu+\nu_0)^2/2(\Delta\nu)^2} \right]. \tag{7.71}$$

Calculate and sketch the autocorrelation function $R(\tau)$ after the filtering.

7.3 Synchronous detection is a technique by which modulation of known frequency and phase is imposed on a weak signal, sometimes by a mechanical "chopper wheel" which alternately passes and blocks incoming radiation. The signal out of the detector is given by

$$x_0(t) = s(t)F(t) + n(t), \tag{7.72}$$

where s(t) is the incident signal, F(t) is the imposed modulation, and the noise n(t) is assumed to be intrinsic to the detector (and therefore unmodulated). The detector output is passed through a prefilter

$$x_1(t) = x_0(t) * h(t), \tag{7.73}$$

where $|\tilde{h}(f)|^2$ is a narrow bandwidth filter centered at the modulation frequency f_0 with a width Δf_0. The filtered output is then demodulated by the operation

$$x_2(t) = x_1(t) \cos(2\pi f_0 t + \phi') \tag{7.74}$$

[3] Adapted from Lèna *et al.* (1998).

and low pass filtered by taking a running mean

$$x_3(t) = x_2(t) * \frac{1}{T}\Pi(t/T). \tag{7.75}$$

The integration time T is typically chosen to be much greater than the reciprocal of the prefilter bandwidth ($T \gg 1/\Delta f_0$). The effect of this signal processing is to largely recover the original signal while removing those components of the noise with frequencies or phases different than that of the imposed modulation.

a. Consider first just the noise $n(t)$. If $n(t)$ has a white spectrum with power density S_N, what are the effects of the various stages of the signal processing? Determine the variance of x_3 in terms of S_N.

b. If $F(t)$ is strictly periodic, but not necessarily sinusoidal, what is the most general form of $\tilde{F}(f)$?

c. Assuming the signal $s(t)$ is just a constant s, what are the effects of the various stages of the processing? What is the optimum value of ϕ'? Calculate the loss of signal for square wave modulation and co-sinusoidal demodulation.[4]

7.4 Consider a stationary Gaussian random process $x(t)$ with zero mean, described by an autocorrelation function (before filtering) of $R_x(\tau)$. This process is filtered by a simple RC filter with a *power* transfer function of

$$H(v) = \frac{1}{1 + (2\pi RCv)^2}. \tag{7.76}$$

Find the autocorrelation function after filtering, $R(\tau)$, in terms of $R_x(\tau)$ and $\tilde{H}(v)$. You will need the Fourier transform pair

$$e^{-|x|} \rightleftharpoons \frac{2}{1 + (2\pi s)^2}. \tag{7.77}$$

Your answer should be in the form of an integral containing R_x.[5]

7.5 Derive the result in Equation 7.38.

7.6 Derive the result in Equation 7.42.

---

[4] Adapted from Lèna *et al.* (1998).
[5] Adapted from Lèna *et al.* (1998).

8

Optics

We begin our treatment of optics by first considering geometrical optics. In the limit of small wavelengths, geometrical optics describes the direction in which light travels through space as it encounters materials with different indices of refraction. Initially the refractive index will simply be assumed to be a property of a material which describes the speed at which light propagates in that material. That will be sufficient to allow us to treat the theory of aberrations and to look into some basic aspects of telescope design. Next we will look at the physical origins of the refractive index and at the Fresnel coefficients, which are important in a number of contexts including the design of various spectroscopic devices. Then we will consider physical optics, the behavior of light in the regime of finite wavelengths where diffractive effects become important. This will include a look at the Airy pattern. Finally we will introduce the concepts of the point spread function and the modulation transfer function and use them to consider some general properties of imaging.

8.1 Geometrical optics

The properties of light propagation can often usefully be described by *geometrical optics*, an approximation which is valid in the limit of small wavelengths. The wavelength λ is assumed to be small compared with all relevant length scales, including the dimensions of any physical objects present. The media of propagation are described by various values of the refractive index n, which in general is wavelength dependent. In this approximation it is possible to visualize the individual paths followed by narrow pencils of light, a process known as ray tracing. These rays are considered to have small cross sectional area A and small divergence Ω and therefore small étendue. Ray tracing depends on two basic laws: the law of reflection and Snell's law.

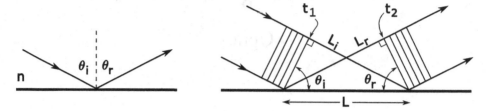

Figure 8.1 The law of reflection in geometrical optics from the ray viewpoint (left) and the wavefront viewpoint (right).

The law of reflection describes the behavior of a ray of light which encounters an interface between different media, as shown in Figure 8.1. The angles of incidence (θ_i) and reflection (θ_r) are defined with respect to the normal to the surface. To determine the relationship between these angles, it is easiest to consider a wavefront picture. To do so we need only two properties: (1) that the direction of propagation is perpendicular to the wavefront and (2) that the velocity of propagation (the phase velocity) is c/n. Incoming plane-parallel light in a medium of refractive index n encounters some interface, and at least part of that light is reflected. Considering a finite section of the wavefront bounded by two limiting rays, a particular wavefront first encounters the interface at time t_1. The remainder of that section of wavefront continues until it all has reached the surface, at time t_2. That last portion of the wave travels an extra distance

$$L_i = \frac{c}{n}(t_2 - t_1), \tag{8.1}$$

and by geometrical construction

$$\sin\theta_i = \frac{L_i}{L}. \tag{8.2}$$

Similarly for the outgoing wave,

$$L_r = \frac{c}{n}(t_2 - t_1), \tag{8.3}$$

$$\sin\theta_r = \frac{L_r}{L}. \tag{8.4}$$

Combining these results we readily see that

$$\theta_i = \theta_r, \tag{8.5}$$

dependent only on geometry and the constancy of the speed of propagation.

Snell's law, the law of refraction, considers the situation in which some portion of the light propagates from a medium of index n_i into a medium of index n_t, as

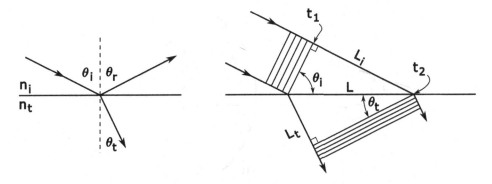

Figure 8.2 Snell's law from the ray viewpoint (left) and the wavefront viewpoint (right).

shown in Figure 8.2. In each material the velocity of propagation is c/n. As before, converting to a wavefront picture we get

$$L_i = \frac{c}{n_i}(t_2 - t_1) = L \sin \theta_i, \tag{8.6}$$

$$L_t = \frac{c}{n_t}(t_2 - t_1) = L \sin \theta_t, \tag{8.7}$$

from which it is easy to see that

$$n_i \sin \theta_i = n_t \sin \theta_t. \tag{8.8}$$

Both of these results are independent of how much power is reflected or transmitted in each case. The possible existence of a transmitted wave does not affect the law of reflection, and the existence of a reflected wave does not affect Snell's law.

8.1.1 Paraxial optics (a first order theory)

Paraxial optics refers to cases in which rays encounter interfaces at small angles of incidence (near the optical axis). Figure 8.3 illustrates the case of a single refractive interface, with the angles exaggerated for clarity. We mostly follow the notation of Schroeder (2000) except for differences in sign conventions. The situation shown is similar to the case of a lens, although a lens has a second refractive interface. Consider rays emerging from the point P in a medium of index n, hitting a spherical refracting surface of radius R, and being bent towards P′ in a medium of index n′. The dashed line is drawn from the center of curvature out through the surface and therefore is normal to the surface. Snell's law gives

$$n \sin \theta = n' \sin \theta', \tag{8.9}$$

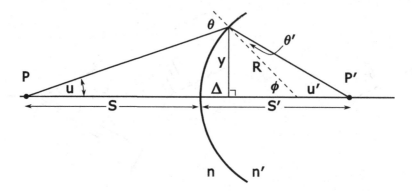

Figure 8.3 Paraxial optics illustrated for a single refracting surface.

which for small angles is

$$n\,\theta \approx n'\,\theta', \tag{8.10}$$

since $\sin\theta \approx \theta$. Now by geometrical construction,

$$\theta = 180° - (90° - u) - (90° - \phi) = u + \phi, \tag{8.11}$$

$$\theta' = (90° - u') - (90° - \phi) = -u' + \phi. \tag{8.12}$$

Plugging into the small-angle version of Snell's law, we get

$$n(u + \phi) \approx n'(-u' + \phi). \tag{8.13}$$

Reexpressing the angles in terms of distances,

$$\phi = \sin^{-1}\frac{y}{R} \approx \frac{y}{R}, \tag{8.14}$$

$$u = \tan^{-1}\frac{y}{S + \Delta} \approx \frac{y}{S + \Delta} = \frac{y}{S + R - R\cos\phi} \approx \frac{y}{S}, \tag{8.15}$$

$$u' = \tan^{-1}\frac{y}{S' - \Delta} \approx \frac{y}{S' - \Delta} = \frac{y}{S' - R + R\cos\phi} \approx \frac{y}{S'}. \tag{8.16}$$

Putting it all together,

$$n\left(\frac{y}{S} + \frac{y}{R}\right) \approx n'\left(-\frac{y}{S'} + \frac{y}{R}\right), \tag{8.17}$$

$$\frac{n}{S} + \frac{n'}{S'} = \frac{n' - n}{R}, \tag{8.18}$$

which holds independent of y. The important conclusion is that *all* paraxial rays leaving the point P (i.e. at small angles with respect to the optical axis) will be focussed onto P'. P and P' are referred to as conjugate points.

The paraxial theory does not apply just to refractive optics (Snell's law). Consider the case of reflective optics, as in Figure 8.4.

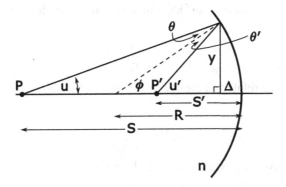

Figure 8.4 Paraxial optics illustrated for a reflecting surface.

$$\theta = \theta', \tag{8.19}$$

$$\theta = (90° - u) - (90° - \phi) = \phi - u, \tag{8.20}$$

$$\theta' = (90° - \phi) - (90° - u') = u' - \phi, \tag{8.21}$$

$$2\phi = u + u'. \tag{8.22}$$

As before, expressing the angles in terms of distances,

$$\phi = \sin^{-1}\frac{y}{R} \approx \frac{y}{R}, \tag{8.23}$$

$$u = \tan^{-1}\frac{y}{S - \Delta} \approx \frac{y}{S - \Delta} = \frac{y}{S - R + R\cos\phi} \approx \frac{y}{S}, \tag{8.24}$$

$$u' = \tan^{-1}\frac{y}{S' - \Delta} \approx \frac{y}{S' - \Delta} = \frac{y}{S' - R + R\cos\phi} \approx \frac{y}{S'}. \tag{8.25}$$

Combining these results we get

$$2\frac{y}{R} = \frac{y}{S} + \frac{y}{S'}, \tag{8.26}$$

$$\frac{2}{R} = \frac{1}{S} + \frac{1}{S'}. \tag{8.27}$$

So in this case also, all rays leaving point P will be focussed onto P', as long as the angles are small.

Within the context of this paraxial theory there are various quantities often useful in optical design. The *power* of a refractive surface is defined by

$$P = \frac{n' - n}{R}, \tag{8.28}$$

where the power is positive for the situation illustrated in Figure 8.3: a surface which is convex to the left with $n' > n$. The power can be made negative either by reversing the sense of curvature (to concave to the left) or by making $n' < n$

(but not both). For a reflective surface which is concave to the left, as in Figure 8.4, the power is

$$P = \frac{2n}{R}.$$ (8.29)

The power is closely related to the focal length. For the refractive case with $s = \infty$ and $s' = f'$,

$$P = \frac{n'}{f'},$$ (8.30)

and for the refractive case with $s' = \infty$ and $s = f$,

$$P = \frac{n}{f}.$$ (8.31)

For the reflective case,

$$P = \frac{n}{f}.$$ (8.32)

The transverse (lateral) magnification, illustrated in Figure 8.5, is defined by

$$m = \frac{h'}{h} = -\frac{S' - R}{S + R} = -\frac{S'}{S}\frac{1 - R/S'}{1 + R/S},$$ (8.33)

which for the refractive case gives

$$m = -\frac{n\,S'}{n'\,S},$$ (8.34)

and for the reflective case gives

$$m = -\frac{S'}{S}.$$ (8.35)

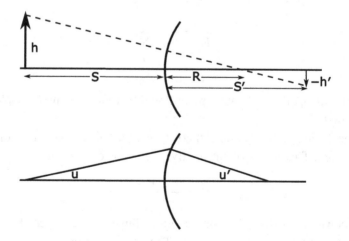

Figure 8.5 Transverse and angular magnification.

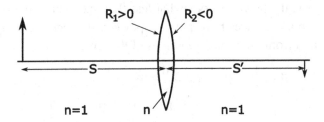

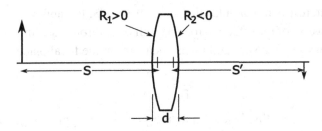

Figure 8.6 Thin and thick lenses.

Similarly, the angular magnification[1] for the refractive case is

$$M = \frac{\tan u'}{\tan u} \approx \frac{S}{S'} = \frac{n}{n' m},$$

(8.36)

and for the reflective case is

$$M = \frac{1}{m}.$$

(8.37)

Note that lateral magnification is accompanied by angular demagnification and vice versa. The Lagrange invariant

$$H = n h u = n' h' u'$$

(8.38)

is closely related to the conservation of étendue. For a "thin" lens, the object and image distances are measured from the center of the lens, as illustrated in Figure 8.6.

$$P = \frac{1}{f} = (n - 1) \left(\frac{1}{R_1} - \frac{1}{R_2} \right)$$

(8.39)

$$= \frac{1}{S} + \frac{1}{S'}.$$

(8.40)

[1] Along with Born & Wolf (1999) we use the term *angular magnification* as synonymous with *convergence ratio*, illustrated in Figure 8.5. Others (e.g. Hecht, 2002) use the term as synonymous with *magnifying power*.

For a "thick" lens, the power is somewhat less than the sum of the powers of the two surfaces, if both surfaces have positive powers. Object and image distances are now measured from the *principal planes* of the lens.

$$P = \frac{1}{f} = P_1 + P_2 - \frac{d}{n}P_1 P_2 \tag{8.41}$$

$$= (n-1)\left[\frac{1}{R_1} - \frac{1}{R_2} + \frac{(n-1)d}{nR_1R_2}\right]. \tag{8.42}$$

For astronomical telescopes the object distance, S, is large so $S' = f$, and a quantity of interest is the angular measure, $\Phi = h/S$, the angle subtended by the object. The inverse of the effective focal length is therefore a scaling factor between angular measure (on the sky) and linear measure (in the focal plane)

$$\frac{1}{f} = \frac{\Phi}{|h'|}. \tag{8.43}$$

For historical reasons this is known as the plate scale, typically given in units of arcsec mm^{-1}.

8.1.2 Seidel aberrations (a third order theory)

In the paraxial approximation we assumed $\sin x \approx \tan x \approx x$, which is the lowest order term in the full series expansions:

$$\sin x = x - \frac{x^3}{3!} + \frac{x^5}{5!} - \cdots, \tag{8.44}$$

$$\tan x = x + \frac{x^3}{3!} + \frac{x^5}{5!} + \cdots. \tag{8.45}$$

As we consider rays farther off axis, these higher order terms become important (first x^3, then $x^5,\ldots$). We will consider here only terms of order x^3, which give rise to the five Seidel (primary) aberrations. Note that these are monochromatic aberrations; they are present even in the absence of any wavelength variation of the refractive index.

Spherical aberration

According to Fermat's principle, a parabolic mirror gives perfect (stigmatic) imaging for an object at infinity. A spherical mirror, therefore, does not do so. This is essentially because the focal length varies with distance from the optical axis, as illustrated in Figure 8.7. A sphere is more strongly curved than a parabola, therefore rays farther from the optical axis will have shorter focal lengths. Rays which fall at the edges (margins) of an optical element are called marginal rays. For a

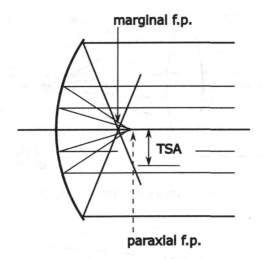

marginal f.p.

TSA

paraxial f.p.

Figure 8.7 Spherical aberration for a mirror, illustrating the marginal rays coming to a focus before the paraxial rays. TSA is the amount of transverse spherical aberration.

spherical mirror the marginal rays come to a focus faster than the paraxial rays. Similar behavior is seen for lenses with spherical surfaces. In spherical aberration the size of the illuminated region is large in the paraxial focal plane because the marginal rays are out of focus. Likewise the illuminated region is large in the marginal focal plane because the paraxial rays are out of focus. There is an optimal distance at which the spot size is smallest. This smallest possible spot is known as the circle of least confusion.

Knowing the radii of curvature, indices of refraction, object distance, and locations of any limiting apertures, it is possible to quantify the amount of spherical aberration present. One measure is the transverse spherical aberration (TSA), the displacement of a transverse ray in the paraxial focal plane. Another is the size of the circle of least confusion, illustrated in Figure 8.8. For the case of a spherical mirror and an object at infinity (collimated light), the diameter of the circle of least confusion, in angular units, is

$$\phi_{\text{CLC}} = \frac{1}{128f^3},$$ \hfill (8.46)

where f is the focal ratio $f = f/D$ and D is the diameter of the mirror. To convert ϕ_{CLC} from angular units to physical (length) units in the focal plane, multiply by the focal length f. Clearly spherical aberration is most significant for fast optics (small f).

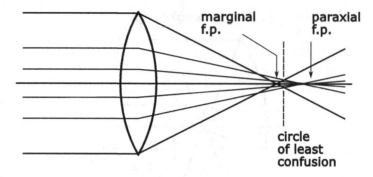

Figure 8.8 Spherical aberration for a lens, illustrating the circle of least confusion.

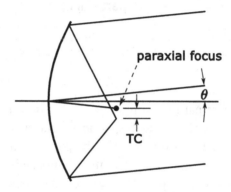

Figure 8.9 Coma produced by rays off axis by an angle θ. The tangential plane is illustrated, and TC is the amount of tangential coma.

Coma

Consider a parabolic mirror in collimated light, a case in which no spherical aberration is present. If the rays hit the mirror at an angle θ with respect to the optical axis, as shown in Figure 8.9, the aberration known as coma will be produced. Rays hitting different parts of the mirror will cross different portions of the focal plane. In the focal plane, the paraxial focus will be displaced off axis. But the marginal foci will be even farther off axis. The asymmetric image produced will have a shape vaguely like that of a comet (whence the name) or an ice cream cone. Coma is important in imaging because it is often fairly easy to remove spherical aberration, usually leaving coma as the largest remaining aberration. Coma is particularly nasty since the degree of image smearing varies across the field of view, depending on the distance from the optical axis.

Note the mapping of rays striking various points on the mirror into various locations in the focal plane, as shown in Figure 8.10. Rays striking the top and bottom

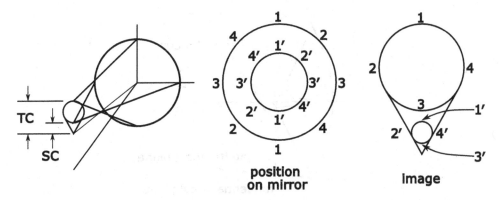

Figure 8.10 Coma in the tangential plane (red in electronic version) and the sagittal plane (blue).

of the mirror (in the plane of incidence) both converge onto the point farthest off axis (the point marked "1"). Rays striking the sides of the mirror (out of the plane of incidence) will come to a point (marked "3") in between the previous point and the paraxial focal point. Rays striking other locations on the periphery of the mirror (e.g. "2" and "4") will be displaced off of the axis of symmetry of the comatic image. One circuit around the mirror corresponds to two circuits in the focal plane.

Coma can be quantified by the size of the tangential coma (TC) or sagittal coma (SC),

$$TC = 3\ SC. \tag{8.47}$$

In angular units the magnitude of the blurring is proportional to θ and $1/f^2$,

$$\phi_{TC} \propto \frac{1}{f^2}\theta. \tag{8.48}$$

The coefficient of proportionality depends on the exact configuration.

Astigmatism

For larger θ (farther off axis), the focal lengths are different for rays in the same plane as the optical axis (the tangential plane) than for the perpendicular plane (the sagittal plane) as shown in Figure 8.11. Therefore a point source produces line images in the tangential and sagittal focal planes (Figure 8.12). For an extended on-axis object, radial features are imaged sharply in the sagittal focal plane and circumferential features are imaged sharply in the tangential focal plane (so don't focus a telescope while looking at a planet). In angular units the amount of astigmatism is proportional to θ^2 and $1/f$:

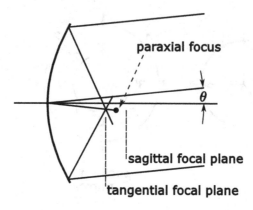

Figure 8.11 Astigmatism gives rise to different focal lengths for the tangential and sagittal rays.

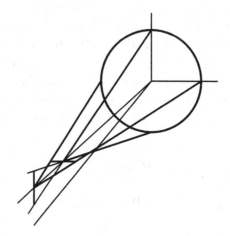

Figure 8.12 Astigmatism produces line foci in the tangential (red in electronic version) and sagittal (blue) planes.

$$\phi_{TAS} \propto \frac{1}{f}\theta^2, \tag{8.49}$$

where TAS refers to the half lengths of the line images at the focal planes S and T. The focal separation of the two line images is proportional to $f\theta^2$.

Distortion

The aberration known as distortion produces stigmatic (point-like) images, but with displaced positions in the focal plane (variations in the transverse magnification), as shown in Figure 8.13. You may have noticed this with photocopy machines; if you enlarge portions of an image and then try to line them up, it usually will

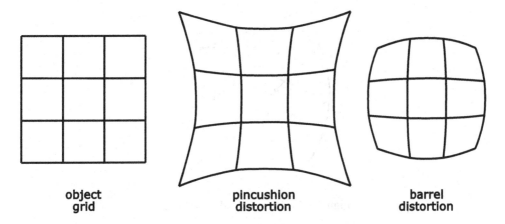

| object
grid | pincushion
distortion | barrel
distortion |

Figure 8.13 Pincushion and barrel distortion.

not work. Under distortion, straight lines are not imaged into straight lines. Distance relationships (e.g. spacings) are not preserved. This is particularly troubling in astrometric applications. The magnitude of the distortion is proportional to θ^3.

Curvature of field

The last of the Seidel aberrations also produces stigmatic images. But in this case, the foci for different values of θ are located on a curved surface, which is to say that the displacement in the focus is proportional to θ^2. If no astigmatism is present, the focal surface is known as the Petzval surface and this aberration is known as Petzval field curvature. If astigmatism is present, there are separate surfaces for the sagittal and tangential foci. The sagittal focal surface is located one third of the way from the Petzval surface to the tangential surface.

As shown in Figure 8.14, for a single mirror the magnitude of the field curvature depends on the radius of curvature of the mirror. In the days of glass photographic plates, the plates were bent (!) before being placed on the telescope. One cannot bend CCD chips. One can, however, tile the focus of a telescope with an array of CCDs in such a way as to approximate a spherical focal surface.

8.1.3 Higher order terms

To some extent, optical design is a tradeoff among the Seidel aberrations. An improvement in one term is often accompanied by deterioration in other terms. But even when a good combination of Seidel aberrations has been achieved, there exist fifth and higher order terms which may be significant, especially for fast optical systems. Numerical optimization using ray tracing is then generally more practical than an analytical approach.

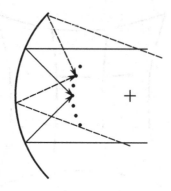

Figure 8.14 Petzval field curvature.

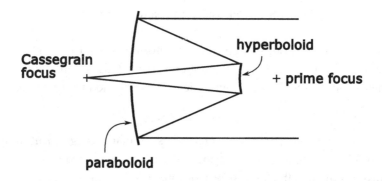

Figure 8.15 Classical Cassegrain telescope.

8.1.4 Telescope design

A classical Cassegrain telescope consists of a parabolic primary mirror and a hyperbolic secondary mirror, as shown in Figure 8.15. Its features include compactness, a long effective focal length, and easy access to the focal region. A classical Cassegrain telescope has zero spherical aberration, and its performance therefore is limited by coma. In angular units its first three Seidel coefficients are given by

$$\phi_{TSA} = 0, \tag{8.50}$$

$$\phi_{TC} = \frac{1}{16f^2}\theta, \tag{8.51}$$

$$\phi_{TAS} \sim \frac{1}{2f}\theta^2. \tag{8.52}$$

Fermat's principle leads to the choice of conic sections as the ideal shapes for stigmatic imaging. A paraboloidal reflector images a point at infinity to a point at a finite distance. An ellipsoidal mirror images a point at finite distance onto another such point. And a hyperboloid images a point at a finite distance into a virtual focus behind the mirror.

In the classical Cassegrain design, the primary mirror was parabolic in order to ensure that the prime focus was free of spherical aberration. However, if the telescope is used *only* in Cassegrain configuration this focus is unused (and inaccessible). The only necessary requirement is that the Cassegrain focus be free of spherical aberration. Therefore the primary mirror is allowed to introduce spherical aberration, as long as that aberration is cancelled by the secondary mirror.[2] This introduces an extra degree of freedom, which may be used to cancel another aberration. In the Ritchey–Chrétien design the shape of the primary mirror is changed (so it is no longer parabolic) and the shape of the secondary mirror is also changed (it is a different hyperbola). This *eliminates* coma while retaining zero spherical aberration, a situation known as the *aplanatic condition*. The cost of eliminating coma, in optical terms, is about a 10% increase in the amount of astigmatism over the classical Cassegrain design. The Ritchey–Chrétien design was used in designing the HST.[3] Both the primary and secondary mirrors are hyperboloids. The disadvantage of this design is that these aspheric surfaces are difficult and expensive to make. For such telescopes the usable field of view is limited by astigmatism.

The above designs illustrate the use of aspherical surfaces to eliminate various types of aberrations. This is an outgrowth of work begun in 1905 by Karl Schwarzschild. In modern terms the generalized Schwarzschild theorem can be stated as:

For a telescope system containing n aspheric elements in any geometry with reasonable separation, it is possible to correct n Seidel conditions (Wilson & Delabre, 1995).

The 8.4 meter Large Synoptic Survey Telescope (LSST) was rated first among large ground-based projects by the 2010 US National Research Council decadal review of astronomy and astrophysics. The LSST uses three large aspheric mirrors to control aberrations across a wide field of view. Cleverly, they have managed to do this with only two pieces of glass. The outer, annular portion of the largest glass blank is ground to the desired figure for the primary mirror, while the inner 5 meter portion (5 meters in diameter) is ground to a different figure for the tertiary mirror. The LSST will have about 50 times the étendue of the Sloan Digital Sky Survey telescope. See Wilson (1996) and Schroeder (2000) for details of some three- and four-mirror telescope designs.

An older approach to wide field imaging is the Schmidt telescope, illustrated in Figure 8.16. A spherical reflector, with an aperture stop at the center of curvature,

---

[2] Expressed as departures from ideal spherical wavefronts, the aberration effects of successive optical surfaces are additive.

[3] However, an error in manufacturing gave the HST significant spherical aberration.

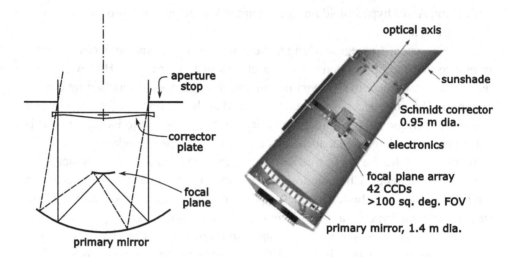

Figure 8.16 (Left) Schmidt telescope. Center of curvature of the mirror is marked by the plus sign, and the axis of symmetry of the corrector plate by the dot-dashed line. (Right) Kepler spacecraft. Credit: NASA.

has *no* off-axis aberrations (no coma, no astigmatism), but *lots* of spherical aberration. This spherical aberration can be cancelled with a thin, aspherical corrector plate. Since the corrector plate has an axis of symmetry, *it* introduces some off-axis aberrations. Fields of view of several square degrees are possible. A disadvantage is that chromatic aberrations are introduced in the corrector plate. The aperture of a Schmidt telescope is determined by the size of the corrector plate; the mirror must be significantly oversized.

The Kepler spacecraft is an important current example of a Schmidt telescope. Its mission is to search for transits of Earth-sized planets around solar type stars. The telescope has a 0.95 meter aperture (a 1.4 meter diameter primary mirror) and a field of view of 105 square degrees. The camera is an array of 42 CCDs each with 2200 × 1024 pixels (95 megapixels in all). A planet is considered confirmed when a transit repeats, in other words, when the period is determined. As of this writing, Kepler has 15 confirmed planets and 1235 candidates. The smallest planet confirmed so far by Kepler (Kepler-10b) has a radius of 1.416 Earth radii, and the least massive (Kepler-11f) has a mass of 2.3 Earth masses. As the mission proceeds through its scheduled 3.5 year lifetime (possibly longer), planets with longer orbital periods will be confirmed.

8.1.5 *Other aspects of telescope design*

Telescope mirrors deform under their own weight. As a simple analogous case of mechanical deformation, consider a *simply supported* uniform rectangular beam (supported at both ends). In this textbook problem, the gravitational deflection at the center is given by

$$\delta = \frac{5}{32} \frac{1}{E} g\rho \frac{L^4}{t^2}, \tag{8.53}$$

where E is Young's modulus, g is the acceleration of gravity, ρ is the density, L is the length, and t is the thickness. The numerical coefficient will vary due to the circular geometry, and can be much smaller depending on the location of the support points, but the functional dependences on E, L, and t are quite general. For an edge-supported disk of diameter D and thickness t,

$$\delta = 0.050 \frac{1}{E} g\rho \frac{D^4}{t^2}. \tag{8.54}$$

For glass, characteristic values are $E = 8 \times 10^{10}$ Pa $= 8 \times 10^{10}$ N m^{-2} and $\rho = 2500$–4500 kg m^{-3}. Taking D = 8.4 m and t = 0.92 m, values characteristic of the LSST primary mirror, one can see that gravitational deformation of a solid piece of glass would be impossibly large (tens of micrometers). For this reason the LSST primary mirror, and that of many other large telescopes, is made of a honeycomb structure,[4] reducing the mass and the resulting gravitational deformation. The Hubble Space Telescope primary mirror, even though not subject to gravitational deformation while in orbit, was fabricated in a similar manner with a honeycomb core fused between 25 mm thick front and back glass plates.

Thermal properties are also important. The thermal conductance of glass characteristically is $K = 0.8$ W m^{-1} K^{-1}. Specific heats are in the range $C_p = 400$–800 J kg^{-1} K^{-1}. For heating and cooling the relevant ratio is $K C_p^{-1} \rho^{-1}$, which is of order 4×10^{-7} m^2 s^{-1}, meaning that glass several centimeters in thickness can have a thermal time constant of many hours. Some form of active thermal control such as forced air circulation is common in large telescope designs. Of greatest importance is preventing any distortion due to non-uniform expansion. Fortunately, glass ceramics such as Zerodur, Cervit, and other ultra-low-expansion (ULE) glasses have been developed with expansion coefficients of $\lesssim 3 \times 10^{-8}$ K^{-1}.

8.1.6 Gravitational lensing

Massive objects affect the direction of propagation of light in their vicinity. In the weak field limit, the space around such objects can be thought of as having an effective index of refraction

$$n = 1 - \frac{2\phi(\vec{r})}{c^2}, \tag{8.55}$$

---

[4] The first use of a large, reduced-weight, ribbed primary was for the 200 inch Hale telescope.

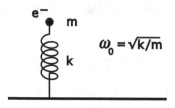

Figure 8.17 Drude–Lorentz model of the refractive index, treating each electron as a mass on a spring.

where ϕ is the Newtonian potential. A massive galaxy cluster can act in this way, forming images from the light from background galaxies. Since neither the lens potential nor the true brightness distribution of the distant galaxy are known *a priori*, the analysis of data from such a system is a difficult *inverse problem*.

8.2 Dispersion

So far we have treated the refractive index as simply an empirical parameter describing the propagation velocity of light in matter. Now we will turn to the physical origin of this parameter and what it tells us about the optical properties of refractive materials. Then we will look at the Fresnel coefficients, which describe the intensities of reflected and refracted waves at boundaries between regions of different refractive indices.

8.2.1 Origin of the refractive index

We will derive a microscopic theory of dispersion, that is, a description of the wavelength dependent (dispersive) nature of the refractive index based on a model at the atomic level. We will use the Drude–Lorentz model, which, although a rather naive, semi-classical picture of the atomic physics, contains all the features necessary to explain the essential properties of the refractive index. The Drude–Lorentz model, as shown in Figure 8.17, considers the motion of bound electrons in matter, subject to a driving force eE_m, a harmonic restoring force, and damping. The subscript m in E_m indicates that we are referring to the microscopic value of the electric field, that is, its value at an individual atomic site instead of its bulk value. The electronic displacement, x, is described by the differential equation

$$e\,E_m - m\,\omega_0^2\,x - m\gamma\,\frac{dx}{dt} = m\,\frac{d^2x}{dt^2}, \tag{8.56}$$

which can be rewritten as

$$\frac{d^2x}{dt^2} + \gamma\,\frac{dx}{dt} + \omega_0^2\,x = \frac{e}{m}\,E_m. \tag{8.57}$$

We assume a time-dependent microscopic field $E_m = \tilde{E}_m\, e^{-i\omega t}$ and seek solutions of the form $x(t) = \tilde{x}\, e^{-i\omega t}$. Plugging in this trial solution we get

$$-\omega^2 \tilde{x} - i\omega\gamma\,\tilde{x} + \omega_0^2\,\tilde{x} = \frac{e}{m}\,\tilde{E}_m, \tag{8.58}$$

$$\tilde{x} = \frac{e/m}{\omega_0^2 - \omega^2 - i\omega\gamma}\,\tilde{E}_m. \tag{8.59}$$

That is, we find a solution which describes simple harmonic motion with peak amplitude at $\omega = \omega_0$ and a resonance width determined by the parameter γ.

To determine the complex index of refraction, we first define the polarization $\vec{P} = N e\, \tilde{x}$. By applying local field theory (not shown) we can get the microscopic field $\vec{E}_m$ in terms of the applied field $\vec{E}$:

$$\vec{E}_m \approx \vec{E} + \frac{\vec{P}}{3\epsilon_0}. \tag{8.60}$$

Combining this with our solution for the electron displacement we can eliminate $\vec{E}_m$. By relating the polarization to the susceptibility χ ($\vec{P} = \chi\vec{E}$) and by relating the susceptibility to the index of refraction ($n^2 = 1 + \chi/\epsilon_0$), we get

$$\frac{n^2 - 1}{n^2 + 2} \approx \frac{Ne^2}{3\epsilon_0 m}\frac{1}{\omega_0^2 - \omega^2 - i\omega\gamma}. \tag{8.61}$$

At this point we need to allow for the fact that not all electrons in the material are bound in the same fashion. Semi-classically we can consider multiple resonant frequencies ω_{0i}, each with an associated damping term γ_i. A certain fraction of the electrons, f_i, is attributed to each resonance. Therefore,

$$\frac{n^2 - 1}{n^2 + 2} \approx \frac{\omega_p^2}{3}\sum_i \frac{f_i}{\omega_{0i}^2 - \omega^2 - i\omega\gamma_i}, \tag{8.62}$$

where $\omega_p^2 = Ne^2/(\epsilon_0 m)$ and $\sum f_i = 1$. This equation can then be solved for the complex index of refraction. Figure 8.18 shows an example of the real and imaginary parts of the complex index of refraction, which in turn is related to the complex wavenumber $\tilde{k} = \omega\tilde{n}/c$.

8.2.2 Fresnel coefficients

Previously when we considered a wave striking a planar dielectric boundary, we considered only the angles of the reflected and transmitted waves. Now we will consider the intensities of reflected and transmitted waves at dielectric interfaces. First consider the case of normal incidence, as in Figure 8.19. It is easiest to treat

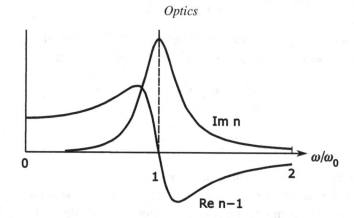

Figure 8.18 Dispersion relation, with $\gamma = 0.3\,\omega_0$. Note the presence of normal dispersion (Re n increasing with ω), a region of anomalous dispersion (Re n decreasing with ω) near resonance, and the offset in Re n−1 well below and well above resonance.

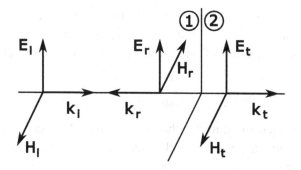

Figure 8.19 Reflected and transmitted waves for a dielectric interface at normal incidence.

this as an electromagnetic boundary value problem. At normal incidence both $\vec{E}$ and $\vec{H}$ are parallel to the surface:

$$\vec{E}_{1,\parallel} = \vec{E}_{2,\parallel}, \tag{8.63}$$

$$\vec{H}_{1,\parallel} = \vec{H}_{2,\parallel}. \tag{8.64}$$

Decomposing the total electric and magnetic fields into incident, transmitted, and reflected waves, the first boundary condition gives

$$E_i + E_r = E_t. \tag{8.65}$$

For the second boundary condition, take the Maxwell equation $\vec{\nabla} \times \vec{E} = -\partial\vec{B}/\partial t$ applied to a plane wave to relate $\vec{H}$ to $\vec{E}$:

$$\omega \vec{B} = \vec{k} \times \vec{E}, \tag{8.66}$$

$$\vec{H} = \frac{1}{\mu_0} \frac{1}{\omega} \vec{k} \times \vec{E}. \tag{8.67}$$

And since $|k|/\omega = n/c$,

$$|\vec{H}| = \frac{1}{\mu_0} \frac{n}{c} |\vec{E}|. \tag{8.68}$$

For the reflected wave there is an implied sign reversal in order to keep $\vec{E} \times \vec{B}$ in the direction of $\vec{k}$. Thus the magnetic boundary condition gives

$$\frac{n_1}{c} (E_i - E_r) = \frac{n_2}{c} E_t. \tag{8.69}$$

Combining the two boundary conditions we get for the amplitudes of the reflected and transmitted waves,

$$\frac{E_r}{E_i} = \frac{n_1 - n_2}{n_1 + n_2}, \tag{8.70}$$

$$\frac{E_t}{E_i} = \frac{2n_1}{n_1 + n_2}. \tag{8.71}$$

This result is closely related to that obtained for impedance changes in waveguides and coaxial cables.

For non-normal incidence we need to consider two polarization cases. S-polarization is where $\vec{E}$ is perpendicular to the plane of incidence, as in Figure 8.20. It turns out to be sufficient to consider boundary conditions only on components tangential to the surface,

$$E_i + E_r = E_t, \tag{8.72}$$

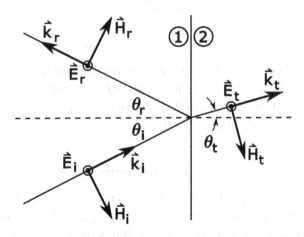

Figure 8.20 S-polarization.

since the E-field is entirely tangential. And for the H-field,

$$H_i \cos \theta_i - H_r \cos \theta_r = H_t \cos \theta_t, \tag{8.73}$$

$$n_1 (\cos \theta_i \, E_i - \cos \theta_r \, E_r) = n_2 \cos \theta_t \, E_t. \tag{8.74}$$

Setting $\theta_i = \theta_r$ and combining equations we get the amplitude transmission and reflection coefficients:

$$r_s = \left(\frac{E_r}{E_i}\right)_\perp = \frac{n_1 \cos \theta_i - n_2 \cos \theta_t}{n_1 \cos \theta_i + n_2 \cos \theta_t}, \tag{8.75}$$

$$t_s = \left(\frac{E_t}{E_i}\right)_\perp = \frac{2 \, n_1 \cos \theta_i}{n_1 \cos \theta_i + n_2 \cos \theta_t}. \tag{8.76}$$

For $\vec{E}$ parallel to the plane of incidence (the case of P-polarization, not shown),

$$r_p = \left(\frac{E_r}{E_i}\right)_\parallel = \frac{n_2 \cos \theta_i - n_1 \cos \theta_t}{n_2 \cos \theta_i + n_1 \cos \theta_t}, \tag{8.77}$$

$$t_p = \left(\frac{E_t}{E_i}\right)_\parallel = \frac{2 \, n_1 \cos \theta_i}{n_2 \cos \theta_i + n_1 \cos \theta_t}. \tag{8.78}$$

Be careful! The differences between these and the formulae for S-polarization are subtle but important. Using Snell's law we can rewrite the entire set as

$$r_s = \frac{\sin(\theta_t - \theta_i)}{\sin(\theta_t + \theta_i)}, \tag{8.79}$$

$$t_s = \frac{2 \cos \theta_i \sin \theta_t}{\sin(\theta_t + \theta_i)}, \tag{8.80}$$

$$r_p = \frac{\tan(\theta_i - \theta_t)}{\tan(\theta_t + \theta_i)}, \tag{8.81}$$

$$t_p = \frac{2 \cos \theta_i \, \sin \theta_t}{\sin(\theta_t + \theta_i) \cos(\theta_i - \theta_t)}. \tag{8.82}$$

In this form the reflection coefficients have been written just in terms of angles. The previous forms contained redundant information in that both angles and indices of refraction were used. Note also that these formulae have the desirable property of $r_s \to r_p$, and $t_s \to t_p$ in the limit $\theta_i \to 0$.

8.3 Physical optics

Geometrical optics was useful in the regime where the wavelength was small compared to any other length scales present. But more generally we need to consider the situation where the wavelength is significant, which is the regime of physical optics.

8.3.1 Vector and scalar diffraction

Optics is, after all, the study of certain aspects of electromagnetic theory. So let us ask what happens when an electromagnetic wave encounters obstacles. The physics of the situation is determined by the vector wave equations,

$$\nabla^2 \vec{E} - \epsilon \mu \frac{\partial^2 \vec{E}}{\partial t^2} = 0, \tag{8.83}$$

$$\nabla^2 \vec{B} - \epsilon \mu \frac{\partial^2 \vec{B}}{\partial t^2} = 0, \tag{8.84}$$

which we must solve subject to appropriate boundary conditions. In effect, we have six second-order partial differential equations (one each for E_x, E_y, E_z, B_x, B_y, and B_z). Furthermore, these equations are coupled, via Maxwell's equations. So this is certainly a formidable problem! Numerical techniques are effective in solving these equations under certain circumstances. But for all practical purposes, exact analytical solutions are difficult to obtain without some further simplifying assumptions.[5]

Perhaps we can solve a simpler problem. Can we describe a wave by a *scalar* function $\psi(\vec{r}) e^{-i\omega t}$ which satisfies the *scalar* wave equation? If so, we are asserting that all polarizations are treated the same, which in detail is certainly false. Nevertheless, this approach captures a surprisingly large share of the physics of diffraction except at very small scales (aperture sizes of order λ or smaller).

8.3.2 Kirchhoff diffraction theory

First we will derive a formula known as Kirchhoff's integral. Green's theorem states that

$$\int_V \left(\psi \nabla^2 \phi - \phi \nabla^2 \psi \right) dV = - \int_S \left(\psi \vec{\nabla} \phi - \phi \vec{\nabla} \psi \right) \cdot \hat{n} \, dA, \tag{8.85}$$

where we adopt the convention that the normal $\hat{n}$ is pointing inward (into the volume V). Choose $\phi(\vec{r}) = e^{ikr}/r$, which is a solution of the wave equation,

$$\nabla^2 \phi + k^2 \phi = 0. \tag{8.86}$$

Let $\psi(\vec{r})$ be some unknown solution of the same wave equation,

$$\nabla^2 \psi + k^2 \psi = 0. \tag{8.87}$$

---

[5] More rigorous treatment of some diffraction problems is given in standard works such as Jackson (1998) and Born & Wolf (1999).

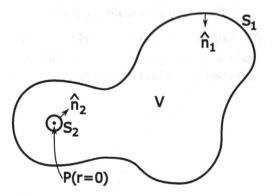

Figure 8.21 Application of Green's theorem to Kirchhoff diffraction.

Since both ϕ and ψ satisfy the wave equation, the volume integral in Green's theorem vanishes. Therefore, the surface integral must also vanish,

$$-\int_S \left(\psi \vec{\nabla} \phi - \phi \vec{\nabla} \psi \right) \cdot \hat{n} \, dA = 0. \tag{8.88}$$

Now let the surface consist of two parts, an outer surface S_1 and an inner surface S_2 excluding the singularity at the origin, as shown in Figure 8.21.

$$\int_{S_2} \left(\psi \vec{\nabla} \phi - \phi \vec{\nabla} \psi \right) \cdot \hat{n}_2 \, dA = \int_{S_2} \left[\psi \left(\frac{ik}{r} - \frac{1}{r^2} \right) e^{ikr} \hat{r} - \frac{e^{ikr}}{r} \vec{\nabla} \psi \right] \cdot \hat{n}_2 \, r^2 \, d\Omega. \tag{8.89}$$

In the limit $r \to 0$,

$$\int_{S_2} \left(\psi \vec{\nabla} \phi - \phi \vec{\nabla} \psi \right) \cdot \hat{n}_2 \, dA = -\int_{S_2} \psi \, d\Omega \tag{8.90}$$

$$= -4\pi \, \psi(P). \tag{8.91}$$

This must equal the negative of the integral over S_1,

$$\psi(P) = \frac{1}{4\pi} \int_{S_1} \left[\psi \vec{\nabla} \frac{e^{ikr}}{r} - \frac{e^{ikr}}{r} \vec{\nabla} \psi \right] \cdot \hat{n} \, dA. \tag{8.92}$$

Thus the value of a scalar function ψ at some point in the interior is related to the values of ψ and $\vec{\nabla}\psi$ on a bounding surface. This is the Kirchhoff integral.

Consider a typical problem in diffraction theory, that of an opaque screen with holes. Apply Kirchhoff's integral to the surface $S = \sigma + \sigma' + \sigma''$, as shown in Figure 8.22. We will take each of the three parts in turn. On σ, *assume* that ψ and $\vec{\nabla}\psi$ are the same as the values would be for the incident wave if no screen were present. This is probably reasonable, except near the edges of the hole where the screen may have some effect. On σ', *assume* $\psi = 0$ and $\vec{\nabla}\psi = 0$. This also is

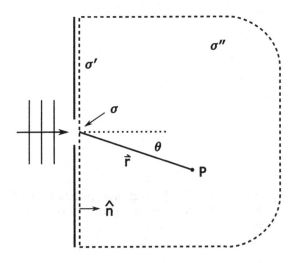

Figure 8.22 Kirchhoff diffraction for an opaque screen with a hole.

reasonable if we are talking about an "opaque" screen, again not near the edge. For σ'' we will play a bit of a trick. Since light travels at a finite speed, pick R sufficiently large that the wave could not have arrived yet (so $\psi = 0$ and $\vec{\nabla}\psi = 0$). The contributions to Kirchhoff's integral from σ' and σ'' vanish, so all that remains is the integral over σ. If we are considering an incident plane wave normal to the screen, then

$$\psi = \psi_0, \tag{8.93}$$

$$\vec{\nabla}\psi = ik\,\psi_0\,\hat{n}. \tag{8.94}$$

$$\psi(P) = \frac{1}{4\pi} \int_\sigma \left[\psi_0 \left(\frac{ik}{r} - \frac{1}{r^2} \right) e^{ikr}\hat{r} - \frac{e^{ikr}}{r}\,ik\,\psi_0\,\hat{n} \right] \cdot \hat{n}\,dA \tag{8.95}$$

$$\approx \frac{1}{4\pi} \int_\sigma \psi_0 \frac{ik}{r}\,e^{ikr}(\hat{r}\cdot\hat{n} - 1)\,dA \tag{8.96}$$

for $r \gg \lambda$. Therefore

$$\psi(P) = -i\frac{\psi_0}{2\lambda} \int_\sigma \frac{e^{ikr}}{r} (1 + \cos\theta)\,dA \tag{8.97}$$

$$\approx -i\frac{\psi_0}{\lambda} \int_\sigma \frac{e^{ikr}}{r}\,dA \tag{8.98}$$

for small θ.

The result of the Kirchhoff diffraction theory is equivalent to saying that each point on a primary wavefront may be viewed as a source of secondary

Figure 8.23 Huygens–Fresnel principle. Note the flatness of the wavefront to the right of the aperture, except near the top and bottom.

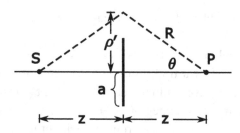

Figure 8.24 Geometry of the Poisson spot.

spherical waves. This is the Huygens–Fresnel principle, illustrated in Figure 8.23. When these waves are superimposed in amplitude and phase, they give the wave amplitude at some later point. For the case of a hole in a screen, the points in the opening contribute as sources of secondary waves while the radiation from points outside the opening does not contribute. The Huygens–Fresnel principle is not proven – in fact, it is clearly wrong at some level since it ignores polarization (the vectorial nature of light). But the Kirchhoff diffraction theory justifies it as a useful approximation.

The Poisson spot

Diffraction theory can produce some amazing results. Consider an opaque circular disk, and consider points on the symmetry axis of the disk, as in Figure 8.24. The Kirchhoff formula, appropriately modified for a finite source distance, is

$$\psi(P) = -i\,\frac{\psi_0}{\lambda}\int_\sigma \frac{e^{i2kR}}{R^2}\cos\theta\,dA \tag{8.99}$$

$$= -i\,\frac{\psi_0}{\lambda}\int_0^{2\pi} d\phi' \int_a^\infty \rho'\,d\rho'\,\frac{e^{i2kR}}{R^2}\cos\theta, \tag{8.100}$$

where the extra factor of $1/R$ is related to the definition of ψ_0 for a spherical wave. Now $\cos\theta = z/R$ and $R^2 = \rho'^2 + z^2$, implying that $R\,dR = \rho'\,d\rho'$.

$$\psi(P) = -i\,\frac{2\pi\,\psi_0}{\lambda}\,z\int_{\sqrt{z^2+a^2}}^\infty \frac{e^{i2kR}}{R^2}\,dR. \tag{8.101}$$

Integrate by parts to get

$$\psi(P) = -ik\,\psi_0\,z\left[\frac{1}{i2k}\,\frac{e^{i2kR}}{R^2}\Big|_{\sqrt{z^2+a^2}}^\infty + \frac{1}{ik}\int_{\sqrt{z^2+a^2}}^\infty \frac{e^{i2kR}}{R^3}\,dR\right]. \tag{8.102}$$

The second term may be neglected with respect to the integral as a whole (the prior equation). So we can write that

$$\psi(P) \approx \frac{\psi_0 z}{2}\,\frac{e^{i2k\sqrt{z^2+a^2}}}{z^2+a^2}, \tag{8.103}$$

for $R \gg \lambda$. One concludes that there is significant intensity on axis, as shown in Figure 8.25, even though P is well inside the geometrical shadow. This is the result of constructive interference of diffracted waves. The fact that it is called the Poisson spot is ironic since Poisson made this calculation in order to debunk Fresnel's theory by demonstrating that it gave such an "absurd" result.

8.3.3 Fresnel and Fraunhofer approximations

In Kirchhoff diffraction theory we used r to denote the distance between our point of interest, P, and an arbitrary point in the aperture, σ. Here we will switch notation

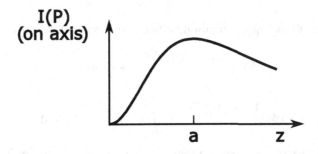

Figure 8.25 Intensity distribution along the axis for the Poisson spot.

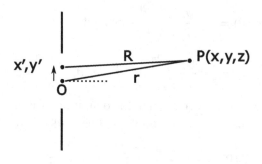

Figure 8.26 Geometry for the Fresnel approximation.

and call that distance R, as in Figure 8.26, letting r denote the distance to the *center* of the aperture. We will use primed Cartesian coordinates to describe locations in the aperture plane and unprimed coordinates for the point P. Consider points far enough from the screen that $R \approx r$.

$$\psi(P) = -i\frac{\psi_0}{\lambda} \int_\sigma \frac{e^{ikR}}{R} \, dA \tag{8.104}$$

$$\approx -i\,\psi_0 \frac{1}{r\lambda} \int_\sigma e^{ikR} \, dA. \tag{8.105}$$

The exponential has an oscillatory part which we cannot take outside the integral.

$$R = \sqrt{(x - x')^2 + (y - y')^2 + z^2} \tag{8.106}$$

$$\approx z \left[1 + \frac{1}{2}\frac{(x - x')^2}{z^2} + \frac{1}{2}\frac{(y - y')^2}{z^2} \right]. \tag{8.107}$$

This is the Fresnel approximation, valid in the *near field*. If we further neglect the x'^2 and y'^2 terms ($k(x'^2 + y'^2)/2z \ll 1$), we get

$$R \approx z \left(1 + \frac{1}{2}\frac{x^2 + y^2}{z^2} \right) - \frac{xx' + yy'}{z}. \tag{8.108}$$

This is the Fraunhofer approximation, which gives the *far field* diffraction pattern,

$$e^{ikR} \approx e^{ikz\left(1 + \frac{1}{2}\frac{x^2+y^2}{z^2}\right)} e^{-ik(xx'+yy')/z}, \tag{8.109}$$

$$\psi(P) \approx -i\psi_0 \frac{1}{r\lambda} e^{ikz\left(1 + \frac{1}{2}\frac{x^2+y^2}{z^2}\right)} \int_\sigma e^{-ik(xx'+yy')/z} \, dx' \, dy'. \tag{8.110}$$

Note the similarity of this aperture integral to a 2-dimensional Fourier transform! It looks like a Fourier transform of unity, but it is restricted to the aperture region. If we define an *aperture function* to be 1 everywhere inside the aperture and zero

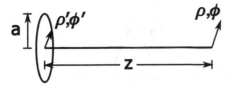

Figure 8.27 Cylindrical coordinates for calculating diffraction from a circular aperture of radius a.

everywhere else, then the far field pattern is the Fourier transform of this aperture function.

Circular apertures and the Airy pattern

A situation commonly encountered is diffraction by a circular aperture, as in Figure 8.27. Letting C be an appropriate collection of constants, the Fraunhofer theory gives the following result for a circular aperture of radius a,

$$\psi(P) = C \int_0^a \rho' \, d\rho' \int_0^{2\pi} d\phi' \, e^{-ik(xx'+yy')/z} \tag{8.111}$$

$$= C \int_0^a \rho' \, d\rho' \int_0^{2\pi} d\phi' \, e^{-ik\rho\rho'(\cos\phi\cos\phi'+\sin\phi\sin\phi')/z} \tag{8.112}$$

$$= C \int_0^a \rho' \, d\rho' \int_0^{2\pi} d\phi' \, e^{-ik\rho\rho'\cos\phi'/z}. \tag{8.113}$$

In the last step we have chosen to set $\phi = 0$, since by symmetry all choices are equivalent. These integrals are related to the Bessel functions J_0 and J_1.

$$J_0(x) = J_0(-x) = \frac{1}{2\pi} \int_0^{2\pi} e^{ix\cos\phi} \, d\phi, \tag{8.114}$$

$$x J_1(x) = \int x J_0(x) \, dx. \tag{8.115}$$

Therefore,

$$\psi(P) = 2\pi C \int_0^a J_0\left(\frac{k\rho\rho'}{z}\right) \rho' \, d\rho' \tag{8.116}$$

$$= 2\pi C \left(\frac{z}{k\rho}\right)^2 \int_0^{k\rho a/z} J_0(x) \, x \, dx \tag{8.117}$$

$$= 2\pi C \left(\frac{z}{k\rho}\right)^2 \frac{k\rho a}{z} J_1\left(\frac{k\rho a}{z}\right) \tag{8.118}$$

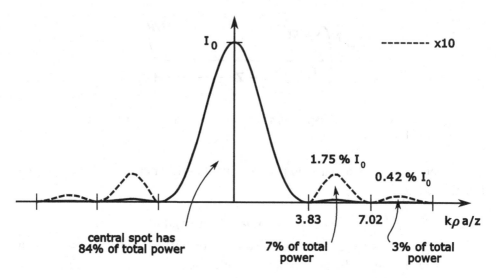

Figure 8.28 The Airy pattern, for diffraction from a circular aperture of radius a.

$$= 2\pi C\, a^2 \, \frac{J_1(k\rho a/z)}{k\rho a/z}. \tag{8.119}$$

The intensity is the square of ψ,

$$I(P) = \pi^2 C^2 a^4 \left[\frac{2J_1(k\rho a/z)}{k\rho a/z} \right]^2, \tag{8.120}$$

which is the Airy pattern, shown in Figure 8.28. The first null of J_1 occurs at $x = 3.83$, so the angular radius of the central bright spot is

$$\theta = \frac{\rho_{\text{null}}}{z} = \frac{1}{ka}\left(\frac{k\rho a}{z} \right)_{\text{null}} = \frac{1}{ka}\, 3.83 = \frac{3.83\lambda}{2\pi a} = 1.22\,\frac{\lambda}{2a}. \tag{8.121}$$

This angular size is a measure of the spatial resolution of the imaging system. The Airy pattern result will be somewhat modified for many astronomical telescopes due to blockage at the center of the aperture (e.g. Cassegrain telescopes).

Diffraction by a straight edge

The problem of Fresnel diffraction by a straight edge, shown in Figure 8.29, is of some relevance in astronomy, so we will treat that next. Here the aperture extends to infinity, so we cannot use the Fraunhofer approximation. Strictly speaking we should not use the Fresnel approximation either.

$$R = |\vec{r} - \vec{r}'| \approx z\left[1 + \frac{1}{2}\frac{(x - x')^2}{z^2} + \frac{1}{2}\frac{(y - y')^2}{z^2} \right]. \tag{8.122}$$

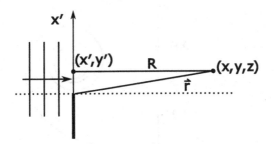

Figure 8.29 Diffraction by a straight edge.

However, it turns out that neglecting terms of order $(x'/z)^4$, $(y'/z)^4$, and higher does not affect the results (Born & Wolf, 1999). Therefore,

$$\psi(P) = -i\frac{\psi_0}{\lambda}\int_\sigma \frac{e^{ikR}}{R} \tag{8.123}$$

$$= -i\frac{\psi_0 e^{ikz}}{\lambda z}\int_\sigma \exp\left[i\frac{k}{2z}\{(x-x')^2 + (y-y')^2\}\right]dx'\,dy'. \tag{8.124}$$

From translational symmetry, the result cannot depend on y, and its integral over y' will give some constant.

$$\psi(P) = C\int_0^\infty \exp\left[i\frac{k}{2z}(x-x')^2\right]dx' \tag{8.125}$$

$$= C\sqrt{\frac{\pi z}{k}}\int_{-\infty}^{u_0} e^{i\pi u^2/2}\,du, \tag{8.126}$$

where we made the change of variables $u = (x-x')\sqrt{k/\pi z}$. Far from the edge $(u_0 \to \infty)$, ψ must approach ψ_0.

$$\psi_0 = C\sqrt{\frac{\pi z}{k}}\int_{-\infty}^\infty e^{i\pi u^2/2}\,du = C\sqrt{\frac{\pi z}{k}}(1+i). \tag{8.127}$$

So, in general,

$$\psi(P) = \frac{\psi_0}{1+i}\int_{-\infty}^{u_0} e^{i\pi u^2/2}\,du, \tag{8.128}$$

$$I(P) = \frac{I_0}{2}\left|\int_{-\infty}^{u_0} e^{i\pi u^2/2}\,du\right|^2. \tag{8.129}$$

Now express the intensity in terms of sine and cosine components,

$$I(P) = \frac{I_0}{2}\left|\int_{-\infty}^{u_0}\cos\frac{\pi u^2}{2}\,du + i\int_{-\infty}^{u_0}\sin\frac{\pi u^2}{2}\,du\right|^2 \tag{8.130}$$

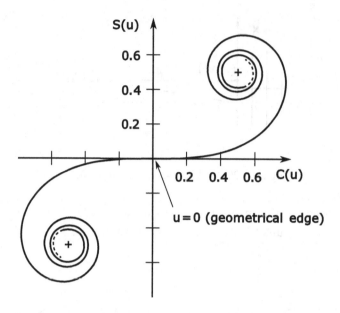

Figure 8.30 The Cornu spiral: a plot of the Fresnel integrals S(u) and C(u).

$$= \frac{I_0}{2} \left[\int_{-\infty}^{u_0} \cos \frac{\pi u^2}{2} \, du \right]^2 + \frac{I_0}{2} \left[\int_{-\infty}^{u_0} \sin \frac{\pi u^2}{2} \, du \right]^2. \tag{8.131}$$

These are related to special functions known as the Fresnel integrals,

$$C(u) = \int_0^u \cos \frac{\pi u^2}{2} \, du, \tag{8.132}$$

$$S(u) = \int_0^u \sin \frac{\pi u^2}{2} \, du. \tag{8.133}$$

These functions are graphically represented by the Cornu spiral in Figure 8.30. The light intensity is proportional to the *square* of the length from the lower left spiral center to the point in question:

$$I(P) = \frac{I_0}{2} \left\{ \left[C(u_0) - C(-\infty) \right]^2 + \left[S(u_0) - S(-\infty) \right]^2 \right\}. \tag{8.134}$$

As one moves through the geometrical shadow towards the edge, the intensity increases monotonically, as shown in Figure 8.31, reaching a value of $I_0/4$ at the edge. Past the geometrical edge the intensity oscillates, eventually settling down on I_0.

This straight edge problem is relevant to lunar occulations. The lunar limb acts as a "knife edge" cutting across a distant source, allowing measurement of source size, structure, and position. The first maximum in the light from a distant star

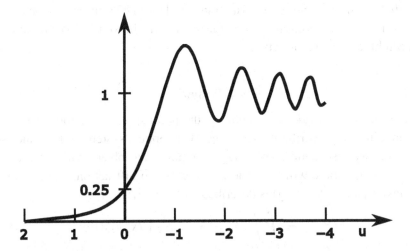

Figure 8.31 Diffraction pattern for a point source of unit flux density as it emerges from the geometrical shadow of a straight edge.

occulted by the Moon occurs at u = 1.22 (where the geometrical edge is at u = 0). Therefore the resolution (at 500 nm) is given by

$$\theta = \frac{x}{z} = 1.22 \sqrt{\frac{\pi}{kz}} = 1.22 \sqrt{\frac{\lambda}{2z}} = 3.1 \times 10^{-8} \text{ rad} = 0.0064''. \qquad (8.135)$$

However, the lunar limb moves at a rate of $\lesssim 0.6'' \text{ s}^{-1}$, so one needs a fast detector and good signal to noise! Also one needs a narrow filter since different colors have different fringe spacings. This technique has also been used at radio wavelengths (e.g. to determine the size of the source 3C273).

8.3.4 Diffraction with aberrations

So far we have treated aberrations and diffraction separately: aberrations in the limit $\lambda \to 0$ and diffraction for "ideal" optical systems. When both are present, they interact and the patterns get very complex. Examples are given in Born & Wolf (1999) in the form of both images and isophotes.

Let σ denote the RMS wavefront distortion in the pupil plane. When the wavefront errors are small, the effects of the aberrations can be calculated. The maximum intensity of the image, relative to the pure diffraction case, is decreased by an amount

$$\frac{I}{I_0} = e^{-(2\pi\sigma/\lambda)^2} \approx 1 - \left(\frac{2\pi}{\lambda}\right)^2 \sigma^2, \qquad (8.136)$$

a quantity known as the Strehl intensity ratio. Wilson (1996) discusses the tolerable amounts of various aberrations, that is, the maximum allowed RMS error consistent with a Strehl ratio of 0.8 or better.

8.4 Imaging

A quantity in widespread use in optics is the point spread function (PSF). Given some input intensity distribution (the object), an optical system will produce some output intensity distribution (the image). A point-like object will *not* produce a point-like image; there will be some blurring due to diffraction, aberrations, etc. This transformation (blurring) is described by the PSF,

$$I_i(X, Y) = \int\int PSF(x, y; X, Y)\, I_o(x, y)\, dx\, dy \qquad (8.137)$$

(Figure 8.32), where we have also assumed that the various points on the object radiate incoherently. For a point-like object,

$$I_o(x, y) = C\,\delta(x - x_0)\,\delta(y - y_0), \qquad (8.138)$$
$$I_i(X, Y) = C\,PSF(x_0, y_0; X, Y). \qquad (8.139)$$

Strictly speaking, I_o and I_i are flux densities.

The Airy pattern is an example of a PSF. In general, the PSF will vary with position (e.g. in the presence of coma). However, if the object is small enough, the PSF may be considered approximately constant (the *isoplanatic condition*), except for displacement in the image plane. For unit magnification we can write

$$PSF(x, y; X, Y) = PSF(X - x, Y - y), \qquad (8.140)$$

$$I_i(X, Y) = \int\int PSF(X - x, Y - y)\, I_o(x, y)\, dx\, dy, \qquad (8.141)$$

which is a 2-dimensional convolution.

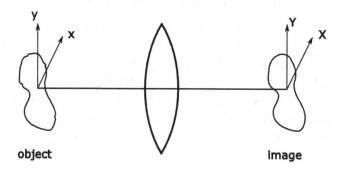

Figure 8.32 The point spread function relates the intensity distribution of the object being observed to the intensity distribution of the image.

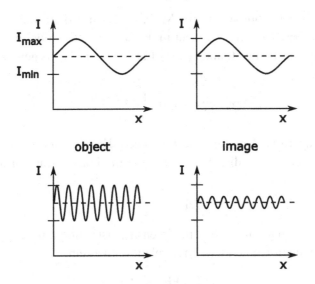

object image

Figure 8.33 The modulation transfer function measures the change in contrast of various spatial frequencies. Low frequencies will retain most of their contrast. High frequencies will suffer loss of contrast.

For many applications it is more convenient to deal with the Fourier transform of the PSF, a complex quantity known as the optical transfer function (OTF). For a symmetric (i.e. even) PSF, the OTF is purely real, in which case it is known as the modulation transfer function (MTF). The MTF may be visualized by considering an object with a sinusoidal intensity variation at some period P_0 (spatial frequency $v_0 = 1/P_0$) as shown in Figure 8.33. If P_0 is large (v_0 is small) the pattern will be imaged accurately. If P_0 is small (v_0 is large) the pattern will be washed out by the image blurring. If we define the contrast as

$$C = \frac{I_{max} - I_{min}}{I_{max} + I_{min}}, \tag{8.142}$$

the MTF describes the change in contrast:

$$MTF(v) = \frac{C_{image}(v)}{C_{object}(v)}. \tag{8.143}$$

Properties of the MTF include:

$$0 \leq MTF(v) \leq 1, \tag{8.144}$$

$$\lim_{v \to 0} MTF(v) = 1, \tag{8.145}$$

and that there exists some cutoff frequency v_c such that

$$MTF(v \geq v_c) = 0. \tag{8.146}$$

Consider an alternate approach to the MTF. As discussed above, the PSF and the MTF are a Fourier transform pair. But in the case of diffraction, we also saw that the Airy pattern was related to the Fourier transform of what is generally called the pupil function,

$$PF(\rho', \theta') = \begin{cases} 1 & \rho' \leq a \\ 0 & \rho' > a \end{cases} \tag{8.147}$$

for a circular aperture. How are these related? The Fraunhofer diffraction integral says the electric field distribution $\psi(P)$ is the Fourier transform of the pupil function,

$$PF \rightleftharpoons \psi(P). \tag{8.148}$$

But the absolute square of $\psi(P)$ is the intensity distribution, the Airy pattern (the PSF). By the Wiener–Khinchin (autocorrelation) theorem,

$$PF \star PF \rightleftharpoons PSF. \tag{8.149}$$

Since we also have the fact that

$$MTF \rightleftharpoons PSF, \tag{8.150}$$

this implies that[6]

$$MTF = PF \star PF = AC(PF), \tag{8.151}$$

or to be precise,

$$MTF(u, v) = \lambda^2 C^2 \, PF\left(\frac{x'}{\lambda}, \frac{y'}{\lambda}\right) \star PF\left(\frac{x'}{\lambda}, \frac{y'}{\lambda}\right). \tag{8.152}$$

This works for aberrations as well as diffraction if we use a complex pupil function, in which the phase of the pupil function is the wavefront phase error introduced.

Consider the MTF of a single-pixel detector. For a rectangular detector of dimensions $a \times b$, as in Figure 8.34, the PSF of the detector aperture is given by

$$PSF(x, y) = \begin{cases} 1 & |x| \leq a/2, |y| \leq b/2 \\ 0 & \text{otherwise} \end{cases} \tag{8.153}$$

$$= {}^2\Pi\left(\frac{x}{a}, \frac{y}{b}\right) = \Pi\left(\frac{x}{a}\right) \Pi\left(\frac{y}{b}\right). \tag{8.154}$$

The MTF then, normalized to unity at $s = 0$, is

$$MTF(s_x, s_y) = \text{sinc}(a \, s_x) \, \text{sinc}(b \, s_y) . \tag{8.155}$$

---

[6] This discussion has been rather informal, since it was meant to convey the general functional relationships between the PSF, MTF, and PF. See the next section for a more careful derivation.

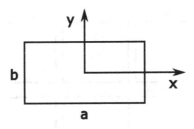

Figure 8.34 A rectangular pixel.

The total MTF is the product of the detector (pixel) MTF and the MTF of the optics. To avoid significant degradation of the MTF of the optics, choose

$$a \approx b \approx \frac{\lambda}{2} \frac{f}{D}. \tag{8.156}$$

Note this size in relation to the radius of the Airy disk, $1.22 \, \lambda f/D$.

What about the MTF of multiple-pixel detectors, such as CCD arrays? Assume some image plane intensity distribution $I(x,y)$, produced by an optical system with some spatial frequency cutoff ν_c,

$$\nu_c = \frac{D}{\lambda} \frac{1}{f} \tag{8.157}$$

for the Airy pattern. The pixel response modifies the MTF but leaves ν_c unchanged. According to Shannon's sampling theorem, we can fully reconstruct $I(x, y)$ if we sample at an interval

$$p = \frac{1}{2\nu_c} = \frac{\lambda f}{2D}. \tag{8.158}$$

Therefore pick p to fully sample, as in Figure 8.35. The pixel dimensions a and b can be somewhat smaller than p without loss of spatial information, but one loses signal intensity if the pixels are too small.

8.5 Addendum

The following is a more careful and complete derivation of the relationships between the PF, MTF, and PSF. We saw that under the Fraunhofer approximation

$$\psi(P) = \frac{C}{\lambda} \int \int PF(x', y') \, e^{-i2\pi (xx'+yy')/\lambda z} \, dx' \, dy'. \tag{8.159}$$

Rewrite this formula by describing the location of the point P in terms of its angle from the optical axis and by redefining the PF in units of wavelength:

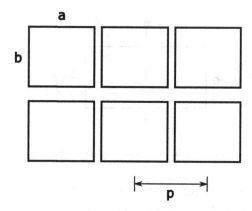

Figure 8.35 Nyquist sampling with a grid of pixels.

$$\psi\left(\frac{x}{z}, \frac{y}{z}\right) = \frac{C}{\lambda} \int \int PF\left(\frac{x'}{\lambda}, \frac{y'}{\lambda}\right) e^{-i2\pi(xx'+yy')/\lambda z} \, dx' \, dy' \tag{8.160}$$

$$= C\lambda \int \int PF\left(\frac{x'}{\lambda}, \frac{y'}{\lambda}\right) e^{-i2\pi(xx'+yy')/\lambda z} \, d\frac{x'}{\lambda} \, d\frac{y'}{\lambda}. \tag{8.161}$$

So we can see that the Fourier transform pair is really

$$f = C\lambda \, PF\left(\frac{x'}{\lambda}, \frac{y'}{\lambda}\right), \tag{8.162}$$

$$\mathcal{F}f = \psi\left(\frac{x}{z}, \frac{y}{z}\right). \tag{8.163}$$

Applying the Wiener–Khinchin (autocorrelation) theorem,

$$PSF\left(\frac{x}{z}, \frac{y}{z}\right) = \left|\psi\left(\frac{x}{z}, \frac{y}{z}\right)\right|^2 \tag{8.164}$$

$$= C^2\lambda^2 \int \int PF\left(\frac{x'}{\lambda}, \frac{y'}{\lambda}\right) \star PF\left(\frac{x'}{\lambda}, \frac{y'}{\lambda}\right) e^{-i2\pi(xx'+yy')/\lambda z} \, d\frac{x'}{\lambda} \, d\frac{y'}{\lambda}. \tag{8.165}$$

By definition the PSF and MTF form a Fourier transform pair,

$$PSF\left(\frac{x}{z}, \frac{y}{z}\right) = \int \int MTF(u, v) \, e^{-i2\pi(ux+vy)/z} \, du \, dv. \tag{8.166}$$

Therefore, taking $u = x'/\lambda$ and $v = y'/\lambda$, we have

$$MTF(u, v) = \lambda^2 C^2 \, PF\left(\frac{x'}{\lambda}, \frac{y'}{\lambda}\right) \star PF\left(\frac{x'}{\lambda}, \frac{y'}{\lambda}\right). \tag{8.167}$$

Exercises

8.1 An $f/0.5$ beam comes to a focus at some point O (the origin of our coordinate system). A glass plate (n = 1.51 at λ = 589.29 nm, the wavelength of the Na D lines) is now introduced into the beam, as shown in Figure 8.36. The plate is 10 mm thick and it is placed with its back surface 10 mm from the origin.

   a. What is the displacement of the paraxial focal plane (at λ = 589.29 nm)? Calculate to the nearest 0.01 mm.
   b. What is the position of the marginal focal plane?
   c. Illustrate the ray paths with a diagram (make it large enough to shown the requisite detail). What aberration does the plate introduce?

8.2 For a spherical mirror with radius of curvature R and an object at infinity, calculate the amount of field curvature (the radius of curvature of the paraxial focal surface).

8.3 We derived formulae for dispersion based on the Lorentz local field

$$\vec{E}_m = \vec{E} + \frac{\vec{P}}{3\epsilon_0}. \tag{8.168}$$

Assume the applicability of those formulae for the following.[7]

   a. Diamond has a static dielectric constant of 5.50 (K = n^2 = 5.50 at $\nu = 0$) and an index of refraction of 2.417 at λ = 589 nm. Assuming that these properties can be described in terms of a single narrow ($\gamma = 0$) resonance in the ultraviolet, what is the wavelength λ_0 of that resonance?
   b. The borosilicate crown glass known as BK1 has an index of refraction of 1.507 63 at 656.2816 nm and 1.515 66 at 486.1327 nm. Using a similar single-resonance model, calculate the index of refraction at 587.5618 nm.

8.4 Calculate and plot the Fresnel coefficients r_s, t_s, r_p, and t_p for the following cases. Comment on any nodes you find and comment on the signs of the coefficients.

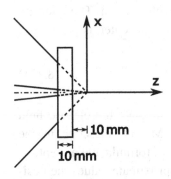

Figure 8.36 Marginal and paraxial rays of an $f/0.5$ beam being intercepted by a glass plate 10 mm thick.

[7] Adapted from Reitz *et al.* (1979).

a. $n_1 = 1$, $n_2 = 2$.

b. $n_1 = 2$, $n_2 = 1$.

c. For both of these cases, verify that

$$n_1 |r|^2 \cos \theta_i + n_2 |t|^2 \cos \theta_t = n_1 \cos \theta_i, \qquad (8.169)$$

and explain the significance of this formula in terms of the Poynting flux.

8.5   A plane wave is incident on a circular aperture in an opaque screen. A lens just after the screen focuses the diffracted light onto a second screen placed at the focal plane of the lens. Show that the Fraunhofer diffraction *pattern* is produced on the screen even though the Fraunhofer *approximation* has not been applied. That is, show the lens cancels the Fresnel terms in the diffraction integral.

8.6   Starting from the pupil function for a uniform circular aperture of radius a, calculate a trigonometric expression for the MTF. Make sure your result has the proper normalization and that you have treated the spatial frequency scale correctly. Plot your result (it should be somewhat triangular in shape).

8.7   Show that any Fraunhofer diffraction pattern will have a center of symmetry as long as the aperture is illuminated with uniform phase. That is, show that $I(x, y) = I(-x, -y)$.

8.8   The 4 meter Mayall telescope at Kitt Peak has its aperture limited to 3.8 meters diameter since the mirror surface is poor near the edge. When the f/8 Cassegrain secondary mirror is used, it blocks a 1.65 meter diameter region at the center of the aperture. The remaining aperture is annular in shape.

   We have seen that for Fraunhofer diffraction by a circular aperture of radius a, as in Figure 8.27, the Kirchhoff diffraction integral gives a field distribution

$$\psi (\rho, z) = C\pi a^2 \, \frac{2J_1 (k\rho a/z)}{k\rho a/z}, \qquad (8.170)$$

where C is a constant and J_1 is the first order Bessel function. The intensity distribution is the Airy pattern which, normalized at the center, is given by

$$\frac{I(\rho, z)}{I(0, z)} = \left(\frac{2J_1 (k\rho a/z)}{k\rho a/z} \right)^2. \qquad (8.171)$$

a. Remembering the principle of superposition, calculate the electric field distribution for the annular aperture of the Mayall telescope in the Fraunhofer limit as a function of $k\rho a/z$. Provide a formula and a graph.

b. Graph the intensity distribution. Label with approximate values the positions of the nodes and the size of the secondary maximum.

c. The spatial resolution is given by the width of the central maximum of the diffraction pattern. Is the resolution of the annular aperture better (narrower) or worse (broader) than that of the circular aperture?

d. The contrast is the ratio of the principal and secondary maxima. Is the contrast better or worse for the annular aperture?

9

Interference

9.1 Mutual coherence function and complex degree of coherence

The *mutual coherence function* describes the relationship between electric fields measured at two points separated in both space and time. For now we will ignore questions of polarization. Let's call the points 1 and 2 as shown in Figure 9.1. Assuming a stationary radiation field,

$$\tilde{\Gamma}_{12}(\tau) = \langle \tilde{E}_1(t)\, \tilde{E}_2^*(t+\tau)\rangle, \tag{9.1}$$

which is in the form of a cross correlation.

The *complex degree of coherence* is defined to be the mutual coherence function normalized by its values for zero spatial separation and zero time delay,

$$\tilde{\gamma}_{12}(\tau) = \frac{\Gamma_{12}(\tau)}{\sqrt{\Gamma_{11}(0)\Gamma_{22}(0)}}. \tag{9.2}$$

We tend to think of lasers as sources of coherent radiation and blackbodies as incoherent sources. Taking the above definition, we have the idealized limiting cases of $|\tilde{\gamma}_{12}| = 1$ for perfectly coherent radiation, and $|\tilde{\gamma}_{12}| = 0$ for perfectly incoherent radiation. Everything in the real world lies between these limits and can be said to be partially coherent:

$$0 < |\tilde{\gamma}_{12}| < 1. \tag{9.3}$$

Our treatment here combines the concepts of spatial and temporal coherence.

9.2 Quasi-monochromatic radiation

Consider a narrow-band light source, such as a mercury lamp. It will have a finite bandwidth associated with either Doppler or pressure broadening. Its waveform will be approximately as shown in Figure 9.2; at different times the various frequency components will combine in and out of phase. The Wiener–Khinchin

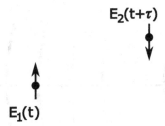

Figure 9.1 Electric field strength measured at points separated in both time and space.

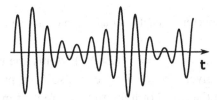

Figure 9.2 Quasi-monochromatic radiation.

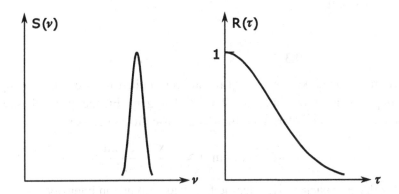

Figure 9.3 Gaussian power spectral density and autocorrelation functions.

theorem says that the power spectral density is the Fourier transform of the auto-correlation function. And we know that the Fourier transform of a Gaussian is a Gaussian. So if the power spectral density of the line is Gaussian, which we write in our conventional form

$$S(\nu) \propto e^{-(\nu-\nu_0)^2/2(\Delta\nu)^2} \tag{9.4}$$

and show in Figure 9.3, then so is its autocorrelation function,

$$R(\tau) \propto e^{-\tau^2/2(\Delta\tau_c)^2}, \tag{9.5}$$

Interference

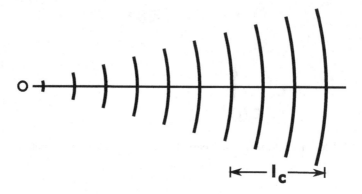

Figure 9.4 Coherence length of quasi-monochromatic radiation.

where the coherence time is $\Delta\tau_c = 1/(2\pi\Delta\nu)$.[1] Only the envelope of the auto-correlation function is given above; there can also be a phase factor. Since the waveform propagates at the speed of light, this coherence time is readily translated into a (longitudinal) *coherence length*, $l_c = c\,\Delta\tau_c$. For typical low pressure gas lamps, $l_c \lesssim 50$ cm, as shown in Figure 9.4. For broadband sources the coherence length is very small.

9.3 Young's two-slit experiment

Consider an idealized point source of radiation illuminating two slits in an opaque screen (in phase) as shown in Figure 9.5. The Huygens–Fresnel principle says that the slits then act as secondary radiators. The phase difference at x is

$$\phi = k\Delta \approx kd\,\sin\theta \approx \frac{kdx}{L} \approx \frac{2\pi xd}{\lambda L},\tag{9.6}$$

where d is the slit spacing. The electric fields add, giving an intensity

$$I = \epsilon_0 c\,\langle E^2\rangle \tag{9.7}$$

$$= \epsilon_0 c\left[\langle E_1^2\rangle + \langle E_2^2\rangle + 2\langle \vec{E}_1\cdot\vec{E}_2\rangle\right] \tag{9.8}$$

$$= I_1 + I_2 + 2\sqrt{I_1 I_2}\,\cos\phi. \tag{9.9}$$

If $I_1 = I_2$, then the fringes are fully modulated (I goes from zero to $4I_1$), as shown in Figure 9.6. Define the "visibility" V:

$$V = \frac{I_{max} - I_{min}}{I_{max} + I_{min}},\tag{9.10}$$

[1] Other conventions for the definitions of bandwidth and coherence length give $\Delta\tau_c = 1/(\Delta\nu)$.

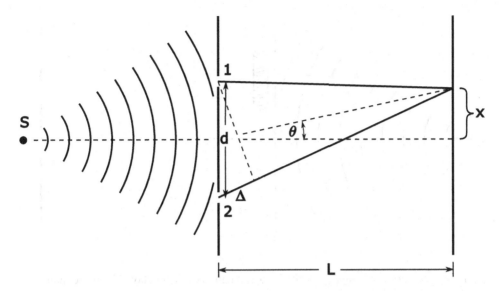

Figure 9.5 Young's two-slit experiment. Quasi-monochromatic radiation from source S illuminates slits 1 and 2, separated by a distance d. Radiation from the slits interferes a distance x from the optical axis.

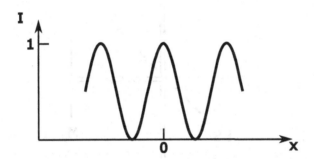

Figure 9.6 Fully modulated fringes.

which is unity for the case shown. There are situations under which the fringes will "wash out." For example, if $\Delta \gtrsim l_c$, as it may be for large x, we get $V < 1$. Similarly, if a plate of glass is placed over slit 1 and the source has a coherence length short compared to the optical delay of the plate, the fringes disappear, as shown in Figure 9.7. A finite size of the source S will also make the fringes disappear. In general, the interference pattern can be written as

$$I = I_1 + I_2 + 2\sqrt{I_1 I_2} \, \text{Re} \, \tilde{\gamma}_{12}(\tau), \tag{9.11}$$

where $\text{Re} \, \tilde{\gamma}_{12}(\tau) = |\tilde{\gamma}_{12}(\tau)| \cos{(\alpha_{12}(\tau) - \phi)}$ and $\alpha_{12}(\tau)$ is the phase offset of slits 1 and 2 with respect to the source.

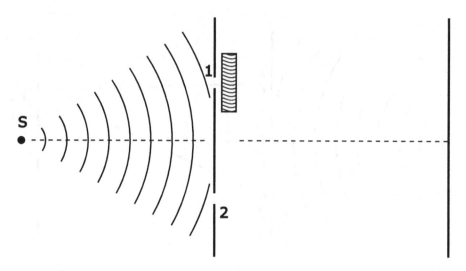

Figure 9.7 Fringes disappear if the light through one slit is delayed by more than its coherence length.

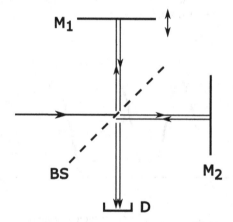

Figure 9.8 Michelson interferometer consisting of a beam splitter (BS), a movable mirror (M_1), a fixed mirror (M_2), and a detector (D).

$$V = \frac{2\sqrt{I_1 I_2}}{I_1 + I_2} |\tilde{\gamma}_{12}(\tau)|, \qquad (9.12)$$

which just equals $|\tilde{\gamma}_{12}(\tau)|$ if $I_1 = I_2$.

9.4 Michelson interferometer

Instead of interfering two *portions* of a wavefront, as in Young's experiment, consider a single wavefront which is split in amplitude (using a partially reflective mirror) as in Figure 9.8. Assume equal intensity division. If M_1 and M_2 are equidistant, the beams will be in phase at the detector. As M_1 is moved, the relative phase will change, modulating the intensity. Fringes will occur only so long as the difference in distances is smaller than the coherence length. Again, the fringe visibility measures the coherence,

$$V = |\tilde{\gamma}_{11}(\tau)|. \tag{9.13}$$

9.5 Michelson stellar interferometer

Consider stellar radiation which is incident on two small, spatially separate apertures, as shown in Figure 9.9. Assume quasi-monochromatic light (some sort of narrow-band filter). For a single point source, fringes will be formed as in Young's experiment (with unit visibility). If there is a second compact source of equal intensity, separated by an angle θ, it also will produce fringes. But because of the extra path length difference from the source to the two apertures (and the resulting phase shift), this second set of fringes will be spatially displaced from the first set. If $h\theta = \lambda/2$, the phase shift will be 180° and the fringe pattern will wash out ($V = 0$), as shown in Figure 9.10. There will be a similar loss of visibility for a single extended source. If we have a circular source of diameter θ with uniform brightness,

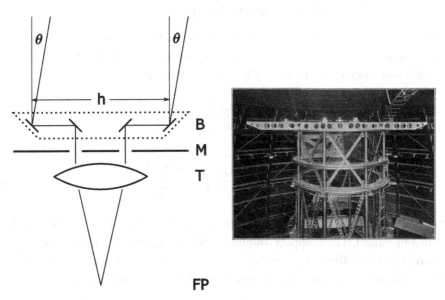

Figure 9.9 (Left) Schematic of Michelson's stellar interferometer showing the rigid support beam B, four small pickoff mirrors, the effective aperture mask M, the telescope T (represented here by a lens), and the focal plane FP. Two sources separated by an angle θ are being observed using a baseline length h. (Right) Photograph of the interferometer attached to the upper end of the Mt. Wilson Hooker telescope.

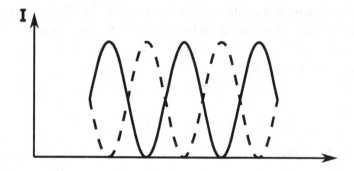

Figure 9.10 Fringes from two point sources being washed out.

$$V = |\tilde{\gamma}_{12}(0)| = \left| \frac{2J_1(\pi h\theta/\lambda)}{\pi h\theta/\lambda} \right|. \qquad (9.14)$$

Compare this result with the Airy pattern. There is clearly a close relationship between coherence and diffraction theory.

9.6 Van Cittert–Zernike theorem

Consider a screen illuminated by quasi-monochromatic radiation from a source of finite extent as in Figure 9.11. Decompose the source into small elements of area, $d\sigma_m$. Each $d\sigma_m$ contributes to the fields at points 1 and 2,

$$E_{m1}(t) = \tilde{A}_m\left(t - \frac{R_{m1}}{c}\right) \frac{e^{-i2\pi \nu(t-R_{m1}/c)}}{R_{m1}}, \qquad (9.15)$$

$$E_{m2}(t) = \tilde{A}_m\left(t - \frac{R_{m2}}{c}\right) \frac{e^{-i2\pi \nu(t-R_{m2}/c)}}{R_{m2}}. \qquad (9.16)$$

Each contribution has a phase term in the numerator and a R^{-2} intensity falloff (R^{-1} in amplitude). The complex amplitude of the emission from σ_m, evaluated at the correct retarded time, is given by $\tilde{A}_m(t - R_{m1}/c)$. The total fields at points 1 and 2 are given by superposition of the contributions from all the σ_m,

$$E_1(t) = \sum_m E_{m1}(t), \qquad (9.17)$$

$$E_2(t) = \sum_m E_{m2}(t). \qquad (9.18)$$

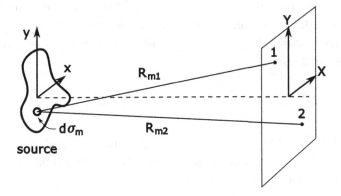

Figure 9.11 A source of radiation (red in electronic version) in the (x, y) plane is decomposed into differential elements of area (blue). These illuminate points 1 and 2 in the (X, Y) plane. Adapted from Born & Wolf (1999).

Evaluate the correlation of the fields at points 1 and 2,

$$\Gamma_{12}(0) = \langle E_1(t)\, E_2^*(t)\rangle \tag{9.19}$$

$$= \sum_m \langle E_{m1}(t)\, E_{m2}^*(t)\rangle + \sum_{m\neq n}\sum \langle E_{m1}(t)\, E_{n2}^*(t)\rangle \tag{9.20}$$

$$= \sum_m \langle E_{m1}(t)\, E_{m2}^*(t)\rangle, \tag{9.21}$$

since m and n are independent radiators. Plugging in the expressions for $E_{m1}(t)$ and $E_{m2}(t)$,

$$\Gamma_{12}(0) = \sum_m \left\langle A_m\left(t - \frac{R_{m1}}{c}\right) A_m^*\left(t - \frac{R_{m2}}{c}\right)\right\rangle \frac{e^{i2\pi v(R_{m1}-R_{m2})/c}}{R_{m1}\,R_{m2}}. \tag{9.22}$$

Assuming temporal coherence (quasi-monochromatic radiation), we can neglect the difference in the arguments of A_m, that is, $l_c \gg R_{m1} - R_{m2}$. Assuming further that the process is stationary,

$$\Gamma_{12}(0) = \sum_m \left\langle A_m(t)\, A_m^*(t)\right\rangle \frac{e^{i2\pi v(R_{m1}-R_{m2})/c}}{R_{m1}\,R_{m2}} \tag{9.23}$$

$$= \int_\sigma I(\vec{r})\, \frac{e^{ik(R_1-R_2)}}{R_1 R_2}\, d\sigma. \tag{9.24}$$

Expressed as the complex degree of coherence,

$$\gamma_{12}(0) = \frac{1}{\sqrt{I_1 I_2}}\int_\sigma I(\vec{r})\, \frac{e^{ik(R_1-R_2)}}{R_1 R_2}\, d\sigma, \tag{9.25}$$

where $I_1 = \int_\sigma d\sigma\, I(\vec{r})/R_1^2$, and $I_2 = \int_\sigma d\sigma\, I(\vec{r})/R_2^2$. Note the similarity with the Kirchhoff diffraction integral. Now let $R_1 \approx R_2 \approx R$ in the denominator. If the screen–source distance is large compared with the source size and the displacements on the screen,

$$R_1 - R_2 \approx \frac{(X_1^2 + Y_1^2) - (X_2^2 + Y_2^2)}{2R} - \frac{(X_1 - X_2)x + (Y_1 - Y_2)y}{R}. \tag{9.26}$$

Terms of order x^2 and y^2 appear in the expansions of R_1 and R_2 but cancel in the subtraction. The first part of the above expression may be taken outside of the integral as a phase factor,

$$\gamma_{12}(0) \approx e^{i\phi} \frac{\int_\sigma I(x, y)\, e^{-ik(px+qy)}\, dx\, dy}{\int_\sigma I(x, y)\, dx\, dy}, \tag{9.27}$$

where $p = (X_1 - X_2)/R$ and $q = (Y_1 - Y_2)/R$. The modulus of the degree of coherence is equal to the Fourier transform of the source intensity distribution, normalized by the total intensity.

9.7 Étendue of coherence

9.7.1 One approach

Consider a uniformly bright circular source of radius r_0. We know that

$$|\gamma_{12}(0)| = \left| \frac{2J_1(u)}{u} \right|, \tag{9.28}$$

where

$$u = 2\pi \frac{r_0}{R\lambda} \sqrt{(X_1 - X_2)^2 + (Y_1 - Y_2)^2}. \tag{9.29}$$

This source subtends a solid angle at the screen of

$$\Omega = \frac{\pi\, r_0^2}{R^2}. \tag{9.30}$$

Let point 1 on the screen be fixed (e.g. $X_1 = Y_1 = 0$), and let point 2 extend out to a radius ρ from point 1. Over how large a region on the screen will the radiation at these two points remain coherent? They will be more coherent if closer together and less coherent if farther apart. Let us *adopt* a criterion of $|\gamma_{12}(0)| \gtrsim 0.577$ as a reasonable dividing line between coherent and incoherent. If $|\gamma_{12}(0)| \gtrsim 0.577$, then the Bessel function argument $u \lesssim 2$. Therefore the largest value of u is $u_{max} = 2$, in which case

$$\rho = \sqrt{X_2^2 + Y_2^2} = \frac{R\lambda}{2\pi r_0} u_{max} = \frac{R\lambda}{\pi r_0}, \tag{9.31}$$

$$A = \pi \rho^2 = \frac{R^2 \lambda^2}{\pi r_0^2}. \tag{9.32}$$

The étendue is then equal to

$$A\Omega = \frac{R^2 \lambda^2}{\pi r_0^2} \frac{\pi r_0^2}{R^2} = \lambda^2. \tag{9.33}$$

9.7.2 An alternate approach

The preceding was a bit unsatisfactory due to the lack of a precise boundary to the coherence area and due to the seemingly arbitrary choice of $|\gamma_{12}| \gtrsim 0.577$. Since A and Ω are related by Fourier transforms, they cannot both have sharp edges, but we can make the derivation seem a bit less arbitrary if we consider a circular aperture with an Airy pattern beam.

Consider a radio telescope of radius a ($A = \pi a^2$) with a single-mode detector at the focus as in Figure 9.12. Assume the aperture is uniformly illuminated with uniform phase and intensity. The sensitivity pattern will be

$$P(\theta, \phi) = P_0 \left[\frac{2J_1(ka\theta)}{ka\theta} \right]^2. \tag{9.34}$$

Define the solid angle of this beam by

$$\Omega_A = \int \frac{P(\theta, \phi)}{P_0} d\Omega \tag{9.35}$$

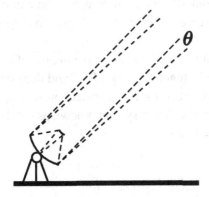

Figure 9.12 In general the radiation pattern of a radio telescope will be a function of the distance θ from the optical axis and an azimuthal angle ϕ.

$$= \int_0^\pi \left[\frac{2J_1(ka\theta)}{ka\theta} \right]^2 2\pi \sin\theta \, d\theta \tag{9.36}$$

$$\approx 4 \int_0^\infty \frac{J_1^2(x)}{x^2} \left(\frac{\lambda}{2\pi a} \right)^2 2\pi \, x \, dx, \tag{9.37}$$

where $k = 2\pi/\lambda$ and $x = ka\theta$.

$$\Omega_A = \frac{2}{\pi} \frac{\lambda^2}{a^2} \int_0^\infty \frac{J_1^2(x)}{x} dx \tag{9.38}$$

$$= \frac{\lambda^2}{\pi a^2}. \tag{9.39}$$

This gives an étendue of precisely λ^2,

$$A\Omega_A = \pi a^2 \frac{\lambda^2}{\pi a^2} = \lambda^2. \tag{9.40}$$

This may seem to lack the arbitrary step necessary in the first approach, but it actually does have one. That step is the definition of the solid angle of the beam. Since the beam lacks sharp edges, there is no unique definition, although the one used is certainly reasonable.

9.8 Aperture synthesis

The van Cittert–Zernike theorem gives us a procedure for determining the intensity distribution of a source: measure the complex degree of coherence for points at various separations and perform a Fourier transform. This is exactly what is done with a Michelson stellar interferometer and with radio interferometers. Using various pairs of apertures (baselines) one samples various spatial frequency components of the source. If enough components are measured, one can reconstruct the appearance of the source as if it were imaged by a single large aperture (hence the name aperture synthesis).

At radio frequencies it is easy to measure the degree of coherence; signals from two telescopes are brought together, multiplied, and then integrated to get $\langle E_1 E_2 \rangle$ as in Figure 9.13. Note the need for linear (phase-preserving) amplifiers. For finite bandwidth there is also a need for a system, known as a delay line, to equalize the time delays ($\Delta\tau \ll \tau_c = 1/\Delta\nu$).

9.8.1 Arrays of antennas

Another way to view aperture synthesis is by considering an array of *transmitting* antennas, as in Figure 9.14. There exists a reciprocal relationship between the

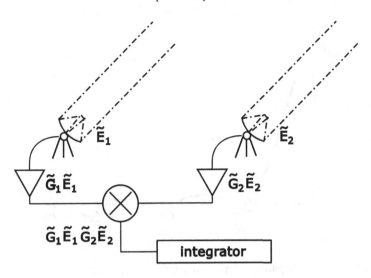

Figure 9.13 In a radio interferometer the electric fields at two points are amplified and then brought together and multiplied. The result is integrated (low pass filtered) to give the visibility.

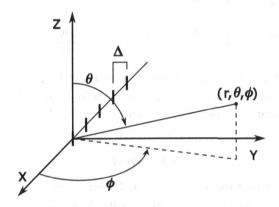

Figure 9.14 A linear array of dipole antennas (red) broadcasting in phase.

power pattern for transmitting and the sensitivity pattern for reception. For simplicity, consider a linear array of N pure dipoles. A single dipole is not very directional. What happens when you have an array of them? Add the electric fields at large distances, where Δ is the spacing of the dipoles:

$$\alpha = k\Delta \sin \theta \cos \phi, \qquad (9.41)$$

$$\vec{E} = \hat{\theta} A_0 \frac{\sin \theta}{r} e^{i(kr - \omega t)} \left\{ 1 + e^{i\alpha} + e^{i2\alpha} + \cdots + e^{i(N-1)\alpha} \right\}. \qquad (9.42)$$

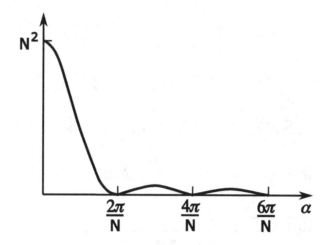

Figure 9.15 Antenna pattern of a linear array relative to that of a single dipole.

Since $1 + x + x^2 + \cdots + x^{N-1} = (x^N - 1)/(x - 1)$,

$$\vec{E} = \hat{\theta}\, A_0\, \frac{\sin\theta}{r}\, e^{i(kr-\omega t)}\, \frac{e^{iN\alpha} - 1}{e^{i\alpha} - 1} \tag{9.43}$$

$$= \hat{\theta}\, A_0\, \frac{\sin\theta}{r}\, e^{i(kr-\omega t)}\, e^{i(N-1)\alpha/2}\, \frac{\sin(N\alpha/2)}{\sin\alpha/2}. \tag{9.44}$$

The radiation pattern of a single dipole is modulated by $\sin^2(N\alpha/2)/\sin^2(\alpha/2)$. This gives a narrow beam, strongly peaked in the $\alpha = 0$ direction, as shown in Figure 9.15. The beam is stronger and narrower for large N.

9.9 Caveat

Second order coherence theory is, in fact, far richer than the treatment in this chapter may suggest. For further details on the interrelationships between spatial and spectral coherence for partially coherent radiation, one may refer to Wolf (2007) or other references on the topic.

9.10 Fourth order coherence

Some phenomena require consideration of higher order coherence functions such as the fourth order quantity

$$\tilde{\Gamma}^{(2,2)}(\vec{r}_1, t_1; \vec{r}_2, t_2; \vec{r}_3, t_3; \vec{r}_4, t_4) = \langle \tilde{E}_1(t_1)\, \tilde{E}_2^*(t_2)\, \tilde{E}_3(t_3)\, \tilde{E}_4^*(t_4) \rangle, \tag{9.45}$$

where, as in the beginning of this chapter, the subscripts refer to separate spatial points. The choice of which fields are complex conjugated is arbitrary, as long as two are conjugated and two are not. Our argument follows that of Wolf (2007), although in a different order and with somewhat different notation. Multivariate Gaussian random processes are completely determined by their mean and their second moments. For fields with zero means, the fourth order moment given above is determined by second order moments in the sense

$$\tilde{\Gamma}^{(2,2)}(\vec{r}_1, t_1; \vec{r}_2, t_2; \vec{r}_3, t_3; \vec{r}_4, t_4) = \tilde{\Gamma}^{(1,1)}(\vec{r}_1, t_1; \vec{r}_2, t_2)\tilde{\Gamma}^{(1,1)}(\vec{r}_3, t_3; \vec{r}_4, t_4)$$

$$+ \tilde{\Gamma}^{(1,1)}(\vec{r}_1, t_1; \vec{r}_4, t_4)\tilde{\Gamma}^{(1,1)}(\vec{r}_2, t_2; \vec{r}_3, t_3) \quad (9.46)$$

$$= \tilde{\Gamma}_{12}(t_1, t_2)\tilde{\Gamma}_{34}(t_3, t_4) + \tilde{\Gamma}_{14}(t_1, t_4)\tilde{\Gamma}_{23}(t_2, t_3). \quad (9.47)$$

The second version corresponds to the notation used at the beginning of this chapter.

9.10.1 Intensity interferometry

One of the first things we are taught in electromagnetic theory is the principle of superposition, by which two electric field components are added and then squared to give the intensity. The squaring operation gives a term proportional to the product of the two fields. The magnitude of this term depends on the phase difference, leading to interference phenomena as in Young's experiment. This is not the whole story. Take the expression for fourth order coherence and let $\vec{r}_1 = \vec{r}_2$ and $\vec{r}_3 = \vec{r}_4$. Assume further that we are looking at a stationary Gaussian random field so that we can write $t = t_1 = t_2$ and $t + \tau = t_3 = t_4$. Therefore we can write

$$\langle \tilde{E}_1(t)\tilde{E}_1^*(t)\tilde{E}_3(t+\tau)\tilde{E}_3^*(t+\tau) \rangle = \tilde{\Gamma}_{11}(0)\tilde{\Gamma}_{33}(0) + \tilde{\Gamma}_{13}(\tau)\tilde{\Gamma}_{31}(-\tau) \quad (9.48)$$

$$= \tilde{\Gamma}_{11}(0)\tilde{\Gamma}_{33}(0) + \tilde{\Gamma}_{13}(\tau)\tilde{\Gamma}_{13}^*(\tau). \quad (9.49)$$

Hanbury Brown realized that this implied there would be correlations between *intensity* fluctuations in two separate antennas and that these would be related to the complex degree of coherence (Hanbury Brown et al., 1952). Many did not initially believe this claim. Hanbury Brown's reasoning goes as follows. For simplicity we will rename point 3 as 2 so that points 1 and 2 correspond to the two separate antennas. Consider intensity fluctuations ΔI_1 and ΔI_2. If one correlates these signals,

$$\langle \Delta I_1(t)\Delta I_2(t+\tau) \rangle = \langle [I_1(t) - \langle I_1(t) \rangle][I_2(t+\tau) - \langle I_2(t+\tau) \rangle] \rangle \quad (9.50)$$

$$= \langle I_1(t)I_2(t+\tau) \rangle - \langle I_1(t) \rangle \langle I_2(t+\tau) \rangle \quad (9.51)$$

$$= \langle E_1(t)E_1^*(t)E_2(t+\tau)E_2^*(t+\tau) \rangle$$

$$- \langle E_1(t)E_1^*(t) \rangle \langle E_2(t)E_2^*(t) \rangle. \quad (9.52)$$

Substituting our result for the fourth order correlation function,

$$\langle \Delta I_1(t) \Delta I_2(t+\tau) \rangle = \tilde{\Gamma}_{11}(0)\tilde{\Gamma}_{22}(0) + \tilde{\Gamma}_{12}(\tau)\tilde{\Gamma}_{12}^*(\tau) - \tilde{\Gamma}_{11}(0)\tilde{\Gamma}_{22}(0) \quad (9.53)$$

$$= |\Gamma_{12}(\tau)|^2, \quad (9.54)$$

$$\frac{\langle \Delta I_1(t) \Delta I_2(t+\tau) \rangle}{\langle I_1(t) \rangle \langle I_2(t) \rangle} = |\gamma_{12}(\tau)|^2. \quad (9.55)$$

This technique produced important early results in the visible (using photomultipliers) and at radio wavelengths, although it is significantly less sensitive than modern techniques.

Exercises

9.1 Consider the antenna array in Section 9.8.1. What is the antenna pattern if $\Delta = \lambda/2$ and if the antennas are driven with phases alternating by π (relative to the antenna at the origin, the remaining antennas have phase shifts of π, 0, π, ...)?

9.2 In Young's double slit experiment, approximately how large can the source size get before the interference fringes are lost? Consider a slit spacing of 1 mm, a wavelength of 600 nm, and a distance between the source and the slits of 1 meter. What is the maximum transverse extent of the source?

10

Spectroscopy

Spectroscopic techniques vary from one wavelength band to another. Methods for performing spectroscopic analysis at radio wavelengths will be covered in Chapter 12. Here we will concentrate on a variety of techniques, focussed on the visible and infrared bands, which also have some applicability at shorter wavelengths. Greater detail on these and other spectroscopic devices may be found in Schroeder (2000), Kitchin (2009), Born & Wolf (1999), and elsewhere.

10.1 Multiple beam interference

Consider a pair of planar parallel dielectric interfaces, as shown in Figure 10.1. The reflected and transmitted waves consist of numerous interfering components. The Fresnel coefficients for the individual surfaces r_{12}, t_{12}, r_{23}, and t_{23} are already known. But what are the overall reflection and transmission coefficients for the pair of interfaces together? Consider the first two contributors to the reflected beam. The difference in optical path length is given by

$$\Delta = \frac{2n_2 d}{\cos\theta_2} - h\, n_1 \sin\theta_1 \tag{10.1}$$

$$= \frac{2n_2 d}{\cos\theta_2} - (2d\tan\theta_2)(n_2\sin\theta_2) \tag{10.2}$$

$$= \frac{2n_2 d}{\cos\theta_2}(1 - \sin^2\theta_2) \tag{10.3}$$

$$= 2n_2 d\cos\theta_2. \tag{10.4}$$

The resulting phase difference is

$$\phi = 2n_2 d\, \frac{\omega}{c}\cos\theta_2. \tag{10.5}$$

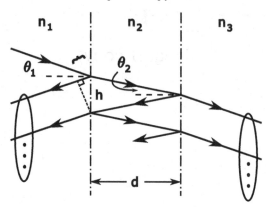

Figure 10.1 Interference between plane-parallel dielectric interfaces. The electric fields of the reflected beams (blue in electronic version) are superimposed, as are the fields of the transmitted beams (green). The brace (orange) shows the optical path length $hn_1 \sin \theta_1$, a contribution to the path length difference between successive reflected beams.

The reflected beam is a superposition of all the individual components

$$r = r_{12} + t_{12}r_{23}t_{21}e^{i\phi} + t_{12}r_{23}r_{21}r_{23}t_{21}e^{i2\phi} + \cdots \qquad (10.6)$$

$$= r_{12} + t_{12}r_{23}t_{21}e^{i\phi} \left[1 + q + q^2 + \cdots \right], \qquad (10.7)$$

where $q = r_{21}r_{23}e^{i\phi}$. Using a well-known series expansion we can write this as

$$r = r_{12} + t_{12}r_{23}t_{21}e^{i\phi} \left[1 - q \right]^{-1} \qquad (10.8)$$

$$= \frac{r_{12} + r_{23}e^{i\phi}}{1 + r_{12}r_{23}e^{i\phi}}, \qquad (10.9)$$

where we have used the Stokes relations $r_{12} = -r_{21}$ and $t_{12}t_{21} = 1 - r_{12}^2$. Similarly for the transmitted wave amplitude, t,

$$t = \frac{t_{12}t_{23}e^{i\phi/2}}{1 + r_{12}r_{23}e^{i\phi}}. \qquad (10.10)$$

There is a variety of alternate approaches to this problem. Treating it as a boundary value problem is cumbersome (see Born & Wolf, 1999). But it can be reduced to an easy and powerful matrix method covered in the exercises at the end of the chapter.

10.1.1 Airy function

In the previous section we found amplitudes of reflected and transmitted waves. We must square these to get intensities. Do not forget that the intensity also depends on the index of refraction. But here, for simplicity, let media 1 and 3 be identical, in which case $r_{12} = -r_{23}$.

$$r = \frac{r_{12} + r_{23}e^{i\phi}}{1 + r_{12}r_{23}e^{i\phi}} \tag{10.11}$$

$$= \frac{r_{12}(1 - e^{i\phi})}{1 - r_{12}^2 e^{i\phi}}, \tag{10.12}$$

$$R = \frac{4r_{12}^2 \sin^2(\phi/2)}{(1 - r_{12}^2)^2 + 4r_{12}^2 \sin^2(\phi/2)}. \tag{10.13}$$

And for the transmitted intensity,

$$T = \frac{(1 - r_{12}^2)^2}{(1 - r_{12}^2)^2 + 4r_{12}^2 \sin^2(\phi/2)}. \tag{10.14}$$

This is known as the Airy function, which is illustrated in Figure 10.2. Note that $R + T = 1$. For large values of r_{12}^2, this gives a good, narrow, well-defined passband.

10.1.2 Anti-reflection coating

An interface between air and a dielectric of index n, at normal incidence, has an intensity reflection coefficient

$$R = \left(\frac{1 - n}{1 + n}\right)^2. \tag{10.15}$$

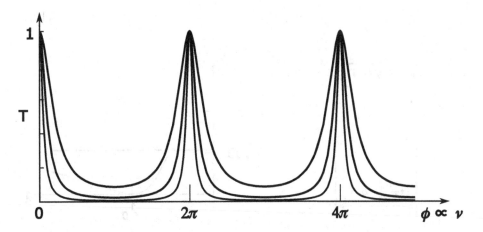

Figure 10.2 The Airy function for a finesse of 5 (upper curve, blue in electronic version), 10 (middle, red), and 20 (lower, black), corresponding to values of $r_{12}^2 = r_{23}^2 = 0.54$, 0.73, and 0.85, respectively. See Section 10.2 for the definition of finesse.

For typical glasses, $n \approx 1.6$, which implies that $R \approx 0.05$. For a typical optical system with multiple surfaces (e.g. several lenses), this can amount to a substantial loss of intensity. It is possible to greatly reduce the reflectivity of such a surface with a thin coating with an intermediate index of refraction. For the general case of $n_1 \neq n_2 \neq n_3$,

$$R = \frac{r_{12}^2 + r_{23}^2 + 2r_{12}r_{23}\cos\phi}{1 + r_{12}^2 r_{23}^2 + 2r_{12}r_{23}\cos\phi},$$ (10.16)

$$\phi = 2n_2 d \frac{\omega}{c}.$$ (10.17)

If we are able to find a material with $n_2 = \sqrt{n_1 n_3}$, then $r_{12} = r_{23}$. If, in addition, $d = \lambda_0/(4n_2)$, then $\cos\phi = -1$ and $R = 0$, giving a perfect anti-reflection coating at λ_0, as in Figure 10.3. One can construct a *broadband* low reflectivity coating using multiple layers, which is also useful if we can't find a material with $n_2 = \sqrt{n_1 n_3}$.

10.1.3 Enhanced reflection coating

It is possible to make coatings which enhance reflectivity in a similar way, but to do so it is necessary to enhance the discontinuities in the refractive index. For example, pick $n_2 \gg n_1, n_3$. Going from low index to high index gives

$$r_{12} = \frac{n_1 - n_2}{n_1 + n_2} < 0,$$ (10.18)

$$|r_{12}| \approx 1.$$ (10.19)

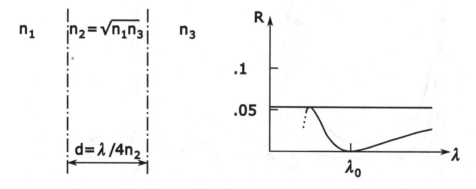

Figure 10.3 (Left) Anti-reflection coat a piece of optics by applying a thin dielectric layer of appropriate index and thickness. (Right) Reflectivity of a material with constant index $n = 1.6$ before (black) and after (red in electronic version) anti-reflection coating.

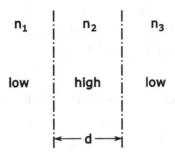

Figure 10.4 Enhanced reflectivity coating.

Going from high to low gives

$$r_{23} = \frac{n_2 - n_3}{n_2 + n_3} > 0, \qquad (10.20)$$

$$|r_{23}| \approx 1. \qquad (10.21)$$

For a resonant thickness $d = \lambda_0/(4n_2)$, as in Figure 10.4, this implies that $\cos\phi = -1$,

$$R = \frac{r_{12}^2 + r_{23}^2 + 2r_{12}r_{23}\cos\phi}{1 + r_{12}^2 r_{23}^2 + 2r_{12}r_{23}\cos\phi}. \qquad (10.22)$$

If we let $r \approx |r_{12}| \approx |r_{23}|$,

$$R \approx \frac{4r^2}{(1 + r^2)^2} \approx 1 \qquad (10.23)$$

since $r^2 \approx 1$.

Mirrors made of glass substrates coated with thin layers of metals such as aluminum, silver, or gold have high – but far from perfect – reflectivity. Aluminum coated mirrors are most common in astronomical telescopes, but such mirrors have reflectivities of 85–92% across the visible band, with even poorer efficiency in the ultraviolet. Metallic surfaced mirrors are often overcoated with dielectrics such as SiO_2 or MgF primarily for protection of the soft underlying metal. But it is possible to make multilayer dielectric coatings which enhance mirror reflectivities up to as high as 99.999% at a specific wavelength. Such mirrors are quite important to the laser industry, but it is not generally possible to make such coatings for the large mirrors used in most astronomical telescopes.

10.1.4 Interference filters

With multiple-layer thin films it is possible to tailor regions of high and low reflectivity to fit your particular application. For example, you might want to make a filter which selectively passes Hα emission. The number of layers can be of order 20, giving some 40 variables (20 thicknesses and 20 indices of refraction), all of which may be adjusted to optimize performance.

10.2 Fabry–Perot interferometer (etalon)

Construct an interferometer using a pair of high reflectivity surfaces, as shown in Figure 10.5. The transmission is given by the Airy function,

$$T = \left[1 + \frac{4r^2}{(1 - r^2)^2} \sin^2(\phi/2)\right]^{-1}. \tag{10.24}$$

The transmission peaks are narrow if r^2 is large. At normal incidence, peaks occur when

$$\phi = 2nd \frac{\omega}{c} \cos \theta = 2d \frac{\omega}{c} = m\, 2\pi. \tag{10.25}$$

A particular transmission peak is designated by the *order* m. What is the full-width at half-maximum (γ) of the Airy function peaks?

$$T = \frac{1}{2}, \tag{10.26}$$

$$\frac{4r^2}{(1 - r^2)^2} \sin^2(\phi/2) = 1, \tag{10.27}$$

$$\Delta\phi^2 \approx \frac{(1 - r^2)^2}{r^2}, \tag{10.28}$$

where $\phi = m\, 2\pi + \Delta\phi$ and $\Delta\phi \ll 2\pi$. The FWHM along the ϕ axis is

$$\gamma = 2 \frac{1 - r^2}{r}. \tag{10.29}$$

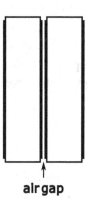

air gap

Figure 10.5 Construction of a Fabry–Perot interferometer using two glass plates, one side of each coated for high reflectivity (red in electronic version) and the other for low reflectivity (blue). The high reflectivity surfaces are flat and parallel with a small, uniform air gap between them.

Define the *finesse*,[1] $\mathcal{F}$, as the ratio of the peak spacing to the width. Under the approximation used above, corresponding to high surface reflectivities and narrow peaks,

$$\mathcal{F} = \frac{2\pi}{\gamma} = \frac{\pi r}{1 - r^2}. \tag{10.30}$$

The resolving power, R, increases with the order, m.

$$R = \frac{\lambda}{\Delta\lambda} = \frac{\nu}{\Delta\nu} = \frac{\phi}{\Delta\phi} = \frac{m2\pi}{\gamma} = m\,\mathcal{F}. \tag{10.31}$$

The transmission pattern shown was for normal incidence. At oblique incidence the resonant peak is shifted,

$$\Delta\phi = 2nd\frac{\omega}{c}(1 - \cos\theta) \approx 2nd\frac{\omega}{c}\frac{\theta^2}{2}, \tag{10.32}$$

for small θ. The angle at which this shift is equal to the peak width is

$$\frac{\theta^2}{2}2d\frac{\omega}{c} = \frac{\gamma}{2}, \tag{10.33}$$

$$\theta^2 = \frac{\gamma}{m\pi} = \frac{2\pi/\mathcal{F}}{\pi R/\mathcal{F}}. \tag{10.34}$$

Therefore the maximum solid angle which can be used without significant loss of resolution is

$$\Omega = \frac{2\pi}{R}. \tag{10.35}$$

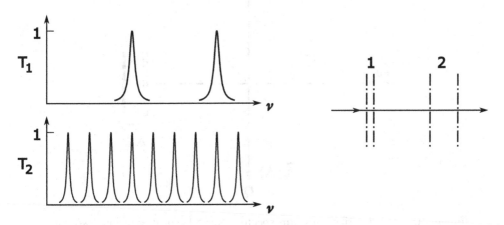

Figure 10.6 A lower order Fabry–Perot interferometer (1) and a high order Fabry–Perot (2) in series, shown along with their transmission patterns.

---

[1] Not to be confused with the *coefficient of finesse*, usually written as F. The exact relationships are $F = 4r^2/(1 - r^2)^2$ and $\mathcal{F} = \pi/[2\arcsin(1/\sqrt{F})]$.

One problem of Fabry–Perots is the transmission of multiple orders.[2] It is possible to use several Fabry–Perots in series, as shown in Figure 10.6, to select the desired order. One can tune to a selected wavelength by varying the spacing(s).

10.3 Fourier transform spectrometer

Consider a Michelson interferometer with a moving mirror, as shown in Figure 10.7. Assume 50/50 division at the beam splitter. For a quasi-monochromatic wave of wavenumber k, the intensity at Q is

$$\tilde{I}_Q(\Delta) = \frac{1}{2} I_0[1 + \cos k\Delta]. \tag{10.36}$$

For multiple frequency components,

$$\tilde{I}_Q(\Delta) = \frac{1}{2} \int_0^\infty I_0(k) \, [1 + \cos k\Delta] \, dk \tag{10.37}$$

$$= \frac{1}{2} \int_0^\infty I_0(k) \, dk + \frac{1}{2} \int_0^\infty I_0(k) \cos k\Delta \, dk, \tag{10.38}$$

$$\tilde{I}_Q(0) = \int_0^\infty I_0(k) \, dk, \tag{10.39}$$

$$\tilde{I}_Q(\Delta) = \frac{1}{2} \tilde{I}_Q(0) + \frac{1}{2} \int_0^\infty I_0(k) \cos k\Delta \, dk. \tag{10.40}$$

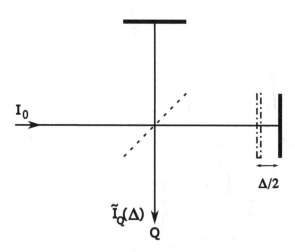

Figure 10.7 In a Fourier transform spectrometer a beam is split into two parts which travel different length paths before recombining. The path difference is varied, and the spectrum is obtained by a Fourier transform of the data.

---

[2] This can be considered as a limitation on the *free spectral range* of the Fabry–Perot. A similar situation arises with diffraction gratings, discussed in Section 10.5.

We can recover the spectrum $I_0(k)$ by an inverse Fourier transform,

$$I_0(k) = \frac{1}{2\pi} \int_0^\infty [2\tilde{I}_Q(\Delta) - \tilde{I}_Q(0)] \cos k\Delta \; d\Delta.$$

(10.41)

The factor of $1/2\pi$ is present since we have deviated from our convention of having 2π present in the kernel of Fourier transforms. Since in practice we can scan only up to some maximum value of Δ, we end up with limited spectral resolution R,

$$R \approx \frac{2\Delta_{max}}{\lambda},$$

(10.42)

but this resolution can be of order 10^6. As with the Fabry–Perot, this is valid for limited solid angle

$$\Omega = \frac{2\pi}{R}.$$

(10.43)

10.4 Prism spectrograph

A light ray entering a prism of index of refraction n and apex angle α, as shown in Figure 10.8, is deflected as shown. After two applications of Snell's law, one finds the total deflection to be

$$\delta = (\theta_{i1} - \theta_{t1}) + (\theta_{t2} - \theta_{i2})$$

(10.44)

$$= \theta_{i1} - \alpha + \sin^{-1}\left[\sin\alpha\sqrt{n^2 - \sin^2\theta_{i1}} - \cos\alpha\sin\theta_{i1}\right].$$

(10.45)

Since δ increases with n, and for normal dispersion n is larger at blue wavelengths, blue light is deflected more than red light.

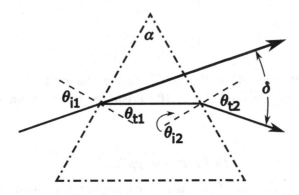

Figure 10.8 A light ray passing through a prism as shown undergoes a total angular deflection of δ. Dashed lines (red in electronic version) show perpendiculars to the surfaces. Notation follows that of Hecht (2002).

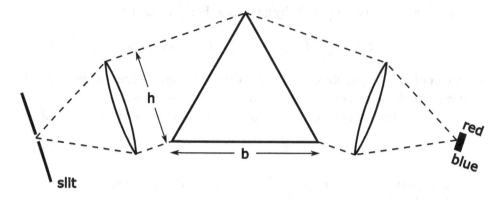

Figure 10.9 A prism spectrograph with slit, collimator lens (left), and camera lens (right).

With prism and grating spectrographs, we distinguish between *dispersion*, the physical separation of different wavelengths, and *resolution*, the ability to distinguish between nearby wavelengths. For a prism, the *angular dispersion*,[3] for the case of symmetrical illumination (illustrated in Figure 10.9; $\theta_{i1} = \theta_{t2}$, $\theta_{i2} = \theta_{t1}$), is given by

$$\frac{d\delta}{d\lambda} = \frac{\sin \alpha}{\cos \theta_{t2} \cos \theta_{t1}} \frac{dn}{d\lambda} \tag{10.46}$$

$$= \frac{b}{h} \frac{dn}{d\lambda}. \tag{10.47}$$

The ray bundle has finite width, h. The resulting diffraction leads to a spread in angle of the deflected beam which limits the resolution. The first minimum of the sinc function gives

$$\Delta \delta = \frac{\lambda}{h}, \tag{10.48}$$

$$R = \frac{\lambda}{\Delta \lambda} = b \left| \frac{dn}{d\lambda} \right|. \tag{10.49}$$

For dense flint glass, $dn/d\lambda \approx 10^{-4}$ nm^{-1}. Taking as a practical limit a prism base thickness of 10 cm, this implies $R \lesssim 10^4$. Prism spectrographs are inherently low resolution.

10.4.1 Prism applications

One important application of prisms is obtaining low resolution spectra of large numbers of stars in *objective prism* surveys. An objective prism is placed at the

---

[3] The linear dispersion is obtained by multiplying the angular dispersion by the focal length of the camera.

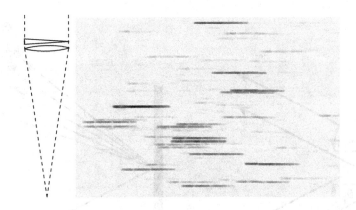

Figure 10.10 (Left) Schematic illustration of an objective prism. (Right) An objective prism image of the Hyades (courtesy of Dean Ketelsen). The Balmer absorption lines can be easily seen. This technique can also be used to find emission lines such as [O III] 495.9, 500.7 nm. For color version of figure, see plate section.

entrance pupil of a telescope, where the light is collimated (near the *objective* lens). In the focal plane each stellar image is replaced by a small image of its spectrum, as shown in Figure 10.10. Note that there is no slit, as would be present in a normal prism spectrograph. This is a type of *slitless* spectrograph. This is useful for spectral classification work and detection of emission-line objects. Typical resolution in such applications is of order $R \approx 1000$. Prisms are also used as pre-dispersers (order-sorting devices) for higher resolution spectrometers such as Fabry–Perots, and in conjunction with gratings (*grism*) for non-objective slitless spectroscopy.

10.5 Diffraction gratings

In astronomy one typically encounters diffraction gratings used in reflection, although transmission gratings are also possible. A diffraction grating is an object with a spatially periodic reflection or transmission function (a complex quantity, affecting the wave amplitude and/or phase). Conceptually, the simplest reflection grating is an array of line scattering centers, as in Figure 10.11. Each line, individually, scatters energy over a wide range of angles. Constructive interference occurs when the net path length difference corresponds to an integral number of wavelengths. The *grating equation* says that

$$m \lambda = a \left(\sin \theta_m - \sin \theta_i \right). \tag{10.50}$$

The integer m gives the spectral *order* of the various maxima. Specular reflection (m = 0), which is non-dispersive, occurs when $\theta_m = \theta_i$. The various orders are

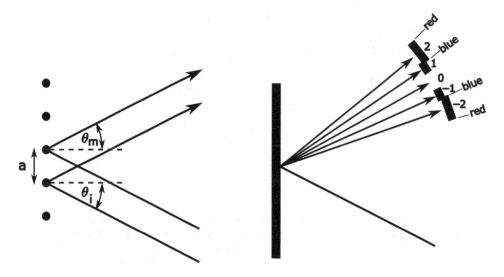

Figure 10.11 (Left) Incident and diffracted light off an array of line scattering centers. Dashed lines (red in electronic version) denote perpendiculars to the plane of the grating. (Right) Arrangement of different orders of diffraction around specular reflection (m = 0).

arrayed on either side of m = 0. Within each order, red light is diffracted more than blue light; it falls farther from m = 0.

10.5.1 *Grating properties*

The dispersion of a grating is obtained by differentiating the grating equation. For fixed θ_i,

$$m \, d\lambda = a \cos \theta_m \, d\theta_m, \tag{10.51}$$

$$\frac{d\theta_m}{d\lambda} = \frac{m}{a \cos \theta_m}. \tag{10.52}$$

The dispersion is larger for higher orders. As with a prism, the size of the grating limits the available resolution. If there are N grooves, the length of the grating is Na. At oblique illumination the ray bundle has a width $Na \cos \theta_m$. Therefore,

$$\Delta\theta_m = \frac{\lambda}{Na \, \cos \theta_m} \tag{10.53}$$

at the first minimum of the sinc function.

$$R = \frac{\lambda}{\Delta\lambda} = m \, N \tag{10.54}$$

$$= Na \, \frac{\sin \theta_m - \sin \theta_i}{\lambda}. \tag{10.55}$$

A typical large astronomical grating might have of order 10^5 grooves (100 mm at 1000 lines per mm), giving a resolving power of 10^5 in first order.

A disadvantage of grating spectrographs is that the various orders will overlap,

$$a \left(\sin \theta_m - \sin \theta_i \right) = (m + 1) \, \lambda = m \, (\lambda + \Delta\lambda), \tag{10.56}$$

$$\Delta\lambda_{FSR} = \frac{\lambda}{m}. \tag{10.57}$$

The *free spectral range* $\Delta\lambda_{FSR}$ is small for large m. Grating spectrographs will also have limited solid angle acceptance. For typical applications, $\Omega \approx 10^{-2}/R$ or $10^{-1}/R$, much smaller than for Fabry–Perot interferometers or Fourier transform spectrometers.

10.5.2 Grating profiles

In practice, one wants a diffraction grating which intercepts a large fraction of the incident light and directs it into the *desired* order. This is done by making triangular *blazed* gratings, as shown in Figure 10.12. The maximum efficiency for order m occurs when

$$\theta_m - \theta_i = 2 \, \gamma, \tag{10.58}$$

which corresponds to specular reflection off each facet of the grating. In essence the grating is acting as an ideal phase transformer, transforming the incident planar wavefront into an outgoing planar wavefront (for some wavelength λ).

10.5.3 Czerny–Turner spectrograph

The *Czerny–Turner* spectrograph is a widely used configuration. Note in Figure 10.13 the presence of both input (collimating) and output (camera) mirrors.

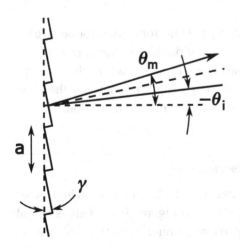

Figure 10.12 Grating blazed at the angle γ. Dashed lines (red in electronic version) show the plane of the grating and its normal. Blue dashed line shows the normal to the facets.

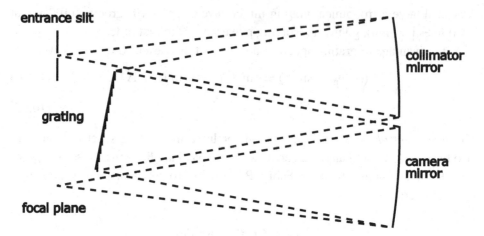

Figure 10.13 A Czerny–Turner spectrograph. Incoming light (red in electronic version) is collimated, diffracted by the grating and then refocussed (blue) onto the focal plane.

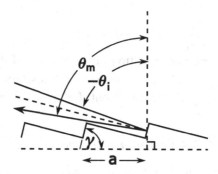

Figure 10.14 An echelle grating, details as in Figure 10.12.

It is possible to design a spectrograph where a single mirror is used for both functions. The input and output beams may use different portions of the mirror. This is called an *Ebert–Fastie* mounting and is considered a special case of the Czerny–Turner design. When the light reflected off of the grating follows nearly the same path as the incoming beam, this is referred to as the *Littrow* configuration.

10.5.4 Echelle spectrograph

An echelle grating is a grating of relatively large period, blazed for use in high order ($m \approx 50$) at a large angle of incidence, as shown in Figure 10.14. This is useful for obtaining high resolution and high dispersion. At high resolution, the echelle has the largest solid angle acceptance and luminosity of all grating spectrometers. There is a disadvantage in that the orders are strongly overlapping. However, these orders may be separated using a dispersing element whose dispersion is in the orthogonal direction (a *cross disperser*). The resulting spectrum appears in a 2-dimensional format, suitable for use with a CCD camera.

10.5.5 Grism spectroscopy

A *grism*, also known as a Carpenter prism, is a combination of a grating and a prism used for slitless, multi-object spectroscopy in much the same way as an objective prism. The grating used is a transmission grating, usually used in first order, and the prism is used to cancel the deflection of the dispersed images at some central wavelength, as shown in Figure 10.15. Since the net result is no deflection, a grism may be inserted into a collimated beam in much the same way as a bandpass filter. With nothing in the beam, a broadband image of the field is obtained. With filters present, narrow-band images are obtained. And with a grism present, each object in the field is dispersed into a spectrum.

10.5.6 Fiber optic spectroscopy

Many modern spectroscopy designs take advantage of the fact that classical long-slit spectrographs can accommodate numerous objects at different heights on the slit. Nature does not put the objects in a line. So a set of optical fibers are placed

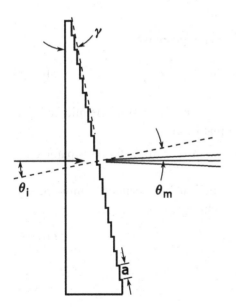

Figure 10.15 A grism consists of a transmission grating in contact with one face of a prism. It may be used for zero deflection at some central wavelength in first order (m = −1).

to collect light from different regions of the focal plane and bring them to different positions on the entrance slit of the spectrograph. An example is the spectrograph of the Sloan Digital Sky Survey.

Exercises

10.1 Consider a set of dielectric layers separated by planar parallel interfaces. The index m designates the various layers, counting from left to right (m = 1, 2, . . .). Let E_m denote the electric field amplitude of the wave traveling left to right in layer m and E'_m the amplitude of the wave traveling right to left. It is claimed that the fields in layers m and m + 1 are related by the matrix equation

$$\begin{pmatrix} E_m \\ E'_m \end{pmatrix} = (C_m) \begin{pmatrix} E_{m+1} \\ E'_{m+1} \end{pmatrix},$$ (10.59)

where

$$(C_m) = \frac{1}{t_{m,m+1}} \begin{pmatrix} e^{-i\beta_m/2} & r_{m,m+1}e^{-i\beta_m/2} \\ r_{m,m+1}e^{i\beta_m/2} & e^{i\beta_m/2} \end{pmatrix}$$ (10.60)

and $r_{m,m+1}$ and $t_{m,m+1}$ are Fresnel coefficients.[4]

a. Show that the results for reflected and transmitted wave amplitudes at a single interface between two dielectrics are given by

$$\begin{pmatrix} E_1 \\ E'_1 \end{pmatrix} = (C_1) \begin{pmatrix} E_2 \\ 0 \end{pmatrix},$$ (10.61)

with $\beta_1 = 0$.

b. Show that the results for two interfaces are given by

$$\begin{pmatrix} E_1 \\ E'_1 \end{pmatrix} = (C_1)(C_2) \begin{pmatrix} E_3 \\ 0 \end{pmatrix},$$ (10.62)

with $\beta_2 = 2d_2 n_2 \omega c^{-1} \cos\theta_2$, where d_2 is the thickness of the middle layer and θ_2 is the angle of propagation in that layer.

This approach may be generalized to a system of multiple layers separated by planar dielectric boundaries.

10.2 Consider a Fourier transform spectrometer which is scanned symmetrically about zero path length difference, with a total excursion Δ_m:

$$-\Delta_m/2 \leq \Delta \leq \Delta_m/2.$$ (10.63)

---

[4] Adapted from Reitz *et al.* (1979).

a. Describe the observed interferogram as the product of the full interferogram (if one scanned from $-\infty$ to ∞) and a boxcar. Calculate the spectral profile of the instrument.

b. By multiplying the observed interferogram by various functions it is possible to modify the instrumental profile (for example, to decrease the wings of the profile). Consider the apodization function

$$1 - \frac{2|\Delta|}{\Delta_m},$$

where $|\Delta| < \Delta_m/2$. Compare the width of the apodized profile to that of the unapodized profile. Compare the heights of the sidelobes.[5]

10.3 The 4 meter Mayall telescope at Kitt Peak has an echelle spectrograph containing a grating with 58 grooves mm^{-1}, blazed at $63°$. The length of the grating is 254 mm. The spectrum is viewed by a CCD camera with a focal length of 590 mm.

a. What is the dispersion of the spectrograph at Hα ($\lambda = 656.3$ nm)?

b. In what order of the grating will Hα fall?

c. What is the free spectral range (FSR) near Hα?

d. What is the limiting (diffraction limited) spectral resolution at this wavelength?

e. In practice, the resolving power is limited by atmospheric turbulence (seeing) to values more like $R = 2 \times 10^4$. Describe the size of the CCD chip (physical size and number of pixels) needed to measure the full FSR of the order containing Hα and to make full use of this resolving power.

10.4 Consider a grism with the geometry shown in Figure 10.15.

a. What is the apex angle, γ, of the prism needed to produce no net deflection at a wavelength λ_c in first order ($m = \pm 1$, depending on the convention used)? Assume the prism is made of glass of index n and the grating has a line spacing of a. Ignore the variation in n with wavelength. (Note that it is necessary to use a modified form of the grating equation.)

b. What is the angular dispersion $d\theta_m/d\lambda$ in first order?

c. Assuming the beam is diffraction limited (ignoring aberrations), what is the spectral resolving power in first order?

d. If placed in a slowly converging beam, what aberrations is this system likely to produce?

---

[5] Adapted from Lèna *et al.* (1998).

11

Ultraviolet, x-ray, and gamma ray astronomy

11.1 Telescopes and imaging

As photon energies increase beyond those of visible and near-ultraviolet light, conventional telescope designs fail. We will look at some variations on conventional optical designs which are effective for x-ray telescopes and then at other approaches to imaging for gamma ray observations. Earth's atmosphere is opaque to radiation beyond the near-ultraviolet, so observations at these energies require space missions.

11.1.1 X-ray telescopes

From our earlier discussion of the complex index of refraction, we can see that for soft x-rays, generally $\omega \gg \omega_{0i}$, and

$$\frac{n^2 - 1}{n^2 + 2} \approx -\frac{1}{3} \frac{\omega_p^2}{\omega^2}. \tag{11.1}$$

When, in addition, $\omega \gg \omega_p$,

$$n^2 \approx 1 - \epsilon. \tag{11.2}$$

The index of refraction is real and slightly less than 1. Materials become transparent – all materials. How then can one hope to build a focussing system, i.e. a telescope? Since materials are transparent, one might first consider making lenses, but the fact that the index of refraction is near unity leads to impractically long focal lengths. Instead, it is better to use mirrors.

Snell's law for refraction is

$$n \sin \theta = n' \sin \theta'. \tag{11.3}$$

If $n = 1$ and $n' < 1$, then for some angles θ there is no allowed value of θ', since $\sin \theta'$ would need to be greater than 1. Thus there can be no refracted ray;

190

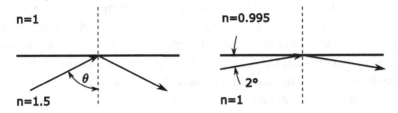

Figure 11.1 Examples of total internal reflection. On the left, visible light is totally internally reflected inside a piece of glass. On the right, an x-ray at grazing incidence is totally "internally" reflected by a metallic surface.

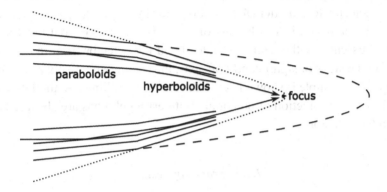

Figure 11.2 Cross section of a Wolter type-I telescope with nested optics such as in ROSAT and Chandra. Paraboloidal shells are shown as solid extensions of dashed lines (red in electronic version). Hyperboloidal shells are extensions of dotted (blue) lines. Not to scale.

there must be total reflection. This is usually called *total internal reflection* and is illustrated in Figure 11.1. The reflected ray carries 100% of the energy and is internal to the higher index material, which for visible optics might be, for example, glass. For x-rays the reflection is still internal to the higher index medium, which in this case is vacuum, but external to the material causing the reflection. To allow the largest possible range of grazing incidence angles, it is desirable to have the index of refraction of the mirror surfaces as much below unity as possible. This requires the largest possible plasma frequency, produced by using high-Z materials.

So to make an x-ray telescope it is necessary for all reflections to occur at grazing incidence, $\theta \approx 89°$. The most common design is known as *Wolter type-I*, a paraboloid followed by a hyperboloid, as shown in Figure 11.2.[1] In designing such a telescope one has the normal optical considerations such as focal length, aberrations, etc., plus the additional complication that all reflections must be at grazing

---

[1] The Wolter type-II design is a paraboloid followed by a hyperboloid placed inside the paraboloid. Wolter type-III consists of a paraboloid followed by an ellipsoid on the outside.

incidence, which means that each reflection can only change the direction of propagation by a small amount. X-ray telescopes are generally not diffraction limited since the diffraction limit is small at such high frequencies and it is very difficult to make and align sufficiently smooth and precise mirrors. Because the grazing-incidence design yields only a small annular collecting area, it is common to use a nested confocal mirror configuration to increase the effective area.

11.1.2 Collimators

A much cruder approach to imaging is the use of *collimators* to limit the field of view of a detector to a particular region on the sky. Such an approach is suitable when one is interested in resolutions of the order of degrees, but not for higher resolution. Essentially the approach is to shield the detector from radiation coming in from the sides or significantly off-axis, thus acting effectively as mechanical "blinders." An example is shown later in the chapter. Collimators, though, are inefficient since a large fraction of potentially interesting photons are absorbed and do not reach the detector.

11.1.3 Tracking designs

High energy photons undergo Compton scattering or pair production events within a detector. Information on the energy of the photon and the direction from which it has arrived must then be derived geometrically from information on the tracks of the resulting energetic electrons and positrons and, if possible, on the characteristics of any scattered photon. We will refer to these very broadly as *tracking designs*, of which there are various implementations. Since these are very detector and mission specific, details are deferred to later in the chapter.

Astronomical sources of x-rays and gamma rays are likely to be highly polarized, but polarimetry is inherently difficult at these energies. One of the more promising approaches is to use the polarization dependence of Compton scattering. A Compton-based detector requires accurate positional detection of both the original Compton scattering event and the direction of the scattered photon (Soffitta *et al.*, 2003).

11.1.4 Coded apertures

Another approach to imaging is one in which directional information is encoded by having an aperture with a variable transmission function located some distance in front of a position-sensitive detector. For best efficiency, the transmission should be either 0 (blocked) or 1 (open) at all positions in the aperture. The transmission

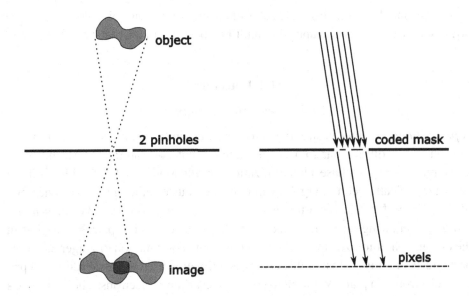

Figure 11.3 Illustration of a pinhole camera and a coded aperture mask.

function must not be patterned in any regular fashion, such as alternating stripes, since that would leave ambiguities in the source position. Instead, the transmission function should be, in some sense, irregular or random.

The technique is analogous to that of a pinhole camera, as illustrated in Figure 11.3. Consider the geometrical optics limit, appropriate for small wavelengths. The only open portion of the aperture plane is the pinhole, which will form a sharp image but with low sensitivity since the aperture is small. How can we increase the sensitivity? Only by opening up more of the aperture plane. So consider adding a second pinhole. In geometrical optics, there will now be a second displaced image whose *intensity* will be added to that of the first image. The intensity distribution of the object being viewed is now encoded, or multiplexed, into the pattern being seen by the detector. Imagine continuing to add pinholes until a significant fraction of the aperture plane is open. It turns out that it is best to have about 50% of the aperture open. A variety of mask patterns are possible, but the principal requirement is that the autocorrelation function of the mask approximate a δ-function. Examples are shown later in the chapter. For further details see In't Zand (1992).

Once the intensity distribution on the detector has been measured, it is necessary to decode it to obtain the original intensity distribution on the sky. A variety of reconstruction algorithms are possible, some of them analogous to those used in radio interferometry. The principal disadvantage of multiplexing systems is that noise associated with a single detector element or Poisson noise due to a single bright source in the sky can appear in multiple locations in the reconstructed image.

11.2 Detectors

11.2.1 Proportional counters

A *proportional counter* is an x-ray detector based on the principle of electron cascade. It is broadly similar to a Geiger counter, but non-saturating. An incoming x-ray induces a gas phase photoionization, as illustrated in Figure 11.4. Filling gases are typically neon or argon, with trace amounts of other gases. The energetic *photoelectron* then ionizes a number of atoms creating some number of *primary electrons*, requiring on average about 25 eV per electron–ion pair. Energy left in the original photoionized atom is released in some combination of Auger electrons or fluorescent x-rays. Most of that energy is ultimately used to create additional primary electron–ion pairs. We want to count these primary electrons. These electrons are accelerated towards a thin wire anode which is held at a moderate potential, of order a few kV. As long as the potential is moderate, the incident photon energy is ultimately converted into a pulse of secondary electrons, where the number of secondary electrons is proportional to the number of primary electrons and hence the initial photon energy. For a 4 keV photon, approximately 4000/25 = 160 primary electons will be generated. If the generation of primary electrons were governed by Poisson statistics,[2] the uncertainty in this number would be about $\sqrt{160}$, yielding a moderate energy resolution, $\Delta E/E \lesssim 0.1$.

This type of detector has good efficiency ($\eta \to 1$), somewhat limited by losses in the thin window material. The window is typically 0.1 mm beryllium or 10 μm mylar, as illustrated in Figure 11.5. An array of wires may be used to make an imaging proportional counter (IPC), as with ROSAT and RXTE, illustrated in Section 11.3.

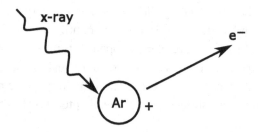

Figure 11.4 Photoionization within a gas proportional counter.

---

[2] The fluctuation statistics are non-Poissonian, as described by something known as the *Fano factor*. The actual fluctuations are smaller than Poisson statistics would suggest, resulting in higher energy resolution.

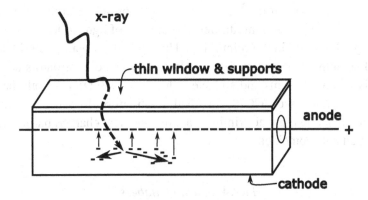

Figure 11.5 Single-pixel gas proportional counter, illustrating photoelectron (red in electronic version) and primary electrons (blue). An Auger electron (green) may also be released and generate primary electrons. As the primary electrons are accelerated towards the anode, they ionize additional atoms, creating a cloud of secondary electrons (not shown).

11.2.2 Solid state detectors

Semiconductor detectors can be operated in a similar mode, as in the ACIS instrument on Chandra. The energy resolution is improved due to the lower threshold for generating pairs, which is a few eV in a semiconductor. Arrays of semiconductors can be used to record the locations of charged particle tracks in tracking detectors.

Superconductors may also be used as detectors. A common application is the *transition edge sensor* (TES) in which superconducting materials are used as microcalorimeters. The absorption of an x-ray photon causes an increase in device temperature, much as in a bolometer. If this temperature change is sufficient to raise the superconductor above its superconducting transition temperature, a large signal can be produced. The energy resolution in this case depends on the heat capacitance of the absorber and the sharpness of the superconducting transition. Resolving powers of order 10^3 have been demonstrated at 10 keV. X-ray detectors based on superconducting tunnel junctions have also been explored.

11.2.3 Scintillators

Scintillators are most commonly used for hard x-rays and gamma rays. The incoming photon interacts within the detector via Compton scattering or pair production, in either case producing energetic electrons. Cross sections are largest for high-Z materials. The energetic electrons then lose energy via bremsstrahlung and ionization. Their energy is transferred to excitons (excited electron states). In inorganic scintillators such as NaI or CsI, these excitons are trapped at *activator sites*, typically thallium (Tl) atoms.[3] The activators fluoresce, emitting a light pulse which is measured by a photomultiplier. The amount of light collected is a measure of the energy deposited in the scintillator. There are also organic scintillators such as crystalline anthracene or a variety of plastics. Organic scintillators consist predominantly of low-Z atoms and so have reduced cross sections. Both the BATSE and OSSE instruments on CGRO use NaI scintillators. In using scintillators as photon detectors, one must discriminate against incident charged particles (cosmic rays) using an anti-coincidence shield.

11.2.4 Spark chambers

Spark chambers are an alternative to scintillators for high energy photons. The initial photon interaction occurs in a high-Z material, such as a sheet of lead or tungsten, producing an e^+e^- pair. This pair, in turn, produce ionization tracks in gas (neon) located between a pair of electrodes. A high voltage pulse, triggered by a scintillator, is applied to the electrodes, producing a visible track of the ionization trail. The EGRET instrument on CGRO uses spark chambers. Since the direction and range of the e^+e^- pair are recorded, one can get the initial energy of the gamma ray. Here also one must use a scintillator as an anti-coincidence shield to discriminate against incident charged particles.

11.3 Recent missions

Although x-ray astronomy dates back to the early 1960s, we begin our discussion with instruments launched in the early 1990s. These are included, not as historical curiosities, but as examples of the wide variety of techniques available to instrumentalists. Of greatest interest, though, will be x-ray and gamma ray instruments launched since the late 1990s and still in operation, such as RXTE, Chandra, XMM-Newton, and Fermi.

11.3.1 ROSAT

ROSAT (ROentgen SATellite) was a German, US, and UK x-ray satellite operating from 1990 to 1999. Originally designed for deployment from the space shuttle, it was reconfigured and launched by a Delta II rocket. The telescope itself was a Wolter type-I design with four nested gold-coated mirror sets. It covered the energy range 0.1–2 keV with approximately 4″ resolution. The peak effective area was $1100\,cm^2$ at about 120 eV, reduced to about $400\,cm^2$ at 1 keV. There were two

---

[3] Or sodium activated cesium iodide, CsI(Na).

principal experiments: PSPC, a multiwire Position Sensitive Proportional Counter, and HRI, a High Resolution Imager which utilized microchannel plates.[4]

The two redundant PSPC detectors were filled with a mix of about 65% argon, 20% xenon, and 15% methane. Photons entered through a window made of Lexan and polypropylene, which was mechanically supported by two grades of wire mesh and by support ribs. Crossed grids of about 180 cathode wires each were used to determine the event location. Quantum efficiency approached 80% at high energies. Spectral (energy) resolving power was a function of photon energy and was of order R ≈ 4.

After the gas supply for the PSPC detectors was nearly exhausted, the HRI was essentially the only operational detector left on ROSAT. The HRI was a microchannel plate detector, similar to that flown on the earlier Einstein satellite but with a CsI photocathode. The spatial resolution of the HRI was 1.7″, which oversamples the telescope PSF. However, the spatial precision of the HRI was somewhat compromised by various problems with the spacecraft pointing control system.

In addition, the main ROSAT telescope carried piggyback a Wide Field Camera (WFC) for the extreme ultraviolet (0.05–0.2 keV). The WFC was also a Wolter type-I design with three nested mirror sets, and was also equipped with a microchannel plate detector with a CsI photocathode.

11.3.2 Compton Gamma Ray Observatory

The Compton Gamma Ray Observatory (CGRO) covered the energy range from 30 keV to 30 GeV and operated from 1991 to 2000. It contained four separate experiments: BATSE, OSSE, COMPTEL, and EGRET.

BATSE

The Burst and Transient Source Experiment (BATSE) was an all-sky monitor to detect gamma rays with energies between 30 keV and 1.9 MeV from a variety of transient sources. It consisted of detector modules located at the eight corners of the satellite. Each detector module contained a NaI scintillator plate 20 inches in diameter and ½ inch thick. The detectors were uncollimated and were oriented like the faces of a regular octahedron. Directional information was based on the variation in projected collecting area and, at high energies, on the incomplete capture of the photon energy. Timing resolution was of order 0.1 seconds, allowing BATSE to rapidly signal detection of gamma ray bursts. The detectors could determine the location of a strong burst to within about 3°. A secondary set of scintillation detectors provided modest spectroscopic capability. One of BATSE's most

---

[4] The properties of microchannel plate detectors were discussed in Chapter 5.

important science results was the measurement of a nearly isotropic distribution for gamma ray bursters.

OSSE

The Oriented Scintillation Spectrometer Experiment (OSSE) consisted of four NaI(Tl) scintillators which could be individually pointed and had spectroscopic capabilities from 50 keV to 10 MeV. Each was surrounded by a variety of other scintillators in anti-coincidence mode to reject cosmic rays, high energy gamma rays, and gamma rays incident from the sides. A passive tungsten collimator in front of the detector defined the $3.8° \times 11.4°$ field of view. One of the principal scientific goals of OSSE was the detection of radioactive nuclei in supernova remnants.

COMPTEL

The Imaging Compton Telescope (COMPTEL) worked from 1 to 30 MeV (Schönfelder *et al.*, 1993). It consisted of two layers of scintillators: an upper layer of liquid scintillators and, 1.5 m below it, a lower layer of NaI scintillators, as illustrated in Figure 11.6. A gamma ray underwent Compton scattering in the upper layer. The scattered photon was then absorbed in the lower layer. Both layers were surrounded by anti-coincidence shields so that only photon interactions were recorded. From the locations and the energies deposited in the two layers it was possible to reconstruct the energy and, to some degree, the direction of the original gamma ray. However, there was an unavoidable positional uncertainty. For a single photon, the source direction could only be said to be confined to somewhere on a circle on the celestial sphere. Multiple photons allowed one to overcome this ambiguity. The required delayed coincidence between the two layers strongly suppressed any background. One might consider this to be a type of tracking detector.

EGRET

The Energetic Gamma Ray Experiment Telescope (EGRET) was designed to observe gamma rays from 20 MeV to 30 GeV. The main part of the instrument was a pair of spark chambers, shown in Figure 11.7. The first of these contained 27 thin tantalum plates. An incoming gamma ray interacted with a tantalum atom and underwent pair production. The electron and positron could be tracked by their ionization trails. These trails had to be seen, in coincidence, in both spark chambers. Typically the electron and positron exited the second spark chamber and deposited the remainder of their energy in a final NaI scintillator acting as a calorimeter. It was possible to reconstruct the direction and approximate energy for each gamma ray.

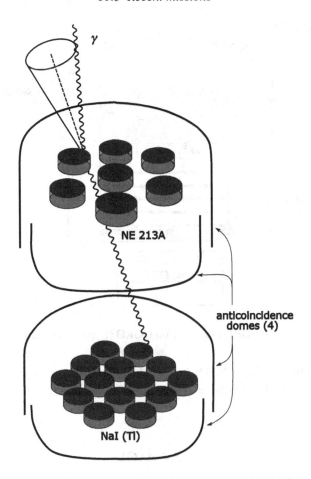

Figure 11.6 COMPTEL detector on CGRO (Schönfelder *et al.*, 1993).

11.3.3 Extreme Ultraviolet Explorer

The Extreme Ultraviolet Explorer (EUVE) was in operation from 1992 to 2001. Generally, the ISM is fairly opaque in the ultraviolet beyond the Lyman limit of 91.2 nm. However, the ISM is inhomogeneous, and numerous extreme ultraviolet sources are detectable. The satellite carried four telescopes, two of which were conventional Wolter type-I designs for the 4 to 36 nm range with microchannel plate detectors. A third telescope was designed for longer wavelengths, 40 to 75 nm, and used a Wolter type-II design with larger grazing angles specifically chosen to block shorter wavelength radiation. The fourth telescope was also Wolter type-II and fed both a microchannel plate imager and three spectrometers (with microchannel plate detectors). The spectrometer gratings provided a third grazing-incidence reflection,

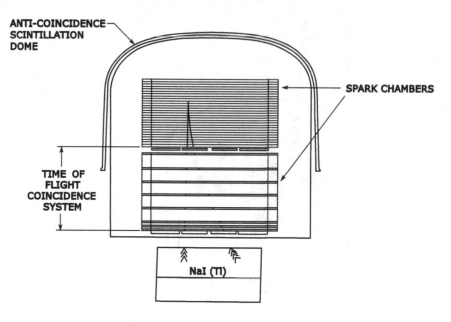

Figure 11.7 EGRET detector on CGRO (Esposito *et al.*, 1999).

leading to the choice of the Wolter type-II design in this instance.[5] The anodes of all these microchannel plates were of a "wedge, strip, and zigzag" design (Vallerga *et al.*, 1989).

11.3.4 ASCA

The Advanced Satellite for Cosmology and Astrophysics (ASCA; formerly Astro-D), was a Japanese-US collaboration operating from 1993 to 2000. It was designed for x-ray observations with four Wolter type-I telescopes. Two telescopes fed Gas Imaging Spectrometers (GIS) based on proportional counters. In this case scintillation light from the xenon gas was used for positional determination. The remaining two telescopes fed Solid-state Imaging Spectrometers (SIS). Each spectrometer detector had four front-illuminated, frame buffered, 420×420 pixel CCDs (Burke *et al.*, 1991), precursors to those in the ACIS instrument on Chandra (below).

11.3.5 Rossi X-ray Timing Explorer

NASA's RXTE satellite, in operation from 1995 to the present, was designed to study time variability of x-ray sources. It did not contain a telescope *per se*;

---

[5] As with the case of XMM-Newton, below, the actual grating design and geometry is complicated since the gratings are placed in the converging telescope beam.

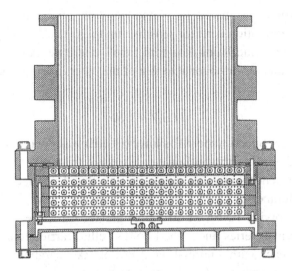

Figure 11.8 Proportional counter unit from RXTE PCA (Jahoda *et al.*, 2006) consisting of a 1° FWHM hexagonal collimator and a three-layer proportional counter array with additional veto layers on top, bottom, and sides. Tin and tantalum shielding not shown.

instead, individual instruments were collimated to provide angular resolution. The PCA (Proportional Counter Array) contained five xenon gas proportional counters covering energies from 2 to 60 keV with a collecting area of 6500 cm^2. The hexagonal BeCu collimators, illustrated in Figure 11.8, provided a field of view of about 1°. Most important to the mission goals was the PCA time resolution of 1 μs. The HEXTE (High Energy X-ray Timing Experiment) contained two sets of four "phoswich" NaI scintillators with CsI anti-coincidence scintillators (as in the CGRO OSSE detectors). These also contained hexagonal collimators but made of $Pb_{0.94}Sb_{0.06}$. HEXTE was sensitive from 20 to 200 keV. There was also an ASM (All Sky Monitor) containing scanning shadow cameras with proportional counter detectors. A shadow camera is a type of coded aperture camera.

11.3.6 BeppoSAX

BeppoSAX was an Italian–Dutch x-ray satellite in operation from 1996 to 2003. Three Wolter type-I telescopes were dedicated for use with medium energy (1.3–10 keV) xenon gas scintillation proportional counter spectrometers and one telescope for use with a low energy (0.1–10 keV) xenon gas scintillation proportional counter spectrometer. A fifth gas scintillation proportional counter was operated at high pressure (5 atm Xe/He) to extend the energy coverage up to 120 keV, although this instrument only had a collimator-defined field of view.

The above instruments shared the same pointing direction as a four-section NaI phoswich scintillation detector sensitive up to 300 keV. This was surrounded by anti-coincidence scintillation detectors. Finally there were two coded mask wide field cameras on an axis orthogonal to that of the other instruments.

11.3.7 FUSE

The Far Ultraviolet Spectroscopic Explorer (FUSE) was operational from 1999 to 2007. It covered a rather narrow wavelength range from 90.5 to 119.5 nm.[6] One of its principal science goals was measuring the abundance ratio of deuterium to hydrogen in a number of sources in order to understand the cosmic evolution of deuterium. A resolving power of 10 000 or greater was necessary to adequately separate their Lyman series lines. The FUSE satellite carried four nearly identical optical systems, each with a normal-incidence off-axis parabolic mirror, a curved grating, and a microchannel plate detector. Two systems were optimized for 90 to 110 nm and two for 100 to 120 nm.

11.3.8 *Chandra*

The Chandra X-ray Observatory (formerly known as AXAF) was launched in 1999 and is still in operation. The x-ray telescope itself, the High Resolution Mirror Assembly (HRMA), is a Wolter type-I design with four nested cylinders, the largest being 1.2 meters in diameter. The result is a large effective collecting area (400 cm^2 at 1 keV).[7] The mirrors are iridium coated. Iridium is a high density material with a correspondingly large plasma frequency, which allows somewhat larger grazing angles at fixed energy and somewhat higher energies at fixed angle of incidence. The accuracy of the HRMA is such that Chandra has an unprecedented angular resolution for high energy detectors of 0.5 arcsec (cf. ROSAT). The focal length of the telescope is 10 meters, and it has two objective transmission gratings which can be used, optionally, for spectroscopy at low or high energy.

Chandra has two focal plane instruments. ACIS, an Advanced CCD Imaging Spectrometer for 0.2–10 keV, is shown in Figure 11.9. It consists of ten CCD chips, each with 1024×1024 pixels and an image scale of 0.5″ per pixel. All of the CCDs can be used for imaging, but six are aligned in a row for use in spectroscopy (ACIS-S) with either of the gratings. Two of these six chips are back-illuminated. The back-illuminated chips have better response at low energy and poorer response

---

[6] Traditional nomenclature for the ultraviolet portion of the spectrum can be confusing, with overlapping terminology including UVA, UVB, UVC, and near-, middle-, far-, vacuum- and extreme-ultraviolet. On a logarithmic scale, FUSE is actually closer to visible wavelengths (400 nm) than to the extreme end of the ultraviolet (10 nm).

[7] For the ACIS-I detector front-illuminated CCDs.

ACIS-S (6 CCDs)

ACIS-I (4 CCDs)

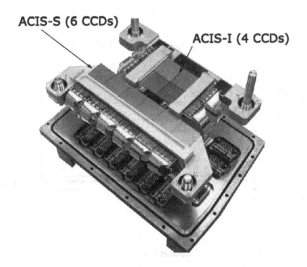

Figure 11.9 Chandra ACIS showing the imaging array ACIS-I and the spectroscopy array ACIS-S. Credit: Lockheed Martin. For color version of figure, see plate section.

at high energy, compared with the front-illuminated chips. The four remaining chips, forming the imaging array (ACIS-I), are tilted to approximate the focal plane curvature. All the CCDs operate in a frame storage mode whereby the accumulated data are rapidly shifted to a frame buffer which can be read out while a new integration is underway. During the Chandra mission the ACIS front-illuminated CCDs experienced significant radiation damage from exposure to low energy protons.

The HRC is a microchannel plate High Resolution Camera for 0.1–10 keV, as shown in Figure 11.10. The imaging part of the instrument, HRC-I, has a $31'' \times 31''$ field of view. The photocathode material is CsI, covered by an aluminized polyimide shield to block visible and ultraviolet radiation as well as electrons and ions. The two microchannel plates have a chevron design with 10 μm pores and a cant of $6°$. The readout is done via a pair of crossed wire grids backed by a reflector plate. The readout time resolution is 16 μs. The spectroscopic detector, HRC-S, is similar but is rectangular with a 30:1 effective aspect ratio. It is designed for use with the low energy transmission grating.

11.3.9 XMM-Newton

The XMM-Newton satellite (Jansen *et al.*, 2001) was launched by the ESA shortly after Chandra. A major design goal was to maximize the effective area, especially at high energy. This was achieved by making three separate co-aligned telescopes, each of which is a Wolter type-I design with 58 [sic] nested thin coaxial mirrors.

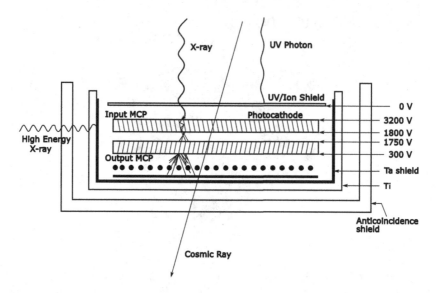

Figure 11.10 Chandra HRC. Credit: NASA/CXC/SAO.

The largest mirror of each set is 70 cm in diameter. The total effective area is of order $3 \times 1550\,\text{cm}^2$ at 1.5 keV. The paraboloidal and hyperboloidal segments were replicated as single units made of gold-plated nickel.[8] The width of the PSF varies with energy, across the field of view, and between telescopes, but is of order 4–6″ FWHM. XMM achieves higher collecting area than Chandra, at the cost of poorer spatial resolution.

The main detectors are the European Photon Imaging Cameras (EPIC), of which there are two types. Two are metal-oxide-semiconductor (MOS) CCD cameras (Turner *et al.*, 2001), each with seven front-illuminated 600×600 pixel CCDs optimized for soft x-rays. Their focal-plane layouts cleverly minimize gaps between CCDs and approximately match the curved focal planes. They are frame-buffered, much like the Chandra ACIS CCDs. The third detector (Strüder *et al.*, 2001) is a set of 12 back-illuminated PN CCDs, optimized for hard x-rays, fabricated on a single piece of silicon with a total of 400×384 pixels, with some gaps. These are not frame-buffered, but have rapid readout.

Two of the Wolter type-I telescopes, those with the EPIC-MOS cameras, are equipped with permanently placed reflection grating arrays (RGA), which decrease

---

[8] *Replication* is a technique whereby one first makes a precise master which is the complement of the desired structure. This is then coated with a release material and then covered with one or more materials by some combination of evaporation, sputtering, electroplating, or electroforming. The replica is then separated from the master at the release layer. The master then may be reused to make multiple replicas, at considerable cost savings since the grinding and polishing need be performed only once. This technique is often used also to make diffraction gratings.

the sensitivity of their main imaging systems but create simultaneous first order spectra with resolving power of 150–800 (den Herder *et al.*, 2001). The grating arrays are sets of 182 grazing-incidence reflection gratings,[9] each with about 650 lines mm^{-1} blazed at 0.7°.[10] The spectrometer cameras contain nine back-illuminated frame-buffered CCDs with a total of 9216 pixels along the direction of dispersion, oversampling the line spread function. The intrinsic energy resolution of the CCDs is used to separate first and second orders. The satellite is also equipped with a smaller telescope for optical and ultraviolet detection with a detector consisting of a microchannel plate followed by a CCD.

11.3.10 INTEGRAL

The International Gamma-Ray Astrophysics Laboratory (INTEGRAL) was launched in 2002 and is still operating. It is in an orbit with a perigee of 10 000 km, which keeps it outside the inner (proton) Van Allen belt, giving stable detector backgrounds and minimizing detector damage. The Spectrometer on INTEGRAL (SPI), has a coded aperture mask with 127 hexagonal cells, shown in Figure 11.11, providing an angular resolution of 2.5°. The 63 opaque cells are fabricated of 3 cm thick tungsten! The SPI detector system is an array of 19 germanium detectors cooled to 85 K, heavily surrounded by anti-coincidence detectors. The germanium detectors are primarily sensitive to Compton scattering events, where in some cases the scattered photon may be seen in an adjacent detector. SPI covers energies from

Figure 11.11 Coded aperture masks for two of the instruments on INTEGRAL: (left) SPI mask (Credit: CAB (INTA-CSIC)) and (right) IBIS mask (Credit: ESA). For color version of figure, see plate section.

---

[9] Also replicated.
[10] The actual grating design and geometry are complicated since the gratings are placed in the converging telescope beam.

20 keV up to 8 MeV, with spectral resolution of order 450 at 1.33 MeV. The Imager on Board the INTEGRAL Satellite (IBIS) also has a tungsten coded aperture mask, also shown in Figure 11.11. Together with a two-layer detector consisting of a plane of 128×128 pixel CdTe detectors followed by a plane of 64×64 CsI(Tl) scintillators, this gives an angular resolution of $12'$. Its energy coverage is similar to that of SPI. At the higher energies, the Compton scattering geometry aids in the reconstruction of the photon energy and in background rejection. INTEGRAL also carries two monitor detectors: the Joint European X-Ray Monitor (JEM-X), which is a 3–35 keV imaging proportional counter with a coded aperture mask, and a refracting telescope silicon CCD Optical Monitor Camera (OMC) for V band.

11.3.11 GALEX

The Galaxy Evolution Explorer (GALEX), launched in 2003, is an ultraviolet telescope operating simultaneously in both far-ultraviolet (134–179 nm) and near-ultraviolet (177–283 nm) using a dichroic beamsplitter. The telescope itself is a 50 cm Ritchey–Chrétien design, equipped for both imaging and slitless (grism) spectroscopy. Each detector contains a photocathode (CsI for far-ultraviolet and Cs_2Te for near-ultraviolet) followed by a series of three microchannel plates and orthogonal anode delay lines for measuring the position of the charge cloud leaving the microchannel plates.

11.3.12 Swift

Swift is a mission launched in 2004 and expected to last until 2011. It is designed primarily for the purpose of detecting and multi-wavelength monitoring of gamma-ray burst sources. The Burst Alert Telescope (BAT) has a coded aperture mask and covers the energy range 15–150 keV over a solid angle of 1.4 sr. Its detector is a 256×128 pixel CdZnTe detector array operating in a photon counting mode. Its resolution is of order $17'$ and it provides burst timing information down to 200 μs. The satellite is designed to rapidly respond to a burst detected in BAT and then to rapidly slew so that the main X-Ray Telescope (XRT) is pointing at the burst position. The XRT is a Wolter type-I design operating from 0.3 to 10 keV with about $18''$ resolution, using a copy of the 600×600 pixel EPIC CCD camera on XMM. The satellite also has an Ultraviolet and Optical Telescope (UVOT), a close copy of the optical/ultraviolet monitor instrument from XMM. The UVOT contains a 2048×2048 pixel intensified CCD array sensitive to the wavelength range 170–650 nm with an angular resolution of order $2''$.

11.3.13 Fermi gamma ray space telescope

The Fermi satellite, formerly known as GLAST, was launched in 2008 to cover the photon energy range from 8 keV to 300 GeV. The Large Area Telescope (LAT) instrument is sensitive over 2 sr. Like a spark chamber, it contains a set of 16 plane-parallel tungsten sheets within which the gamma ray produces e^+e^- pairs. But instead of spark detection, particle tracks are followed using interleaved silicon tracking detectors. Any charged particles leaving the stack are stopped in a CsI tracking calorimeter, allowing a determination of the initial gamma ray energy. The CsI scintillator crystals are long rectangular bars arranged in eight crossed layers of 12 bars each, allowing 3-dimensional positional information on the energy deposition of the showers. There is also an instrument known as the GLAST Burst Monitor (GBM), which covers the 8 keV−1 MeV range using 12 NaI scintillator plates with a 10:1 aspect ratio[11] and covers the range 150 keV–30 MeV using an additional two BiGeO (BGO) scintillation detectors. The BGO detectors provide important spectroscopic information but virtually no positional information.

11.4 Possible future missions

11.4.1 IXO

A new mission under consideration is the International X-ray Observatory (IXO). It is a joint effort of NASA, ESA, and JAXA (Japanese Aerospace Exploration Agency), resulting from a merger of their previously separate concepts named XEUS and Constellation X. Although instrument details are in flux at the present time, it appears that one of the main requirements will be a very large collecting area, of order 30 000 cm^2 at 1.25 keV, from a 3 meter diameter grazing incidence telescope with nested optics. Plans are to place it in an orbit around the L2 Lagrangian point to provide a stable, low-background environment. IXO was ranked fourth in priority among future, large-scale space projects in the 2010 US National Research Council decadal report on astronomy and astrophysics.

11.4.2 MAXIM or BHI

For some time now NASA has considered the possibility of an x-ray interferometry mission under the name of Micro-Arcsecond X-ray Imaging Mission (MAXIM). Some laboratory work towards this design has been accomplished. Concepts along this line were presented for the 2010 decadal review using the name Black Hole Imager (BHI) to emphasize the main science goal. It is too early to predict

---

[11] Like BATSE, these provide directional information by the dependence on projected collecting area.

the prospects for this concept. But the current emphasis seems to be on further technology development for a mission target date of around 2030.

Exercises

11.1 Calculate the angular response of an ideal collimator. Consider only one dimension and assume that the collimator width is 1 cm and its length is 1 m.

11.2 For nickel, gold, and iridium coatings for x-ray mirrors:

    a. Calculate the plasma frequencies, based on bulk densities of 8.91, 19.32, and 22.56 g cm^{-3} and mean atomic weights of 58.69, 196.97, and 192.22 daltons for nickel, gold, and iridium, respectively.

    b. What two factors, unrelated to the plasma frequency, might influence the use of these materials for x-ray mirror coatings?

    c. What third factor, related to the atomic number, might be an additional consideration? (Hint: One assumption used in the beginning of the chapter is only approximately correct.)

    d. What other techniques may be employed in making materials with high x-ray reflectivity?

11.3 Explain why the autocorrelation function of a coded aperture mask should approximate a delta function. For example, what happens if the autocorrelation function is double peaked?

11.4 An energetic photoelectron with kinetic energy E_0 traveling through matter loses energy through atomic interactions which may ionize the atom or induce a non-ionizing excitation. Let W be the average energy loss (by both mechanisms) divided by the average number of ionizations. The average number of ionizations is $\bar{N} = E/W$. Calculate the mean square fluctuations in the number of ionizations N, namely $\langle (N - \bar{N})^2 \rangle = \langle (N - E/\epsilon)^2 \rangle$. (Hint: Consider each atomic interaction to be statistically independent so that the fluctuations *per interaction* may be added in quadrature. Your answer should depend only on $\bar{N}$ and on properties of the material; the number of ionizations per atom encountered and the energy loss per atom.) The fluctuations can be sub-Poissonian because in all cases the total energy loss is fixed. The only variability is related to whether particular events are ionizing or not.

12

Radio receivers, spectrometers, and interferometers

12.1 Astrophysical radio sources

Most astronomical radio sources are fundamentally different than the most common optical sources, stars. Some radio continuum sources exhibit *thermal* emission, in which flux increases with frequency (remember that $S_\nu \propto \nu^2$ at low frequencies for a blackbody). This type of spectrum is characteristic of thermal bremsstrahlung, also known as free–free emission, from a hot electron plasma such as an H II region, as shown in Figure 12.1. At low frequencies such a source is optically thick and the spectrum rises as ν^2. At high frequencies such a source becomes optically thin, and the spectrum is nearly flat. The cosmic microwave background (CMB) is another example of a thermal source. Other continuum sources are *nonthermal*, with flux increasing at longer wavelengths. A typical spectrum from synchrotron radiation varies as $S_\nu \propto \nu^{-0.8}$. The spatial structure of the emitting region is often quite complex and of great importance astrophysically. Spectral line emission at radio wavelengths comes from the 21 cm hyperfine structure line of H I (a tracer of neutral hydrogen), from recombination lines primarily of H and He (useful as probes of ionization conditions), and from molecular rotational lines (probes of dense gas and star forming regions). Some radio sources show rapid temporal variations (pulsars).

12.2 Fundamentals of radio receivers

At radio frequencies ($\lambda \gtrsim 300$ μm; $\nu \lesssim 10^{12}$ Hz) generally the wave picture of electromagnetic radiation is more appropriate than the photon picture. The power contained in radio waves propagating in free space can, at best, be localized on size scales of order λ ($A\Omega = \lambda^2$), which is much larger than typical solid state devices. Therefore the wave needs to be collected out of free space using some combination of conductors and dielectrics and directed into some confined region

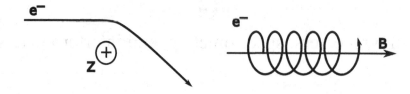

Figure 12.1 Radio emission by thermal bremsstrahlung (left) and synchrotron emission (right).

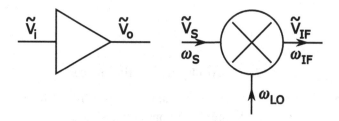

Figure 12.2 Examples of linear systems: an amplifier (left) and a mixer (right).

for further processing. Such a system is called an antenna; we will discuss more about antennas later. The signal may then be: (1) amplified, (2) translated in frequency, (3) filtered, or (4) detected, that is, have its power measured. For best noise performance, and for a variety of other reasons, one generally chooses to do (1) or (2) before doing (3) or (4).

12.2.1 Linear systems

Both amplifiers and mixers (frequency translators), shown in Figure 12.2, are examples of what we call *linear systems*. An amplifier takes some incoming voltage and amplifies it, possibly adding a phase shift. In terms of amplitude and phase we can relate input and output voltages by

$$V_o e^{i(\omega t + \phi_o)} = G_V e^{i\Delta\phi} V_i e^{i(\omega t + \phi_i)}. \qquad (12.1)$$

In operator notation we can write

$$\tilde{V}_o = \tilde{G}_V \tilde{V}_i, \qquad (12.2)$$

which shows that amplification is a linear operation.

A mixer takes an input signal at angular frequency ω_S, combines it with a local oscillator at ω_{LO}, and produces an output at the "intermediate frequency" ω_{IF}. The frequencies are related by

$$\pm \omega_{IF} = \omega_{LO} - \omega_S. \qquad (12.3)$$

In a mixer the signal typically suffers some loss (gain less than unity), hopefully a small loss. In operator notation we can relate the amplitude and phase shift of this operation by

$$\tilde{V}_{IF} = \tilde{G}_V \, \tilde{V}_S. \tag{12.4}$$

So mixers also are linear systems. Both amplifiers and mixers will also add noise. Don't be confused by the fact that linear *systems* may incorporate non-linear *devices* such as diodes.

12.2.2 Quantum noise limit

Quantum mechanics sets a fundamental lower limit on the noise of *any* linear system. There exists a number–phase uncertainty principle for the electromagnetic field, $\Delta N \, \Delta \phi \gtrsim 1$, somewhat analogous to the Heisenberg uncertainty principle, $\Delta x \, \Delta p_x \gtrsim h$. Therefore for a phase preserving system, one with $\Delta \phi \lesssim 1$, there is an uncertainty in photon number of unity or greater. This corresponds to one photon in a time $t = 1/\Delta \nu$, where $\Delta \nu$ is the bandwidth. This noise may also be viewed as being due to zero-point fluctuations in the electromagnetic field. Each mode of the radiation field can be thought of as a simple harmonic oscillator with minimum energy $E = \frac{1}{2} h\nu$. This minimum noise corresponds to a power per unit bandwidth of $h\nu$, equivalent to a thermal noise source at a Rayleigh–Jeans temperature of

$$T_N = \frac{h\nu}{k}. \tag{12.5}$$

12.2.3 Components in series

If linear systems are combined in series, the net gain is the product of the individual gains, and the individual noise temperatures all contribute to the system noise, as illustrated in Figure 12.3.

$$P_1 = G_1(P_S + kT_{N1} \, \Delta \nu), \tag{12.6}$$

$$P_2 = G_2(P_1 + kT_{N2} \, \Delta \nu) \tag{12.7}$$

$$= G_1 G_2 P_S + G_1 G_2 \left(kT_{N1} \, \Delta \nu + \frac{1}{G_1} kT_{N2} \, \Delta \nu \right). \tag{12.8}$$

The net signal gain is

$$G = G_1 G_2, \tag{12.9}$$

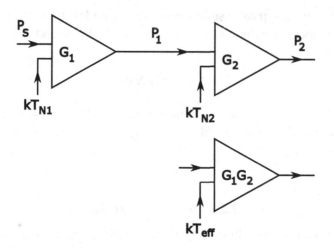

Figure 12.3 Amplifiers connected in series (top) and the equivalent circuit (bottom).

and the effective noise temperature is

$$T_{eff} = T_{N1} + \frac{1}{G_1} T_{N2}. \qquad (12.10)$$

If the gain of the first stage G_1 is large, then the first stage noise dominates. One generally divides a radio receiver into *front end* components, those for which low noise performance is critical, and *back end* components whose noise performance is less critical.

12.2.4 Low noise GaAs FET amplifiers

For frequencies in the range 1–100 GHz, the best noise performance currently is achieved by using field-effect transistor (FET) amplifiers cooled to cryogenic temperatures, $T_{phys} \approx 10$ K. The types of FET are the JFET (junction FET), the MOSFET (metal-oxide-semiconductor FET), and the MESFET (metal-semiconductor FET). Thorough discussions of the physics of these devices may be found in Streetman & Banerjee (2005) and especially Sze & Ng (2006). Some details are also given in Rieke (2002).

For low-noise, high-frequency FET amplifiers, the material of choice is the compound[1] semiconductor gallium arsenide (GaAs), partly because of that material's high electron mobility. The mobility of holes is more than an order of magnitude lower. In a generic FET, as shown in Figure 12.4, the voltage applied to the gate

---

[1] The word *compound* refers to the fact that GaAs is a compound of two chemical elements. Silicon would be called an elemental semiconductor.

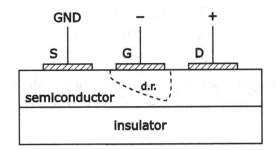

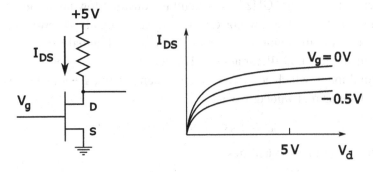

Figure 12.4 (Top) Generic drawing of an FET showing source (S), gate (G), and drain (D) electrodes and depletion region (d.r.). In this case an n-channel depletion-mode MESFET is illustrated. (Bottom left) Schematic circuit of an FET amplifier. (Bottom right) Current–voltage characteristics of an FET.

controls the size of the depletion region, thereby controlling the size (and therefore the resistance) of the channel in which current flows from the drain to the source.

There is a type of FET known as a high electron mobility transistor (HEMT),[2] incorporating a thin layer of n-Al$_x$Ga$_{1-x}$As on top of undoped GaAs. This forms a quantum well, just below the interface, within which electrons (in the undoped GaAs) have a higher than normal mobility in two dimensions. This is often referred to as a 2-dimensional electron gas (2-DEG). These devices are small, inexpensive, and have very low noise, making them good first stage amplifiers. Typical noise performance is

---

[2] A brief note on some of the jargon in this field: MMIC stands for Monolithic Microwave Integrated Circuit. Such devices might include multiple amplification stages, combine amplification and mixing, or even more complicated systems. They are the key elements in cellular phones. Some employ silicon rather than GaAs. Pseudomorphic (as in the term pHEMT) refers to the use of a thin layer of a different compound semiconductor within, say, GaAs. The materials have a lattice mismatch and the thin layer is forced to adopt the lattice constants of the underlying material. Metamorphic (mHEMT) refers to a gradual transition between mismatched lattice structures, allowing for less strain in the material. InP is another compound III-V semiconductor (built from elements in columns III and V of the periodic table). InP may be used as a base material with good lattice matching to InGaAs, which has advantageous materials properties for HEMTs. However, the mHEMT technology may allow use of InGaAs and obviate the need to use the more expensive and difficult InP technology.

$$T_N(\nu \lesssim 8 \text{ GHz}) \approx 3 \text{ K},$$ (12.11)

$$T_N(\nu = 40 \text{ GHz}) \approx 20 \text{ K},$$ (12.12)

$$T_N(\nu = 100 \text{ GHz}) \approx 45 \text{ K},$$ (12.13)

which is approaching limits set at lower frequencies by the 2.7 K CMB and at higher frequencies by the quantum noise limit (0.048 K/GHz).

12.2.5 Radio frequency mixers

At frequencies above 100 GHz it is difficult to construct low noise amplifiers. It is more convenient to make low noise mixers to shift signals to lower frequencies. One combines the incoming signal at ω_S (or a band of signal frequencies) with a monochromatic local oscillator at ω_{LO}. The conceptually simplest type of mixer is a simple multiplier, which acts as a single-sideband mixer. Treating both signal and local oscillator as real valued, we get

$$V_{out} = V_S \cos(\omega_S t) \, V_{LO} \cos(\omega_{LO} t).$$ (12.14)

By simple trigonometric identities

$$V_{out} = \frac{V_S V_{LO}}{2} \{\cos(\omega_S - \omega_{LO})t + \cos(\omega_S + \omega_{LO})t\},$$ (12.15)

indicating output is present at the sum and difference frequencies. In many cases it is more practical to add the signal and local oscillator

$$V_{in} = V_{LO} \cos(\omega_{LO} t) + V_S \cos(\omega_S t),$$ (12.16)

and then impose this voltage on a non-linear device. If the device has a square-law response, which it generally will for small voltage swings, the mixer output will have components at the difference frequency ω_{IF},

$$V_{out} \propto V_{LO} V_S \cos(\omega_{IF} t),$$ (12.17)

plus components at DC, the sum frequency, and the second harmonics of both ω_{LO} and ω_S. If the response departs from a pure square law, there will be higher harmonics and other sum and difference frequencies as well. In practice, these usually can be kept small. Reverting to complex notation, we can see that

$$\tilde{V}_{out} \propto \tilde{V}_{LO} \tilde{V}_S^* \, e^{i[(\omega_{LO} - \omega_S)t - \phi]},$$ (12.18)

indicating that this is a linear phase-preserving system. The mixer may add an instrumental phase shift ϕ, but the phase information present in the original signal is preserved. The signal frequency can be either above (upper sideband, USB) or below (lower sideband, LSB) the local oscillator frequency, as shown in Figure 12.5.

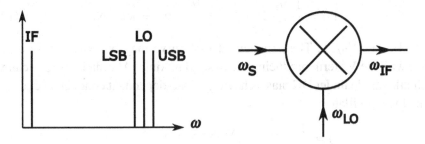

Figure 12.5 Relationship between the USB and LSB signal frequencies and the LO and IF frequencies in a mixer.

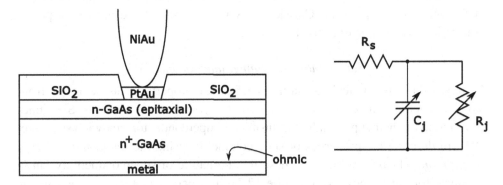

Figure 12.6 A Schottky diode formed between epitaxial n-type GaAs and a PtAu alloy dot. Contact to the metal side of the barrier is made with a NiAu alloy wire. Equivalent circuit is shown on the right.

Schottky diode mixers

A simple device that can perform this mixing function is a Schottky barrier diode, an example of which is shown in Figure 12.6. A Schottky diode is a metal–semiconductor junction with a non-linear conductance:

$$I \approx I_S \, e^{V/V_0}. \tag{12.19}$$

Its high frequency performance is limited by the capacitance C_j and the series resistance R_S. One keeps C_j small by using a small device ($\sim 2\,\mu m$ diameter). R_S is then determined by properties of the materials. There will be a high frequency cutoff at

$$\nu_c = (2\pi \, R_S \, C_j)^{-1} \approx 2 \times 10^{12} \text{ Hz}. \tag{12.20}$$

There is shot noise due to current flow through the dynamic resistance

$$R_j = \left(\frac{dI}{dV}\right)^{-1} = \frac{V_0}{I}, \tag{12.21}$$

$$P_N = \frac{1}{4}\frac{V_0}{I} 2e\, I\, \Delta\nu = \frac{V_0}{2} e\, \Delta\nu = kT_N\, \Delta\nu, \qquad (12.22)$$

where $T_N = eV_0/2k$. The exponential I–V curve may be expanded in a power series, with the V^2 term producing the desired mixing. Alternatively, one can view a Schottky diode under DC bias as having a time-dependent conductance produced by the local oscillator:

$$g(t) = \frac{I_S}{V_0} e^{V_{DC}/V_0} e^{V_{LO}(\cos \omega_{LO}t)/V_0} \qquad (12.23)$$

$$= g_0 + 2g_1 \cos \omega_{LO}t + 2g_2 \cos 2\omega_{LO}t + \cdots. \qquad (12.24)$$

The signal voltage $V_S(t) = V_S \cos \omega_S t$ is applied to this conductance, giving output at frequencies $\pm n\,\nu_{LO} \pm \nu_S$. Classical mixer theory says there must be a power conversion loss of at least 3 dB ($L_M^{min} = 2$).

Superconducting tunnel junction mixers

Mixers with non-classical, quantum mechanical properties can be made from superconducting tunnel junctions. These are commonly known as SIS junctions based on their superconductor–insulator–superconductor sandwich structure. We will describe the properties of superconductors using the *semiconductor representation*, which does not fully describe the nature of superconductivity but is adequate for our purposes (see Figure 12.7). In superconductors there is a superconducting energy gap, a region of no allowed states, which prevents current flow at bias voltages less than the gap ($V < 2\Delta/e$). At bias voltages greater than the gap

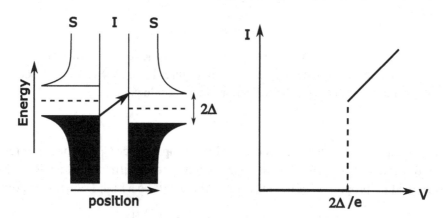

Figure 12.7 (Left) Energy level diagram of an SIS tri-layer. The horizontal width is indicative of the density of states increasing near the energy gap. Filled states are shown in black (green in electronic version). Photon-assisted tunneling allows a quasi-particle to make the transition shown (red). (Right) An idealized current–voltage curve.

(V > 2Δ/e), current is able to flow by tunneling through the insulator. Since the density of states is singular at the edges of the gap, this onset of current flow is sudden. In the idealized T = 0 case the current jumps from zero to a finite value at V = 2Δ/e. This extreme non-linearity, faster than exponential, gives efficient mixing. If the behavior is sufficiently close to ideal (sufficiently sharp), one needs to view this as photon-assisted tunneling. With the junction biassed at V < 2Δ/e, the photon provides the remaining energy needed for the electron to tunnel across the barrier. One electron flows for each photon detected. This is nearly an ideal situation (and reminiscent of many optical photon detectors). In practice, these devices are not used to count photons but as mixers (for which their performance is also nearly ideal). With SIS mixers one can achieve $T_N \approx h\nu/k$, that is, approaching the quantum limit. In principle, one can also achieve conversion gain, which is a quantum effect not allowed for classical mixers.

12.2.6 Detectors and the radiometer equation

Non-linear devices such as Schottky diodes also may be used for detectors, as shown in Figure 12.8. As we have seen, after a square-law device one sees a DC current

$$I \propto |V_S|^2 \propto P_{tot} \propto k(T_{sys} + T_{sig})\Delta\nu. \tag{12.25}$$

Typically one measures the power with and without the signal present. The difference is proportional to T_{sig}. There are also noise fluctuations associated with T_{sys}. For an integration time t, the uncertainty in the signal is given by

$$\Delta T_{sig} = \frac{T_{sys}}{\sqrt{\Delta\nu\, t}}. \tag{12.26}$$

A relatively new type of detector is the Transition Edge Sensor (TES). A TES is a type of bolometer, a class of detector discussed in Chapter 5. Germanium bolometers, a standard for infrared detection, may also be used at millimeter wavelengths. All bolometers work by sensing small temperature changes caused by

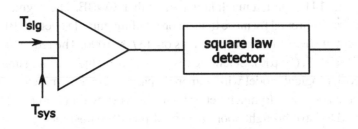

Figure 12.8 A simple radiometer system.

the absorption of radiant energy. In a TES this temperature change is sensed by observing the rapid change in resistance of a superconductor near its superconducting transition (Irwin & Hilton, 2005). TESs can be made from a wide variety of superconducting materials. Typically feedback is required to maintain their temperature near the superconducting transition, and low noise SQUID[3] amplifiers are required to read out the TESs. Other applications of TES technology range from x-ray detectors to searches for dark matter.

12.3 Precision radiometry of the CMB

12.3.1 COBE

The COsmic Background Explorer (COBE) spacecraft, launched in 1989, had two instruments devoted to the study of the CMB. The Differential Microwave Radiometer (DMR) mapped the spatial fluctuations in the CMB with Schottky diode mixers at 31.5, 53, and 90 GHz, which formed differential radiometers pointing at directions separated in the sky by 60°. The results showed a dipole anisotropy of the CMB of 3 mK and residual fluctuations on a 5° spatial scale of order 30 μK. The Far-InfraRed Absolute Spectrophotometer (FIRAS) was a polarizing Michelson interferometer using composite bolometer radiometric detectors. Its scientific results included a determination that the CMB spectrum matches that of a blackbody to better than 0.03%. Both DMR and FIRAS used horn antennas, which limited the spatial resolution.

12.3.2 WMAP

The Wilkinson Microwave Anisotropy Probe (WMAP) was launched in 2001 (Bennett *et al.*, 2003). WMAP repeated the general COBE DMR idea of differential radiometry. A larger variety of frequencies (23, 33, 41, 61, and 93 GHz) were used in order to facilitate the separation of the anisotropy signal from that of galactic foregrounds. The telescope design is discussed below in Section 12.5.3. The receivers employed HEMT amplifier front ends. The fields of view being differenced were set 141° apart, a much larger angle than COBE. The angular resolution of WMAP was improved by more than an order of magnitude over COBE, allowing measurement of CMB multipole moments out to $l \approx 1000$. The required sensitivity was $\Delta T/T \approx 10^{-5}$ (30 μK). Scientific results so far include strong support for the Λ-CDM cosmological model with numerous parameters determined to the level of a few percent and some to much better than 1%, as shown in Figure 12.9. WMAP is considered to have brought about the era of precision cosmology.

---

[3] SQUID stands for Superconducting QUantum Interference Device.

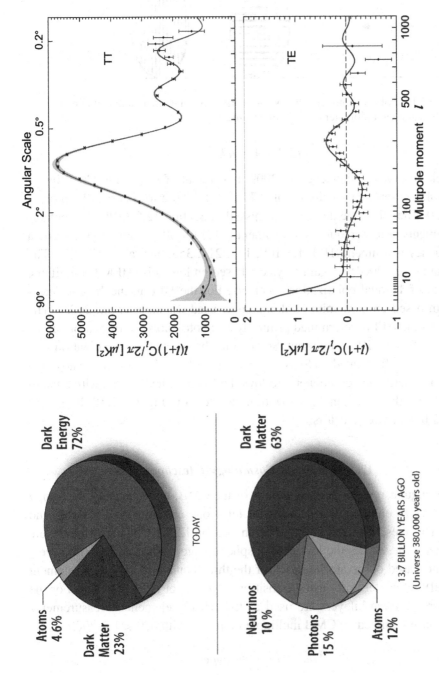

Figure 12.9 (Left) Constituents of the universe today and 13.7 billion years ago, as determined by WMAP. (Right) WMAP multipole moments. TT (red curve) refers to the temperature fluctuation power spectrum, and TE (green curve) refers to the cross correlation of temperature fluctuations with E-mode polarization. Courtesy of WMAP Science Team. For methodology, see Hu & White (1997). For color version of figure, see plate section.

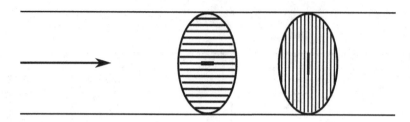

Figure 12.10 Polarization-sensitive bolometers in circular waveguide using resistive wires and Ge bolometers (red in electronic version).

12.3.3 Planck

The Planck observatory, launched in 2009, has a goal of measuring CMB intensity and polarization to a sensitivity of $\Delta T/T \approx 10^{-6}$ out to spherical harmonics of $l \approx 2500$. The telescope design is discussed in Section 12.5.3. Planck contains two instruments, a low frequency instrument (LFI) for 30, 44, and 70 GHz and a high frequency instrument (HFI) for 100, 143, 217, 353, 545, and 857 GHz. The LFI (Cuttaia *et al.*, 2004) is again a system based on low noise HEMT amplifiers, with a total of 11 dual-polarization receivers. For the HFI an incoherent detection system based on germanium bolometers is used (Lamarre *et al.*, 2003). The sensitivity of the HFI is determined primarily by photon noise. Information on the polarization of the CMB is important. Some of the bolometers are polarization sensitive. These were fabricated on Si_3N_4 by using parallel resistive wires to absorb a single linear polarization component, followed by an identical system with crossed wires to absorb the orthogonal polarization, as shown in Figure 12.10. In all, the HFI has 52 bolometric detectors.

12.3.4 Atacama Cosmology Telescope

For higher angular resolution observations of the CMB, the Atacama Cosmology Telescope (ACT) employs a 6 m diameter telescope operating at 145, 215, and 280 GHz. The telescope itself is discussed in Section 12.5.3. The high altitude site in the Atacama desert has the lowest atmospheric water vapor of all easily accessible ground-based observatories. Each of the three bands has a 32×32 element (1024 pixel) bolometer array using transition-edge sensor (TES) devices. For the ACT cameras Mo/Au bilayers are used as the sensing elements. Measurements with ACT have focussed on CMB multipole moments with $600 < l < 8000$.

South Pole Telescope

The South Pole Telescope (SPT) is a 10 m diameter telescope constructed at the Amundsen–Scott South Pole station. Details of the telescope are contained in

Section 12.5.3. One of the primary missions of the SPT is the study of galaxy clusters via the Sunyaev–Zel'dovich (SZ) effect. The SZ effect causes fluctuations in the CMB due to inverse Compton scattering by electrons in hot gas contained in galaxy clusters along the line of sight. These fluctuations will be seen primarily at multipole moments of $l = 2000$ and greater, requiring high angular resolution. A thorough study of the SZ effect is expected to provide information about the nature of dark energy by studying the growth of clusters with time. The polarization of the CMB will also be studied, with the potential of uncovering information about the inflationary epoch in the early universe and other new physics. The SPT employs arrays of TES bolometer detectors.

12.4 Radio spectrometers

12.4.1 Autocorrelation spectrometers

The Wiener–Khinchin theorem says that the power spectrum of a signal is given by the Fourier transform of its autocorrelation function. For IF signals with frequencies of order 100 MHz, it is possible to make use of this relationship by sampling the signal rapidly, at the Nyquist rate, digitally forming the autocorrelation for various time delays, τ, as shown in Figure 12.11. The power spectrum is then recovered by taking the Fourier transform. It is advantageous to weight the autocorrelation function before transforming, to minimize "ringing" in the spectral response. Spectral resolution is determined by the range of the utilized delays. A digitized signal

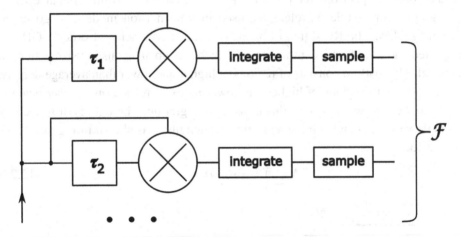

Figure 12.11 Autocorrelation spectrometer. Signals are correlated with time-delayed versions of themselves. The sampled outputs are weighted and then Fourier transformed. Although this diagram suggests an analog system, in practice the original signal is usually first sampled and the delay products formed digitally.

is necessarily quantized, and sometimes this is done rather coarsely, with as few as two quantization levels (1 bit). Quantization always results in a loss of sensitivity (efficiency) but by modest factors that are easily calculable (Thompson *et al.*, 2001). Some of the lost sensitivity may be recovered by oversampling, sampling faster than the Nyquist rate of the unquantized signal.

This technique is also useful in spectral line interferometry, where one forms the *cross* correlation between signals from two different antennas (the cross power spectrum).

12.4.2 Filter banks

A method which is even easier, conceptually, is to split the amplified signal into a number of identical copies and then pass the copies through different individual filters before square-law detection. This can be done, without loss of sensitivity, with linear amplifiers and power splitters, once the signal has been amplified sufficiently in the front end. This is currently less practical than digital techniques due to the need to design and fabricate multiple filters. A filter bank may need to contain thousands of channels.

12.4.3 Acousto-optical spectrometers

A novel approach to obtaining spectral information at radio frequencies is the acousto-optical spectrometer (AOS). This is based on a 3-dimensional diffraction grating for visible wavelengths, used in transmission mode, and governed by Bragg's law. The RF signal to be analyzed, at frequencies of order 1 GHz, is amplified and used to drive a transducer, which sets up an acoustic wave pattern in a crystal. This pattern consists of regions of higher and lower than average density, corresponding to regions of higher and lower index of refraction. A laser beam is sent into the cell, where it sees this transmission grating. The light is diffracted by an angle which depends on the acoustic wavelength Λ as shown in Figures 12.12 and 12.13.

$$\Lambda \, (\sin \theta - \sin \phi) = n \, \lambda. \tag{12.27}$$

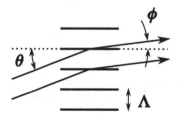

Figure 12.12 Bragg reflection.

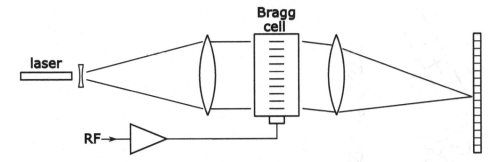

Figure 12.13 Acousto-optical spectrometer deflects laser light onto different pixels of a position-sensitive detector depending on the radio frequency.

In practice θ and the optical wavelength λ are fixed. A 1-dimensional detector array is used to examine the amount of light diffracted into various angles ϕ. This corresponds to the RF power spectrum.

12.5 Radio antennas

Most modern radio antennas are steerable Cassegrain reflectors. A large size (large collecting area) gives good point source sensitivity but, for a given wavelength, there are practical limits on the size imposed by the need for high surface accuracy and mechanical stability. Surface irregularities produce an irregular (non-planar) wavefront.

According to Ruze theory (Figure 12.14), for random irregularities the efficiency (the *gain*) is reduced by the factor $e^{-(4\pi\epsilon/\lambda)^2}$, where ϵ is the RMS surface irregularity.[4] This is due to the fact that waves from different portions of the aperture no longer add coherently (in phase) at the receiver. Clearly, telescopes for shorter wavelengths need more accurate surfaces. Steerable antennas are also subject to a varying gravitational load. Beyond a certain size it is not possible to make a rigid structure. Adding structural elements increases the stiffness but also adds weight. A *homologous* design accepts the existence of gravitational deformations, but insures that the parabolic surface deforms into another parabola.

12.5.1 Antenna patterns

Describe an antenna by the power it would radiate in different directions, if it were used as a transmitter. Let the antenna be pointed in the nominal direction (θ_0, ϕ_0).

---

[4] In optics the corresponding quantity is the Strehl ratio, usually written $S = e^{-(2\pi\sigma/\lambda)^2}$. The apparent difference is due to the convention in optics of referring to the RMS wavefront error σ and in Ruze theory of using the RMS surface error ϵ.

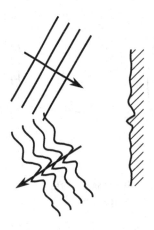

Figure 12.14 Ruze theory describes the loss in antenna gain on reflection off a surface with random irregularities.

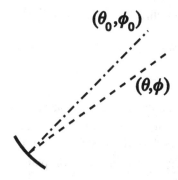

Figure 12.15 The antenna pattern describes the sensitivity in the direction (θ, ϕ), where the nominal pointing direction is (θ_0, ϕ_0).

The beam is described by the function $P(\theta - \theta_0, \phi - \phi_0)$, as shown in Figure 12.15. By reciprocity, this is proportional to its sensitivity for receiving power from a source in those various directions. If the aperture were *illuminated* with uniform amplitude and phase, this pattern would be the Airy pattern. But in practice the feed systems for radio telescopes are designed to provide non-uniform illumination, known as taper. The illumination of the edges of the telescope is less than that nearer the center, in order to minimize sidelobes.

Define the beam solid angle to be

$$\Omega_A = \int \frac{P(\theta - \theta_0, \phi - \phi_0)}{P_0} \, d\Omega. \tag{12.28}$$

This can be related to an *effective area*, the effective collecting area for an on-axis point source, by the van Cittert–Zernike theorem,

$$A_e \Omega_A = \lambda^2. \tag{12.29}$$

For uniform illumination and no surface errors, A_e is just the geometrical area of the telescope. It is smaller than the geometrical area if the illumination is *tapered*, in which case the outer portions of the dish are largely unused. We can define an aperture efficiency as the ratio of effective area to geometrical area:

$$\eta_A = \frac{A_e}{A_g}. \tag{12.30}$$

We can also define a main beam solid angle and main beam efficiency by

$$\Omega_{MB} = \int_{MB} \frac{P(\theta - \theta_0, \phi - \phi_0)}{P_0} d\Omega, \tag{12.31}$$

$$\eta_{MB} = \frac{\Omega_{MB}}{\Omega_A}. \tag{12.32}$$

12.5.2 Antenna temperature

Consider an astronomical source with a brightness distribution $I_\nu(\theta, \phi)$. Remembering that blackbody radiation is described by

$$I_\nu = \frac{2h\nu^3}{c^2} \frac{1}{e^{h\nu/kT} - 1}, \tag{12.33}$$

we can consider the low frequency (Rayleigh–Jeans) limit of this equation and use it to define a Rayleigh–Jeans brightness temperature,

$$T_B(\theta, \phi) = \frac{\lambda^2}{2k} I_\nu(\theta, \phi). \tag{12.34}$$

This Rayleigh–Jeans brightness temperature is the same as the physical temperature for a blackbody in the Rayleigh–Jeans limit. Since the specific intensity is power per unit area per unit solid angle, we get the power by taking the specific intensity times the power pattern and integrating over area and solid angle. Including a factor of ½ to consider a single polarization, the power collected per unit bandwidth is

$$W = \frac{1}{2} A_e \int I_\nu(\theta, \phi) \frac{P(\theta - \theta_0, \phi - \phi_0)}{P_0} d\Omega. \tag{12.35}$$

Equating this to an equivalent thermal source, $W = kT_A$,

$$T_A(\theta_0, \phi_0) = \frac{1}{2k} A_e \int I_\nu(\theta, \phi) \frac{P(\theta - \theta_0, \phi - \phi_0)}{P_0} d\Omega \tag{12.36}$$

$$= \frac{A_e}{\lambda^2} \int T_B(\theta, \phi) \frac{P(\theta - \theta_0, \phi - \phi_0)}{P_0} d\Omega \tag{12.37}$$

$$= \frac{1}{\Omega_A} \int T_B(\theta, \phi) \frac{P(\theta - \theta_0, \phi - \phi_0)}{P_0} d\Omega. \tag{12.38}$$

In single-dish observing we actually measure $T_A(\theta_0, \phi_0)$. From our knowledge of the beam pattern we can attempt to reconstruct $T_B(\theta, \phi)$.

12.5.3 Special antenna designs

As mentioned above, a large number of ground-based radio telescopes have been blocked Cassegrain designs. Examples include the 25 m telescopes of the VLA, the 100 m Effelsburg telescope, and many others. In contrast, the Byrd Green Bank Telescope (GBT) was designed with a primary mirror consisting of an off-axis paraboloid with a 100 m diameter clear aperture. The clear aperture provides exceptionally low sidelobes for precision mapping. The GBT is equipped with both prime and Gregorian foci. Off-axis designs are becoming increasingly common, both in orbit and for ground-based radio telescopes, as the remaining examples from this section show.

WMAP utilized off-axis Gregorian telescopes[5] with an unblocked 1.4 m diameter aperture. This and a 20 dB taper were important in limiting sidelobe response. Additional shields were added to further limit sidelobe response, and the telescope mirrors were specially shaped to minimize aberrations and optimize polarization characteristics. All of the receivers were dual polarization designs with corrugated feed horns and used orthomode transducers to separate the polarizations.

The Planck satellite also has an off-axis Gregorian telescope with an unblocked aperture of 1.5 m, only slightly larger than WMAP. However, there is only a single telescope. Planck radiometry is direct, not differential. The Gregorian design had to be modified to be aplanatic due to the large focal plane (large field of view).

The ACT is a 6 m off-axis Gregorian optimized to be nearly aplanatic. One essential feature is a ground screen, necessary to prevent thermal emission from the nearby landscape from entering the sidelobes of the antenna pattern. The SPT is a 10 m off-axis classical Gregorian (parabolic primary). Like the ACT it is surrounded by an extensive ground screen.

12.6 Radio interferometry

12.6.1 Basic two-element interferometer

Consider two antennas separated by a baseline $\vec{B}$ and consider emission from a small quasi-monochromatic source, as shown in Figure 12.16. The receivers produce output voltages whose phase difference is proportional to the path-length difference $B \sin \theta$,

---

[5] A standard Gregorian telescope has a parabolic primary and an elliptical secondary. This type of Gregorian is not aplanatic. An aplanatic Gregorian would have an elliptical primary and hyperbolic secondary.

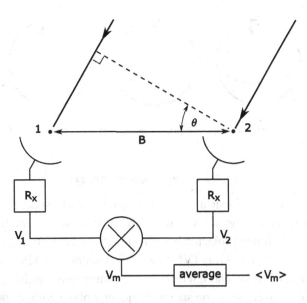

Figure 12.16 In a radio interferometer each antenna pair measure time-delayed versions of the astronomical signal. These are brought together, correlated, and averaged to produce the complex visibility for the baseline $\vec{B}$.

$$V_2 \propto E \cos \omega t, \tag{12.39}$$

$$V_1 \propto E \cos \omega (t - \tau), \tag{12.40}$$

$$\tau = \frac{B \sin \theta}{c}. \tag{12.41}$$

In a correlation interferometer these voltages are multiplied and averaged,

$$\langle V_m \rangle \propto \frac{1}{2} E^2 \cos \omega \tau. \tag{12.42}$$

For a source in a direction normal to the baseline ($\theta = 0$), the fringes are separated by $\Delta \theta = \lambda / B$. This applies, for example, to a source on the meridian when the baseline runs east–west. The sinusoidal pattern measures a single Fourier component of the source intensity distribution.

The complex value of $\langle V_m \rangle$ is referred to as the complex fringe visibility[6] for some particular baseline $\vec{B}$. Its 2-dimensional Fourier transform is the source intensity distribution. One can measure additional Fourier components via (1) using multiple pairs of telescopes, (2) movable baselines, (3) variation of the projected baseline via the Earth's rotation. The last is illustrated in Figure 12.17.

[6] Sometimes visibility refers to a dimensionless quantity normalized to unity at zero spacing. Sometimes it remains unnormalized and has units of flux density. It is necessary to look at the context to determine which convention is being used.

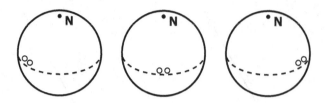

Figure 12.17 Variation in baseline projected length and orientation due to Earth's rotation.

12.6.2 Interferometer arrays

The earliest arrays were linear arrays (often with equally spaced antennas), providing a 1-dimensional view of the source structure. However, the Earth's rotation provides some variation in projected baseline length and orientation (Earth-rotation synthesis). The Very Large Array (VLA) has 27 elements in an inverted-Y configuration. The design provides enough different instantaneous baselines ($27 \times 26/2 = 351$) to give good coverage of the spatial frequency plane without needing to wait for Earth's rotation (snapshot mode). As Earth rotates, a fixed baseline $\vec{B}$ traces out an ellipse in Fourier space, as discussed in Section 12.6.4. At long wavelengths one needs continental or intercontinental baselines for good angular resolution. The general term for this is Very Long Baseline Interferometry (VLBI). A dedicated VLBI instrument is the 10-element Very Long Baseline Array (VLBA).

EVLA

The Expanded Very Large Array (EVLA) project consists mostly of upgrades in receivers, IF electronics, correlator, and software to provide continuous frequency coverage from 1 to 50 GHz with improved sensitivity, bandwidth, and spectral resolution. Additional proposed changes included eight new antennas up to 250 km away from the center of the array, two converted VLBA antennas, and a new "super-compact" configuration. However, these additional changes have not been funded. Central to the improvements which were funded is the new WIDAR correlator, discussed below.

CARMA

The Combined Array for Research in Millimeter-wave Astronomy (CARMA)[7] is a consolidation of the previous BIMA (Hat Creek) instrument with its OVRO (Owens Valley) counterpart at a new higher site at Cedar Flat (7200 feet (2200 m) elevation) in the Inyo mountains, with additional telescopes supplied by

---

[7] The participating institutions are the California Institute of Technology, University of California (Berkeley), University of Illinois, University of Maryland, and University of Chicago with additional funding from the National Science Foundation.

the SZ-array (University of Chicago). In all CARMA has six 10.4-meter telescopes, nine 6.1-meter telescopes, and eight 3.5-meter telescopes. This is an example of a heterogeneous array, in which one must make allowance for the different primary beam patterns of the different sized antennas. The receivers cover the 7 mm, 3 mm, and 1.3 mm wavelength bands. Baselines range up to 2 km.

ALMA

The Atacama Large Millimeter/submillimeter Array (AMLA) was conceived of as an array of 64 12-meter telescopes capable of operating from 30 to 900 GHz in ten frequency bands. The Altiplano de Chajnantor in northern Chile near the borders of Bolivia and Argentina, at 5000 meters elevation, is one of the driest sites on Earth, a feature essential for optimal operation at submillimeter wavelengths. ALMA was initially a US/European collaboration with cost constraints limiting the number of antennas to 50. A Japanese collaboration has added four 12-meter antennas and twelve 7-meter antennas and the capability of having short baselines, essential for discerning large-scale structure. This is often called the Atacama Compact Array (ACA) .

12.6.3 Correlators

The first consideration in designing a spectral correlator is whether one will correlate antennas before Fourier transforming (an XF correlator) or Fourier transform before correlating (an FX correlator). The X part of the correlator is essentially performing the operation illustrated in Figure 12.16. The F part of the correlator is essentially performing the operation illustrated in Figure 12.11. For design considerations, see Thompson *et al.* (2001). The general trend is away from designs incorporating ASICs (Application Specific Integrated Circuits), and towards designs incorporating FPGAs (Field Programmable Gate Arrays).

The WIDAR correlator (Wideband Interferometric Digital ARchitecture) in use at the EVLA is, at its heart, an XF correlator, although frequency separation into sub-bands occurs before the XF correlation.

The initial CARMA correlator is known as the Caltech Owens-Valley Broadband Reconfigurable Array (COBRA). Receiver signals are first digitized at a 1 GHz sampling rate. The correlator is an XF design. Future improvements will increase the number of bits per sample to be correlated, improving the efficiency.

The design for the initial ALMA correlator is for an XF correlator capable of handling 64 antennas with up to 4096 spectral channels. Work is being done on a "European Future Correlator," which will have some similarities with the WIDAR design. The Japanese are working on an FX correlator design which will be needed when there are more than 64 antennas. Although the current funding seems to support only 66 antennas $(50+4+12)$, there is still a possibility that an 80 antenna correlator will be needed $(64+4+12)$.

The correlator for the future Square Kilometer Array (SKA) will be an FX design since that architecture is advantageous for large numbers of antennas.

12.6.4 Fourier inversion

For a particular observation, the coordinates u and v are the components of the interferometer baseline vector, projected onto the plane of the sky in the directions of right ascension and declination, respectively, measured in units of wavelength,

$$\frac{\vec{B}}{\lambda} = \begin{pmatrix} u \\ v \\ w \end{pmatrix}, \tag{12.43}$$

where w is the component of the baseline along the line of sight towards the source. The values of u and v refer to the 2-dimensional Fourier component being measured, and the complex visibility gives the amplitude and phase of that component.

The brightness distribution $I(x, y)$ may be recovered from the visibilities $V(u, v)$ only if we have *complete* visibility data (densely spaced measurements with u and v extending to infinity). Ignoring the beam patterns of the individual telescopes,

$$V(u, v) = \int \int_{-\infty}^{\infty} I(x, y) \, e^{i2\pi (ux+vy)} \, dx \, dy, \tag{12.44}$$

$$I(x, y) = \int \int_{-\infty}^{\infty} V(u, v) \, e^{-i2\pi (ux+vy)} \, du \, dv. \tag{12.45}$$

However, we only know $V(u, v)$ at a finite number of points. Any attempt to recover $I(x, y)$ implicitly requires some assumptions about $V(u, v)$ in regions where it was not measured.

One aspect of this incompleteness is known as the *zero-spacing problem*. Any set of purely interferometric data inevitably has a hole in the center of the uv-plane whose radius corresponds to the minimum projected spacing between antennas. In an inhomogeneous array this will be determined by the average of the diameters of the two smallest telescopes in the array. This information may be obtained by single-dish mapping with a telescope of sufficient diameter (typically at least twice the diameter of the telescopes in the array) or by interferometric mosaic mapping.

The simplest inversion, a direct sum over the available data, is equivalent to assuming $V(u, v) = 0$ where it was not measured. This is the logical equivalent of the myth of an ostrich hiding its head in the sand, thinking "What I cannot see does not exist."

$$I''(x, y) = \sum_k V(u_k, v_k) \, e^{-i2\pi(u_k x + v_k y)} \, \Delta u \, \Delta v \qquad (12.46)$$

$$= \int \int_{-\infty}^{\infty} V(u, v) \sum_k \delta(u - u_k, v - v_k) \, e^{-i2\pi(u_k x + v_k y)} \, du \, dv. \qquad (12.47)$$

Since the Fourier transform of a product is a convolution of Fourier transforms,

$$I''(x, y) = I(x, y) * P_D(x, y), \qquad (12.48)$$

where

$$P_D(x, y) = \int \int_{-\infty}^{\infty} \sum_k \delta(u - u_k, v - v_k) \, e^{-i2\pi(u_k x + v_k y)} \, du \, dv \qquad (12.49)$$

$$= \sum_k e^{-i2\pi(u_k x + v_k y)}. \qquad (12.50)$$

This function $P_D(x, y)$ is referred to as the *dirty* or *synthesized* beam. It is essentially a point spread function. The map we get by direct Fourier inversion is a convolution of the true intensity distribution with this "dirty beam."

Clean algorithm

There is no *unique* way to deconvolve $I(x, y)$,

$$I''(x, y) = I(x, y) * P_D(x, y). \qquad (12.51)$$

However, our dirty map will contain sidelobes which are artifacts of where $V(u, v)$ was sampled. These are clearly unrelated to the true source structure, and we want to eliminate them. Consider, for example, a dirty map which is identical to the dirty beam. It is appealing to say that the intensity distribution is just a single point source. Similarly one might "recognize" the dirty map as appearing to be the sum of two (displaced) copies of the dirty beam, suggesting $I(x, y)$ consists of two point sources. *Cleaning* is an iterative procedure of this sort which attempts to recover $I(x, y)$ by identifying and fitting the strongest features in the map. It is non-linear and will not necessarily converge to a unique result, but it is often quite successful. Its strong non-linearity makes the effects of *Clean* difficult to analyze.

Maximum entropy method (MEM)

Any method of image reconstruction is essentially filling in some assumptions about the unmeasured Fourier components. Ideally, we would like to have a map reflecting the true source structure with as few artifacts as possible relating to the measurement process. We want to minimize false features. If we can quantify the concept of *structure*, we can try various solutions and pick the one with the least structure. Since our map will always reflect the true source structure, this minimum structure map will be the map least contaminated by artifacts. In practice we maximize the opposite of structure, the *entropy* of an image. There can be various definitions of entropy, such as

$$E = - \int \int \frac{I(x, y)}{I_{TOT}} \ln \left(\frac{I(x, y)}{I_{TOT}} \right) dx \, dy, \qquad (12.52)$$

where $I_{TOT} = \int \int I(x, y) \, dx \, dy$. This quantity is then maximized within the constraints imposed by the available data.

In Chapter 13 we discuss a better, statistical approach to this topic. Fourier inversion is what is known as an *inverse problem*, of which there are many examples in statistics.

13

Modern statistical methods

It is ironic that a chapter entitled "Modern statistical methods" begins with Bayes' theorem, since the work of the Reverend Thomas Bayes was published posthumously in 1763. But the acceptance of this approach was largely delayed until the latter half of the twentieth century. This is despite an imposing pedigree containing major contributions from Pierre-Simon Laplace, who in 1814 independently and much more thoroughly developed the field we now call Bayesian statistics. The modern revival of these ideas is largely attributable to Harold Jeffreys, with additional creative contributions from Richard Cox, Claude Shannon (the father of information theory), George Pólya, Edwin Jaynes, and others.

In this chapter we begin with Bayes' theorem, followed by a sampling of some of the statistical issues that should be of concern to astronomical observers. We also cover a few of the powerful statistical tools available to modern astronomers.

13.1 Bayes' theorem

In probability theory one often deals with what are known as conditional probabilities. That is, in addition to the normal probability that something will occur, $p(A)$, one also needs to express the probability of A given B, denoted $p(A|B)$. Now consider the joint probability of both A and B, which we will write as $p(AB)$. Clearly this is equal to the probability of B times the conditional probability of A given B:

$$p(AB) = p(B)\, p(A|B). \tag{13.1}$$

But that joint probability is also equal to the probability of A times the conditional probability of B given A:

$$p(AB) = p(A)\, p(B|A). \tag{13.2}$$

Combining these results and rearranging terms we get Bayes' theorem:

$$p(A|B) = p(A)\,\frac{p(B|A)}{p(B)}.\tag{13.3}$$

Taking a somewhat broader view, we can use the above formula to describe the plausibility of hypotheses. To do so, we need to make all of the probabilities conditional on some set of background assumptions. Letting H_i stand for one hypothesis out of the set $\{H_i\}$ and letting D denote some data,

$$p(H_i|DI) = p(H_i|I)\,\frac{p(D|H_iI)}{p(D|I)}.\tag{13.4}$$

The symbol I represents the information which was known *a priori*, before the data were collected, and we call $p(H_i|I)$ the *prior*. Thus $p(H_i|DI)$ is the probability that the hypothesis is supported by both the prior information and the data. This we call the *posterior* probability. The numerator $p(D|H_iI)$ is known as the *likelihood* function. The denominator is simply a normalization factor known as the *prior predictive probability* or the *evidence*,

$$p(D|I) = \sum_i p(H_i|I)\,p(D|H_iI).\tag{13.5}$$

One might paraphrase Bayes' theorem by saying

The probability that some hypothesis is true depends on what we know *a priori* and the likelihood that that hypothesis would produce the data we observe.

We can also say that Bayes' theorem describes how newly acquired data change our knowledge of the various hypotheses. It is interesting to note that the order in which the data are arranged does not matter. Bayes' theorem may be applied to all of the data at once or successively to various subsets of the data. What is important is that each measurement be included in the analysis once and only once. As long as the data are independent,

$$p(\{D\}|H_iI) = \prod_j p(D_j|H_iI),\tag{13.6}$$

and multiplication is a commutative (and associative) operation.

In the previous paragraph we took a rather large step from the language of probability theory to that of inductive reasoning or inference. That we are entitled to do so was shown by Cox (1946), who proved that the only procedure for what he called *reasonable expectation* that was logically consistent was one which, mathematically, was identical to the rules of probability theory.

13.2 Maximum likelihood

Next we take a diversion to the classical concept of maximum likelihood, a technique with origins dating back to Gauss and Lagrange, contemporaries of Laplace. What happens when $p(H_i|I)$ is a constant, that is, when the models are equally likely *a priori*? In that case the most plausible hypothesis is the one for which the likelihood function $p(D|H_i I)$ is maximum. We just need to find the theory with the highest probability of predicting the actual data,

$$H_{\underset{i}{\mathrm{argmax}}(p(H_i|DI))} = H_{\underset{i}{\mathrm{argmax}}(p(D|H_i I))} , \qquad (13.7)$$

since $p(H_i|I)$ and $p(D|I)$ are both constants.

Suppose we have a set of data containing N values $\{y_j\}$ with known uncertainties $\{\sigma_j\}$. We wish to compare these data with models which predict values $\{\xi_j\}$. These models contain free parameters $\{a_i\}$. In effect, the various values of the $\{a_i\}$ correspond to various hypotheses (alternative theories). We want to find the best theory. If we further assume that the probability distributions of the experimental results are Gaussian,

$$p(D|H_i I) = L(\{a_i\}) = (2\pi)^{-N/2} \left[\prod_{j=1}^{N} \sigma_j^{-1} \right] \exp \left[-\sum_{j=1}^{N} \frac{(y_j - \xi_j)^2}{2\sigma_j^2} \right]. \qquad (13.8)$$

Maximizing the likelihood L (or log L), is then equivalent to *minimizing*

$$\chi^2(\{a_i\}) = \sum_{j=1}^{N} \frac{(y_j - \xi_j)^2}{\sigma_j^2}. \qquad (13.9)$$

This is the justification for the familiar least-squares fitting process: adjusting parameters to give a minimum value for χ^2. The optimum parameters $\{a_i\}$ may be found either numerically or by solving a system of N linear equations. Note that whereas one might always try to seek a least-squares solution, that solution will in general be a maximum likelihood solution only under the conditions outlined above. In other words, the technique of least squares is justified in cases where different values of the parameters $\{a_i\}$ are equally likely *a priori* and the data $\{y_j\}$ are corrupted by uncorrelated Gaussian noise.

The technique of least squares is also frequently used to fit a function $f(x)$ to a set of data pairs $\{x_j, y_j\}$. The same justification applies here as long as the independent variables $\{x_j\}$ are noiseless and the dependent variables $\{y_j\}$ are corrupted by uncorrelated Gaussian noise.

Earlier we introduced the concepts of biassed and unbiassed estimators. Maximum likelihood estimators are in some cases biassed and in some cases unbiassed.

13.3 So what is Bayesian inference?

At the beginning of this chapter we introduced an important distinction between two approaches to statistics. The traditional *frequentist* school of thought, represented by the material in Chapter 6, focusses on the data and provides descriptors which are characteristics of the observed data, such as the mean and the variance. Probability is viewed as the frequency of occurrence of various outcomes if an experiment is repeated (e.g. rolling dice). In practice, the frequentist considers the existence of alternative data sets and considers various operations (e.g. averages) over the ensemble of possible data sets. The *Bayesian* school, in contrast, compares alternative hypotheses in the light of a single observational data set. Probability takes on the meaning of "plausibility of some proposition." In practice, a Bayesian does "arithmetic" on hypotheses.

These approaches often agree, but they do not always agree. The fact that they frequently agree has led to confusion and to the supposition that they are different ways of looking at the same thing. However, they are fundamentally different in concept. Certainly it is true that two valid approaches to a problem should give the same answer. But it can be difficult to pose some problems in language that means the same thing to a frequentist and to a Bayesian, particularly if that language uses the word "probability."

It is interesting to note that Bayesian inference provides an automatic application of Occam's razor. As we will see, hypotheses with additional (unnecessary) free parameters are necessarily less plausible.

In support of the Bayesian approach, one frequently encounters assertions such as:

1. A Bayesian approach yields a unique solution; a frequentist approach does not.
2. A frequentist approach cannot incorporate *a priori* information; a Bayesian approach can.

Such statements are controversial. In my view (1) is true; the lack of a unique solution in the frequentist approach is related to the well-known problem of choosing an appropriate "statistic." In my view (2) is false, although the use of *a priori* information is more difficult and often less transparent in a frequentist approach.

13.3.1 Example 1

Consider the following question: was there a galactic supernova in 2008? There are only two possible hypotheses, T (True) and F (False).

Now consider some *a-priori* information. Hypothetically, let us assume, based on our understanding of galactic structure and stellar evolution, that $p(T|I) = 0.05$ and $p(F|I) = 0.95$. That is, we assume that we live in a galaxy where SN occur on average once every 20 years. Let us also assume that we know enough about galactic extinction and other observational constraints to say that *if* a supernova occurred, we would have a 50% chance of seeing it visually.

Now consider observational data for the year 2008. If a supernova was observed, the probabilities *must* jump to 100%/0%. On the other hand, if a supernova was not observed, the probability of T should drop somewhat below 5%, but not to zero, since a supernova might have occurred without being seen. Let's work through the math:

$$p(T|DI) = p(T|I) \frac{p(D|TI)}{p(D|I)} = 0.05 \frac{p(D|TI)}{p(D|I)}, \tag{13.10}$$

$$p(F|DI) = p(F|I) \frac{p(D|FI)}{p(D|I)} = 0.95 \frac{p(D|FI)}{p(D|I)}, \tag{13.11}$$

$$p(D|I) = p(T|I)p(D|TI) + p(F|I)p(D|FI). \tag{13.12}$$

Now if D = OBS,

$$p(D|I) = 0.05 \times 0.5 + 0.95 \times 0 = 0.025, \tag{13.13}$$

$$p(T|OBS, I) = 0.05 \frac{0.5}{0.025} = 1, \tag{13.14}$$

$$p(F|OBS, I) = 0.95 \frac{0}{0.025} = 0, \tag{13.15}$$

which is the trivial result we expected. For the somewhat less trivial case of D = NOTOBS,

$$p(D|I) = 0.05 \times 0.5 + 0.95 \times 1 = 0.975, \tag{13.16}$$

$$p(T|NOTOBS, I) = 0.05 \frac{0.5}{0.975} \approx 0.025, \tag{13.17}$$

$$p(F|NOTOBS, I) = 0.95 \frac{1}{0.975} \approx 0.975. \tag{13.18}$$

This should give some idea how one could use Bayes' theorem in practice. Obviously things would be more complex in a situation with multiple hypotheses.

13.3.2 Example 2

This example has been adapted from Loredo (1992). Let us assume that we have sufficient (*a priori*) knowledge of supernova explosions to say that the probability of a neutrino detection decays exponentially after the starting time t_0, as shown in Figure 13.1,

$$p(t) = \begin{cases} 0 & t < t_0, \\ \tau^{-1} e^{-(t-t_0)/\tau} & t \geq t_0. \end{cases} \tag{13.19}$$

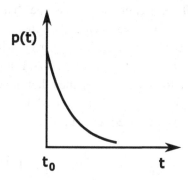

Figure 13.1 In example 2, the probability of a neutrino detection.

Let us also assume that we know the decay constant τ. For simplicity, set $\tau = 1$. For some data set (e.g. for SN 1987A) what is our best estimate of t_0?

A *frequentist* would correctly say that for a data set $\{t_i\}$ with N neutrinos detected, the unbiassed, maximum-likelihood estimator of t_0 is

$$\hat{t} = \langle t_i \rangle - \tau = \frac{1}{N} \sum_{i=1}^{N} (t_i - \tau). \tag{13.20}$$

For a data set consisting of three events at $t = 12, 14, 16$ (a possible, but unlikely, data set), this would give $\hat{t} = 13$. The 90% confidence interval says $12.15 < t_0 < 13.83$. *But* the first event occurs at $t = 12$, and t_0 *cannot* be later than the first event! So we have missed some critical information available in the data. This is a somewhat contrived example. A frequentist actually could have done better with a better choice of statistic (better than $\hat{t}$). But it illustrates some of the difficulties that can emerge in analyzing a data set which may be somewhat improbable.

A *Bayesian*, on the other hand, would say that, given the above model, the *likelihood* of some time-ordered data set $\{t_1, t_2, \ldots\}$, is given by

$$p(\{t_i\}|t_0 I) = \prod_{i=1}^{N} \left[e^{-(t_i - t_0)} H(t_i - t_0) \right]. \tag{13.21}$$

Since t_1 is the earliest detection, the Heaviside function $H(t_1 - t_0)$ dominates, and the likelihood is

$$p(\{t_i\}|t_0 I) = \begin{cases} e^{N t_0} e^{-\sum_{i=1}^{N} t_i} & t_0 \leq t_1, \\ 0 & t_0 > t_1. \end{cases} \tag{13.22}$$

The term $e^{-\sum_{i=1}^{N} t_i}$ is just a constant (independent of t_0). The posterior distribution of probabilities for various t_0, properly normalized, is given by

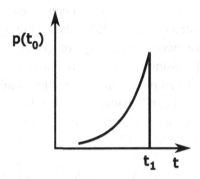

Figure 13.2 The Bayesian posterior for t_0 in example 2.

$$p(t_0|\{t_i\}I) = \begin{cases} N\, e^{N(t_0-t_1)} & t_0 < t_1, \\ 0 & t_0 > t_1, \end{cases} \qquad (13.23)$$

as shown in Figure 13.2. Thus the most probable value for t_0 is just the earliest data point ($t_1 = 12$ in the example above). No information about the times t_2 and t_3 is included. Only start times before t_1 are allowed (i.e. $t_0 \leq t_1$). Start times which precede t_1 become increasing unlikely as the interval between t_0 and t_1 is increased. The width of the posterior is narrower if more neutrinos are detected. In this case the difference between the frequentist and Bayesian results arises from the non-Gaussian shape of the probability distribution.

This is an example of Bayesian *parameter estimation*. We have used Bayes' theorem to discriminate between hypotheses which differ only through the value of a free parameter, in this case t_0. In general, Bayes' theorem may be used in this way for multiple free parameters.

13.4 Maximum entropy

Having so far largely ignored the *a-priori* probability $p(H_i|I)$ in Bayes' theorem, let us now look at that in more detail. Suppose that our set of hypotheses $\{H_i\}$ consists of 100 possibilities, each of them in equally good agreement with the data: $p(D|H_iI)$ is constant. Suppose they are also equally likely *a priori*. And further suppose that 99 of these are *indistinguishable*. What should we adopt as the most likely hypothesis? In fact, according to Bayes' theorem, all are equally likely. But if we ask whether the single distinguishable hypothesis is most likely, the answer is clearly not. In essence, we have collapsed the 99 indistinguishable hypotheses into a single hypothesis (call it A) and are comparing that with the remaining distinguishable hypotheses (call it B). The hypothesis A is 99 times more probable than B, by a counting argument.

This line of reasoning is similar to that encountered in statistical mechanics. A disordered physical state is more probable than a highly ordered physical state simply because there are many more ways of constructing a disordered state. In physics we describe this using the concept of *entropy*, the degree of disorder.

Similar reasoning may be employed in the statistics of image analysis. We would like to maximize p(H$_i$|DI), or equivalently the natural logarithm of p(H$_i$|DI). If each hypothesis is described by some number N$_i$, the number of ways that state can be constructed (N$_i$ taking the place of p(H$_i$|I)), then

$$Max\ (\ln p(H_i|DI)) \propto Max(\ln(N_i \times p(D|H_iI))) \qquad (13.24)$$
$$\propto Max(\ln N_i + \ln p(D|H_iI)) \qquad (13.25)$$
$$\propto Max(\ln N_i - \chi^2/2) \qquad (13.26)$$
$$\propto Max(\lambda S - \chi^2/2), \qquad (13.27)$$

where we have once again assumed Gaussian noise. Essentially N$_i$ represents the number of "microscopic" states per "macroscopic" state. We have ignored here the normalization of the prior and assert only that the entropy, S, is proportional to ln N$_i$. For a more careful and complete statement of the imaging problem see Sutton & Wandelt (2006).

In the previous equation, S takes on the meaning of *entropy*, and our task becomes one of maximizing S while jointly minimizing χ^2. This is a very powerful technique often applied to image analysis in radio astronomy and increasingly in other wavelength regions as well. It is also applicable to a wide range of situations in which information is to be extracted from experimental data (medical tomography, crystallography, NMR spectroscopy, etc.).

There has been much discussion of the appropriate form of the entropy function S. For image analysis, strong arguments have led to the conclusion that

$$S = -\sum_i f_i \ln f_i \qquad (13.28)$$

is the appropriate choice, where f$_i$ is the intensity of the image at pixel i. This is usually referred to as the Shannon or Shannon–Jaynes entropy,[1] and is equivalent to the Gibbs entropy, apart from the absence of Boltzmann's constant.[2] Note that this formulation appropriately ensures non-negativity to the image intensity. Additional details are contained, below, in Section 13.5.3.

[1] As written, the units of entropy are called nits or nats. Changing ln to log$_2$ makes the units bits.
[2] Again, this is a question of units. The quantum extension of this in density matrix theory is known as the von Neumann entropy.

13.5 Uninformative priors

Bayesian analysis requires us to specify our prior information about a problem in a suitable mathematical form. Failure to include *all* relevant prior information means that we will not be able to take full advantage of Bayes' theorem. This may result in a broader (less certain) posterior distribution than one might obtain with full use of prior information. Incorporation of incorrect prior information, on the other hand, is a much more serious error. In the worst cases it may entirely exclude the true answer from the region of finite posterior density. For this reason, it is often useful to consider *uninformative priors*, those which incorporate the least possible amount of information.

13.5.1 Location priors

The symmetries of a problem may lead to the appropriate choice of prior. For example, if there is translational symmetry, then the answer should not depend on location. In one dimension (let us say x), this would be expressed by a uniform prior in x. But a uniform prior in x from $-\infty$ to ∞ is an *improper prior* since it cannot be normalized. A uniform prior may be made proper by applying some lower and upper bounds. In real life our full information usually allows us to set some bounds to the allowed region, enabling us to use a prior which is uniform (uninformative) over some finite region of space.

13.5.2 Scale priors

Under other circumstances the appropriate symmetry may be one of scale. Complete ignorance about the scale of a positive continuous variable, r, is expressed by a *Jeffreys' prior*,[3] a prior which is proportional to 1/r. The Jeffreys' prior is equivalent to a uniform probability density for log(r). In other words, scales of watts, mW, and kW are all equally likely. This is necessary to ensure that the numerical result of any Bayesian posterior calculation be independent of the units chosen to perform the calculation.

This prior in r is also an improper prior since the integral of 1/r is logarithmically divergent at both zero and infinity. As above, one may construct from this a proper prior by prescribing upper and lower cutoffs in r, and the normalization will be only weakly sensitive to the choice of cutoffs.

13.5.3 Positive, additive distributions

An uninformative prior to a positive, additive distribution is often said to be the entropy prior. As motivation, consider an image to be made up of n pixels into

---

[3] The term *Jeffreys' prior* has a broader meaning in statistics, beyond the scope of this text.

which a team of monkeys randomly drops λ elements of luminance. If $\{N_i\}$ describe the number of elements which end up in each of the pixels ($\lambda = \sum N_i$), then combinatorial arguments give the multiplicity of a particular distribution to be

$$W = \frac{\lambda!}{N_1! \, N_2! \, \ldots \, N_n!}. \tag{13.29}$$

Ignoring the normalization of the prior, taking the limit $\lambda \to \infty$, and expanding all of the factorials using Stirling's formula, the leading order term in the natural logarithm of the prior is given by the entropy

$$S = -\sum_i \frac{I_i}{I_t} \ln \frac{I_i}{I_t}, \tag{13.30}$$

where I_i is the intensity of pixel i and $I_t = \sum I_i$. This is essentially the development of the principle of maximum entropy, discussed above. As a practical matter this entropy prior has had much success, although some objections to the method have been noted by Sutton & Wandelt (2006).

13.6 Inverse problems

There is a large class of what are known as *inverse problems* in which one tries to infer, from noisy data, some original state. Let us call this original state the object of observation, O, which for astronomers might be an unknown intensity distribution, an unknown spectrum, or something else entirely. We observe the object via some measurement process M. In matrix notation we could write

$$D = MO + N, \tag{13.31}$$

where N represents noise and D represents our data. We would like to get from D back to O. Of course, we cannot "subtract" the noise N since we do not know the actual realization of the noise, only perhaps its statistical properties. Nor could we then "divide" by M since M will usually contain zeros, meaning that some information is irretrievably lost in the measurement process.

In mathematics when a matrix is not invertable, either because it is not square or it is square but singular, a *pseudoinverse* may be defined. Leaving aside the issue of noise, the Moore–Penrose pseudoinverse may be viewed as a method of solving the set of linear equations given by

$$D = MO \tag{13.32}$$

in such a way as to minimize the Euclidian norm of MO − D in the case of an over-determined set of equations and minimizing the Euclidean norm of O in the under-determined case.

The real world is not this simple. For a measurement process such as that represented by Equation 13.31, one can develop a variety of inversion *procedures*. As we saw for radio astronomy in Chapter 12, each of those procedures may have its own merits. But there is no single, correct inversion procedure because inversion in this case is an ill-formulated problem. The forward problem, however, is well formulated; it is simply Equation 13.31.

The solution is never to try to perform an inverse procedure; it is a lost cause. Instead we should always solve the forward problem, repeatedly if necessary. If the measurement matrix M is known and the noise N is statistically known, then we can try varying the object O until we achieve statistical agreement with the data. And we can characterize the uncertainties by continuing to vary O in such a way as to obtain a statistical representation of our knowledge about O. Stated this way, this is clearly a Bayesian approach, as discussed by Sutton & Wandelt (2006) for radio interferometric imaging.

There are other advantages as well. If M is only partially known, we can parameterize the unknown aspects of M and solve jointly for those parameters and O. We then marginalize over the measurement parameters to obtain information about O. Marginalization is a procedure in statistics in which the dimensionality of a posterior distribution is reduced by integrating over one or more undesired or *nuisance parameters*. For example, if the posterior is a function of parameters x_1 and x_2, marginalizing over x_2 means integrating

$$p(x_1) = \int p(x_1, x_2)\, dx_2. \tag{13.33}$$

In radio interferometry the nuisance parameters might include the antenna-based complex gains which vary, among other things, due to atmospheric effects.

13.7 Sampling the posterior

The Bayesian posterior probability density function (PDF) hardly ever will be a simple analytic function. Sometimes it may be sufficiently well approximated by a Gaussian, in which case what is needed are estimates of the mean and the variance and some tests of Gaussianity. But most commonly, the shape of the PDF is unknown and possibly multimodal. In such cases, rather than seeking a few parameters (such as a mean and a variance) to describe the PDF, a more complete characterization is necessary. This may be provided by obtaining a sampling of the PDF using Monte Carlo techniques. The sample size may be made as large as necessary to provide an accurate description. Assuming that the posterior can be calculated, via Bayes' theorem, there is a variety of ways of sampling the PDF. Samplings may be done for parameter spaces of any dimensionality, although here for simplicity we will mostly illustrate with 1-dimensional distributions. From such samplings one may obtain marginal distributions, moments, credible intervals, etc.

13.7.1 Rejection sampling

Assume $p(x)$ is an unknown PDF, which we will call the target density, and $g(x)$ is a function which, for some constant c, provides an upper bound to $p(x)$ in the sense

$$p(x) \le c\, g(x). \tag{13.34}$$

The function $g(x)$ is known as the proposal density and is chosen to be a distribution from which it is easy to make random samples. The procedure for *rejection sampling*[4] is to draw a random value of x from the distribution $g(x)$. Another random number is then drawn from the uniform distribution $U(0, cg(x))$. The sample is accepted if the second random number is less than $p(x)$ and rejected otherwise. In other words, the sample is accepted with probability $p(x)/[cg(x)]$. This process is then repeated. An advantage of this method is that it produces independent (uncorrelated) samples. A disadvantage is that it can be inefficient.

As an illustration of the potential inefficiency of importance sampling, consider a narrow PDF defined on $[0, 1)$, as shown in Figure 13.3. If little is known about $p(x)$, one might choose the proposal density to be $U(0, 1)$. Let us assume we know the peak value of $p(x)$ well enough so that an efficient choice may be made for c. As the figure shows, most samples of x drawn from $U(0, 1)$ will fall in regions

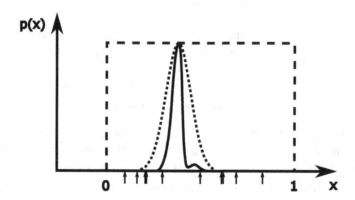

Figure 13.3 An example of rejection sampling, in which eight of the first ten draws from $U(0,1)$ are rejected with certainty, and the remaining two have greater than 95% probability of rejection. The solid line is $p(x)$, and the dashed line is $g(x)$. A better proposal density, in this case, would be a Gaussian such as that shown by the dotted line.

---

[4] Another procedure, known as *importance sampling*, does not actually provide a sampling from $p(x)$ but does allow one to calculate expectation values over the distribution $p(x)$.

of low probability and be rejected. If a more favorable choice of g(x) is possible, the efficiency will improve. But rejection sampling is generally impractical in problems of high dimensionality due to its inefficiency (MacKay, 2003).

13.7.2 Metropolis–Hastings algorithm

A *Markov chain* is a sequence of samples in which each subsequent sample, x_{i+1} is determined by the current sample x_i and by the probability density ratio $p(x_{i+1})/p(x_i)$ but is not explicitly dependent on the values of previous samples $(x_1, \ldots, x_{i-1})$. Since the progress of the chain is determined probabilistically, methods such as these are known as Markov Chain Monte Carlo (MCMC) techniques.

An important example of MCMC techniques is the Metropolis algorithm. First, a random starting point is chosen. Then a symmetric proposal density is chosen in a neighborhood around that point. Typically this might be a Gaussian distribution, whose width can be chosen based on the curvature of the PDF at the current point (if known). A proposed move is selected. If the PDF at the proposed point x' is greater than the current PDF, $p(x') > p(x_i)$, the proposed point is accepted. If not, it is accepted with probability $p(x')/p(x_i)$. Equivalently, one can say that the move is accepted with acceptance probability

$$a = \min\left(1, \frac{p(x')}{p(x_i)}\right). \tag{13.35}$$

Note that if a move is rejected, this implies that the current point is repeated, $x_{i+1} = x_i$.

The Metropolis algorithm generates a type of random walk. It will converge asymptotically to the distribution p(x), although the rate of convergence is difficult to estimate. Disadvantages of this approach include the fact that it produces correlated samples. The initial part of the chain depends on the starting point. So, during an initial *burn-in* period, the samples are not honest draws from the PDF and must be discarded. Since the rate of convergence is unknown, it is difficult to know how long a burn-in period is necessary. Finally, it is possible for the sampler to get trapped for some time in local maxima of the PDF. Many of these features are evident in the example shown in Figure 13.4.

The efficiency of the sampler depends on the width of the proposal density. Too wide a proposal density leads to a low rate of acceptance. Too narrow a proposal density leads to a high rate of acceptance but slow exploration of parameter space. For the case of an approximately Gaussian posterior of width σ, a natural scale for the proposal density would be a width of order σ. Under certain circumstances, in 1-dimension the optimal[5] proposal width is 2.38σ with an acceptance rate of

---

[5] Optimal in the sense of minimizing the autocorrelation of the samples.

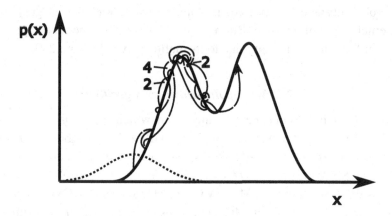

Figure 13.4 An example of the Metropolis algorithm exploring p(x), starting near the lower edge of the distribution, using the Gaussian proposal density shown by the dotted line. The numbers of times points are repeated, due to rejected moves, are shown by the numerals. In this case, the most probable part of the distribution has not been reached in 30 samples.

0.44, and in higher dimensions (N > 6) the optimal acceptance rate approaches 0.234 with an optimal proposal width of less than σ (Gelman *et al.*, 1996; Roberts *et al.*, 1997). In practice, having acceptance rates near these values is sufficient to ensure near optimal performance. For an unknown and potentially multimodal distribution, a non-Gaussian proposal density is a better choice since the tails of a Gaussian fall off too rapidly. Some large trial steps may be necessary to explore parameter space efficiently.

The Metropolis–Hastings algorithm allows use of a proposal density $q(x'; x_i)$ for moving to x' from x_i which is non-symmetrical:

$$q(x'; x_i) \neq q(x_i; x').$$ (13.36)

In this case detailed balance requires that the acceptance probability be changed to

$$a = \min\left(1, \ \frac{p(x') \, q(x_i; x')}{p(x_i) \, q(x'; x_i)}\right).$$ (13.37)

The general term Metropolis–Hastings is sometimes employed whether or not the proposal density is symmetric.

13.7.3 Gibbs sampling

In two or more dimensions, the Gibbs sampler is an MCMC method which relies on conditional probabilities. Let a point in n-dimensional space be written as

$$\vec{x} = (x_1, x_2, x_3, \ldots x_n).$$ (13.38)

The Gibbs sampler is obtained by updating each component successively, conditioned on the values of all the other components. In other words, starting from an initial $\vec{x}^0$, the first step $\vec{x}^1$ is obtained by making successive draws from the conditional posteriors

$$
\begin{aligned}
&p(x_1^1 \mid x_2^0, x_3^0, \ldots, x_n^0)\\
&p(x_2^1 \mid x_1^1, x_3^0, \ldots, x_n^0)\\
&\qquad\qquad \vdots\\
&p(x_n^1 \mid x_1^1, x_2^1, \ldots, x_{n-1}^1).
\end{aligned}
\tag{13.39}
$$

The Gibbs sampler exhibits many of the characteristics of the Metropolis–Hastings method. It undergoes a random walk, generates correlated samples, and requires some burn-in time. However, unlike the Metropolis–Hastings method, no moves are rejected (for continuous variables).

13.7.4 Mixing behavior

A Markov chain is said to be *well mixed* if it rapidly explores parameter space. Frequently Markov chains are not very well mixed, so a variety of techniques have been developed to enhance mixing. One of the simplest is to employ multiple Markov chains started from different regions in parameter space. This is frequently employed for practical reasons, but it is difficult to evaluate how well this technique works. One method is to compare inter-chain and intra-chain variances.

Another technique is based on an analogy with Hamiltonian dynamics. Slow MCMC mixing is due to the diffusive nature of a random walk. If one makes successive steps in the same direction more likely to be chosen, the resulting set of samples is no longer a Markov chain. But such an algorithm may still be used to explore the posterior PDF in a probabilistic (Monte Carlo) fashion. A third method is based on an analogy with thermodynamic systems. By raising the "temperature" of a system it is possible to overcome the barriers between peaks in a multimodal posterior. A fourth method known as "overrelaxation" enhances the mobility of a sampler in multidimensional spaces with highly correlated parameters. These and other techniques are discussed in texts such as MacKay (2003) and Gelman *et al.* (2004).

13.8 Model comparison

Above we discussed the use of Bayes' theorem for parameter estimation. The hypotheses differed by varying one or more parameters common to all of the hypotheses being considered. But Bayes' theorem may also be used for model comparison; for example, determining whether a more complex model with more free parameters is justified by the data. For models labelled A and B, the posterior odds ratio is given by

$$\frac{p(M_A|D)}{p(M_B|D)} = \frac{p(D|M_A)}{p(D|M_B)} \frac{p(M_A)}{p(M_B)}. \tag{13.40}$$

Let us assume the prior odds ratio $p(M_A)/p(M_B)$ is unity. If model A has no free parameters, the meaning of $p(D|M_A)$ is clear. If model B has one free parameter, then $p(D|M_B)$ is obtained by marginalizing over that free parameter, e.g.

$$p(D|M_B) = \int p(D|\lambda M_B)\, p(\lambda|M_B)\, d\lambda. \tag{13.41}$$

Consider the situations illustrated in Figure 13.5. For simplicity consider a uniform prior probability density of the free parameter λ over the range $\Delta\lambda$. Normalization requires then that the probability density be small if the allowed range of λ is large, so

$$p(\lambda|M_B) = \frac{1}{\Delta\lambda} \tag{13.42}$$

within the allowed range. The peak likelihood $p_{max}(D|\lambda M_B)$ usually will be at least as large as $p(D|M_A)$. For simplicity, assume that maximum likelihood of model B is uniform over the range $\delta\lambda$ with a maximum, relative to model A, of $p_{max}(D|\lambda M_B)/p(D|M_A)$. The posterior odds ratio is then

$$\frac{p(M_A|D)}{p(M_B|D)} = \frac{p(D|M_A)}{p_{max}(D|\lambda M_B)} \frac{\Delta\lambda}{\delta\lambda}. \tag{13.43}$$

Therefore model B, the model with the free parameter, is favored if the likelihood of model B (conditional on λ) exceeds the likelihood of model A by a sufficiently large amount. Otherwise the factor $\Delta\lambda/\delta\lambda$ penalizes model B, and the simpler

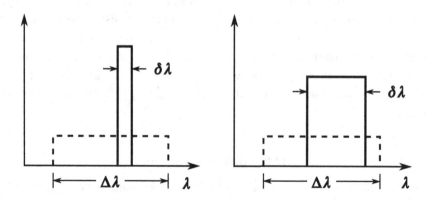

Figure 13.5 Two examples of model comparison. The dashed lines represent the prior probability density of the free parameter λ. The solid lines represent the conditional likelihood of model B.

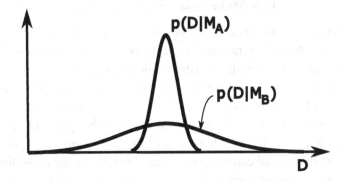

Figure 13.6 Evidence comparison for two models. The simpler theory, A, gives a narrower distribution, and so is favored in the region where it fits the data.

model, model A, is favored. This is the automatic application of Occam's razor in Bayesian analysis.

Model comparison may also be considered using the *evidence* term we encountered in our introduction to Bayes' theorem. Continuing the previous example, if we were to write Bayes equation just for model B, incorporating the model as part of the background information, we would get

$$p(\lambda|DM_B) = p(\lambda|M_B) \frac{p(D|\lambda M_B)}{p(D|M_B)}. \tag{13.44}$$

In this equation the evidence term may be described as the evidence for model B, alternatively written as the marginalized likelihood,

$$p(D|M_B) = \int p(D|\lambda M_B) \, p(\lambda|M_B) \, d\lambda. \tag{13.45}$$

The evidence for model A is $p(D|M_A)$. The model comparison is then given by the posterior odds ratio

$$\frac{p(M_A|D)}{p(M_B|D)} = \frac{p(D|M_A)}{p(D|M_B)} \frac{p(M_A)}{p(M_B)}. \tag{13.46}$$

Assuming no prior preference for either A or B, the posterior odds ratio is just the ratio of the evidences. Since the evidence terms are functions of the data, we end up with a situation like that in Figure 13.6. Where the data are consistent with both theories, the simpler theory is favored. Where the data are more consistent with the more complex theory, the complex theory is favored.

We have given only a few examples of Bayesian model comparison. For further examples, see Sivia (1996) and MacKay (2003).

13.9 Malmquist (truncation) bias

Many astronomical problems involve analyzing data from objects in incomplete samples. Incompleteness can take many forms, but we will concern ourselves with that due to the limited sensitivity of astronomical detection systems. Suppose one can detect objects down to a certain *flux limit*. If one makes a survey for all objects down to that flux limit, one obtains what is called a flux-limited sample. This may seem like a benign selection criterion, but it is not. Subsequent analysis is biased by this flux limit, and such bias is known as *Malmquist* or *truncation bias*. In some cases the biases introduced can be quite severe. In studying the most distant galaxies one sees only the most luminous objects, providing a very biassed view of the constituents in the early universe.

Consider the general truncation problem. One obtains a flux-limited sample of some objects, let's say galaxies. At each redshift there will be a range of luminosities. For low z, one sees the high and low luminosity objects. At high z, only high luminosity objects will be above the flux limit and be seen. Therefore, in a comparison of *any* property of the galaxies (luminosity, color, metallicity, etc.) one will get misleading results. Essentially one is comparing apples (at high z) with apples + oranges (low z).

Let's look at this in the context of the Tully–Fisher method for estimating galactic distances. The Tully–Fisher method is based on the assumption that there is an intrinsic relationship between velocity dispersion and luminosity. Galaxies with larger dispersions are more massive, contain more stars, and are therefore more luminous. But obviously this will not be an exact relationship – there will be some scatter. For simplicity let's assume this relation is of the form

$$L = c \, \sigma + \xi. \tag{13.47}$$

That is, for a given velocity dispersion σ, on average the luminosity will be $c\sigma$, but there will be some galaxies with more or less luminosity, as represented by the random variable ξ. The relationship between σ and L for a complete set of galaxies would look something like that shown in Figure 13.7.

Now consider a flux-limited sample of galaxies at a fixed velocity dispersion σ_0. Those galaxies which are intrinsically more luminous than average ($L > c\sigma_0$) can be seen at larger distances than the less luminous galaxies (at fixed σ_0). Since the number of galaxies seen is proportional to the volume of space surveyed, the sample will contain a greater than usual number of overluminous galaxies.

But we have no way of knowing these are intrinsically overluminous. Instead we naively apply the standard relationship $L = c\sigma$, which assigns them simply the average luminosity. We then use their apparent magnitudes and these assumed luminosities to derive distances, and these distances will be systematically

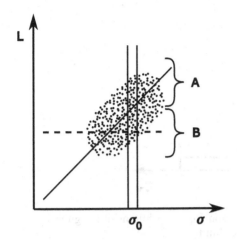

Figure 13.7 A hypothetical plot of luminosity versus velocity dispersion for a set of galaxies. Considering galaxies inside some interval around a dispersion of σ_0, those in group A will be overrepresented in a flux-limited sample, while some of those in group B will fall below the flux limit (dashed line).

underestimated. If we then use these incorrect distances together with the measured redshifts to determine the Hubble constant, the value of H_0 will be overestimated. The issue of Malmquist bias played an important role in earlier debates over the correct value for the Hubble constant. All is not lost. If one knows enough about the astrophysics of the situation (the spread in luminosities) and the exact nature of the sample truncation (the flux limit), one can try to correct for the biases which have been introduced.

More literature is becoming available on the subject of statistical methods for astronomy, but much of it is unknown to a large portion of the astronomical community. As Press (1997) has put it, "...make the choice to live your professional life at a high level of statistical sophistication, and not at the level ...that is the unfortunate common currency of most astronomers."

13.10 Censoring

A related problem is *censoring*. In this case one starts with a known sample of objects and then fails to detect some of them, for example in observations in a different wavelength band. One therefore obtains upper limits on the flux of such objects. Upper limits contain real information, often vital information. There are well-developed statistical techniques for dealing with them. The most powerful of these techniques is known as survival analysis.

The basic idea of survival analysis can be easily introduced by an intuitive argument. Suppose one starts with a known sample of galaxies (perhaps known from an optical survey) and then obtains the x-ray luminosities of some and x-ray upper limits for others. The question is: what is the distribution function of x-ray luminosities? Note that even if all objects are observed in the same fashion, down to

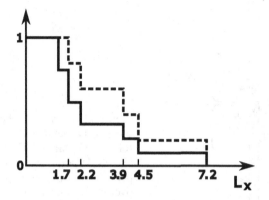

Figure 13.8 Example of a distribution function for x-ray luminosities ignoring (dashed line) and including (solid line) upper limits.

some limit in x-ray flux, this will not correspond to a sharp cutoff in x-ray luminosity (since the galaxies are at different distances). So one obtains various values for luminosity and various upper limits. Let the data be

$$L_x = \{1.3-, 1.7, 1.9-, 2.2, 2.7-, 3.3-, 3.9, 4.5, 5.0-, 7.2\}. \qquad (13.48)$$

Each number represents a luminosity, in arbitrary units, and those followed by "−" are upper limits.

If we plotted a distribution function (akin to a "cumulative" probability function) based on just the five real detections we would get the dashed line in Figure 13.8. But what about the upper limits? Each upper limit represents 1/10 of the data we obtained, so each should somehow be given a "weight" of 1/10 in the final result. But where should that weight be placed? Not *at* the upper limit value. Perhaps spread out evenly over the interval below it. But why that way?

In fact one should use the remaining data to determine how to spread out the weight. Consider the point 5.0−. Redistribute its weight of 0.1 evenly over the eight data points below it. Proceeding down in luminosity, one would then redistribute the weight of 3.3− (now with a weight of 0.1125) over the remaining five points, etc. The result is the solid line, clearly significantly different than the result using only the five true detections.

Statisticians are familiar with this procedure and would call it "redistribution to the left" (although they are more accustomed to using lower limits and therefore usually redistribute to the right). This procedure has theoretical justification, although in practice one must be careful about the pattern of censoring present. The most useful formulation of this method is not the redistribution procedure just described, but what is known as the Kaplan–Meier (product-limit) estimator.

As a final note I will mention the powerful statistical technique known as bootstrap resampling (Efron, 1981; Diaconis & Efron, 1983). That and related techniques allow one to set uncertainties on parameters derived from limited data sets.

13.11 Confidence limits

In one dimension, assuming a Gaussian distribution, we have the readily determined results shown in Table 13.1. For non-Gaussian likelihoods, one should choose a certain value of $\Delta\chi^2$ and determine the likelihood contained between upper and lower values as shown. This will produce the shortest confidence interval containing the specified probability. The same applies to posterior probabilities, although in that case the correct terminology would be *credible interval*.

In two dimensions, we can imagine attempts, for example, to jointly determine effective temperature and stellar luminosity (T_{eff}, L), Hubble parameters (H_0, q_0), or cosmological density parameters (Ω_Λ, Ω_M). In general, such confidence intervals may not have simple shapes, and may not even be simply connected.

But for simplicity, assume a bivariate Gaussian probability distribution, centered at the origin, which can be written

$$p(x, y) = \frac{1}{2\pi ab} \exp\left\{-r^2 \left[b^2 \cos^2(\theta - \theta_0) + a^2 \sin^2(\theta - \theta_0)\right]/2a^2b^2\right\} \quad (13.49)$$

$$= \frac{1}{2\pi ab} \exp\left[-x^2(b^2 \cos^2 \theta_0 + a^2 \sin^2 \theta_0)/2a^2b^2\right]$$
$$\times \exp\left[-y^2(b^2 \sin^2 \theta_0 + a^2 \cos^2 \theta_0)/2a^2b^2\right]$$
$$\times \exp\left[-xy(2b^2 - 2a^2) \sin \theta_0 \cos \theta_0/2a^2b^2\right], \quad (13.50)$$

which is an elliptical distribution with a semi-major axis a (to an $e^{-1/2}$ drop in probability density), semi-minor axis b, and with the major axis at an angle θ_0 measured counterclockwise from the x-axis. Since a line of constant probability density is an ellipse, it is natural to use an ellipse as a confidence limit boundary. For a desired confidence limit CL (e.g. CL = 0.9),

Table 13.1. *One-dimensional confidence limits*

Confidence limit	Uncertainty	$\Delta\chi^2$
68.3%	$1.00\,\sigma$	1.00
90.0%	$1.64\,\sigma$	2.71
95.4%	$2.00\,\sigma$	4.00
99.0%	$2.57\,\sigma$	6.63

Table 13.2. *Two-dimensional*
confidence limits

Confidence limits	$\Delta\chi^2$
68.3%	2.30
90.0%	4.61
95.4%	6.17
99.0%	9.21

$$- \ln(1 - CL) = \frac{1}{2} r^2 \left[\frac{1}{a^2} \cos^2(\theta - \theta_0) + \frac{1}{b^2} \sin^2(\theta - \theta_0) \right]. \tag{13.51}$$

How do we know this? Do the integral! The required value of $\Delta\chi^2$ for different confidence limits is shown in Table 13.2.

What if we have the 2-dimensional distribution, but only want information about one of the parameters, let us say x? To get a particular confidence limit for x by itself, we first need to marginalize (integrate over) y. For a confidence limit CL,

$$x = x_0 \pm \sqrt{2} \, \text{erf}^{-1}(CL) \sqrt{b^2 \sin^2 \theta_0 + a^2 \cos^2 \theta_0}. \tag{13.52}$$

How is this related to χ^2? Well χ^2 is a natural starting point since that is the information we have from the observations. At its minimum, χ^2 is a constant χ_0^2, plus quadratic terms, plus higher order terms. There are no linear terms since this represents a minimum. Neglecting higher order terms, the χ^2 surface can be described by a curvature matrix $\boldsymbol{\alpha}$, also known as the *Hessian matrix*,

$$\Delta\chi^2 = \delta\mathbf{a} \cdot \boldsymbol{\alpha} \cdot \delta\mathbf{a} \tag{13.53}$$
$$= \delta\mathbf{a}^T \boldsymbol{\alpha} \, \delta\mathbf{a}, \tag{13.54}$$

where $\delta\mathbf{a}$ is the vector displacement from the minimum. The matrix $\boldsymbol{\alpha}$ in this case is a 2×2 symmetric matrix with *at most* three independent terms (related to the a, b, and θ_0 used above). The inverse of the curvature matrix is known as the covariance matrix,

$$\mathbf{C} = \boldsymbol{\alpha}^{-1}. \tag{13.55}$$

The diagonal elements of the covariance matrix are the true variances in x and y, after marginalizing over the other variable. This curvature matrix treatment is easily generalized to multiple dimensions.[6]

---

[6] It is easy to make mistakes regarding confidence limits in multiple dimensions. One treatment may be found in §15.6 of *Numerical Recipes*, 3rd edn. (Press *et al.*, 2007). For users of earlier editions, see §14.5 of the 1st edn. (1986) and §15.6 of the 2nd edn. (1995).

Exercises

13.1 Consider a random variable x with a Gaussian probability distribution,

$$p(x) = \frac{1}{\sqrt{2\pi}\sigma} \exp\left[-\frac{(x-a)^2}{2\sigma^2}\right],$$

where the parameters a and σ are unknown. Suppose we make n independent measurements of x and obtain the values $\{x_i\}, i = 1 \to n$.

   a. From the data $\{x_i\}$ calculate $\hat{a}$ and $\hat{\sigma}$, maximum likelihood estimates of the quantities a and σ.
   b. What are the expectation values of $\hat{a}$ and $\hat{\sigma}$? If the expectation value of an estimator equals the underlying parameter, the estimator is said to be unbiassed. Are $\hat{a}$ and $\hat{\sigma}$ unbiassed?
   c. If either estimator in part b is biassed, calculate an unbiassed estimator.

13.2 Prove that Bayesian inference automatically includes Occam's razor. For simplicity, start with one parameter belonging to a finite discrete set. Then try to add a second parameter, also from a finite discrete set.

13.3 Consider the 2-dimensional joint probability density written below in both Cartesian and polar coordinates,

$$p_1(x, y) = \frac{1}{4\pi} e^{-x^2/8} e^{-y^2/2}, \tag{13.56}$$

$$p_1(r, \theta) = \frac{1}{4\pi} e^{-r^2(\cos^2\theta + 4\sin^2\theta)/8} = \frac{1}{4\pi} e^{-r^2(1+3\sin^2\theta)/8}. \tag{13.57}$$

As the Cartesian form shows, this is a bivariate Gaussian distribution centered on the origin (for simplicity) with width $\sigma = 2$ in the x-direction and width $\sigma = 1$ in the y-direction.

Consider also a probability density equivalent in shape to p_1 but rotated by 45°.

$$p_2(x, y) = \frac{1}{4\pi} e^{-5x^2/16} e^{3xy/8} e^{-5y^2/16}, \tag{13.58}$$

$$p_2(r, \theta) = \frac{1}{4\pi} e^{-r^2(5\cos^2\theta - 6\sin\theta\cos\theta + 5\sin^2\theta)/16} = \frac{1}{4\pi} e^{-r^2(1+3\sin^2(\theta - \pi/4))/8}. \tag{13.59}$$

   a. Verify that both p_1 and p_2 are normalized (that their integrals over the xy-plane are unity).
   b. For both p_1 and p_2 find the equation for the curve where the probability drops to 1/e of its peak value.
   c. For both cases, give a curve that corresponds to the 90% confidence limit (that encloses 90% of the probability density).

d. For both cases, give 90% confidence limits for x. In other words, integrating the probability density over y, what is the 90% confidence limit in x?

e. Graph your answers to b and c, and mark your answers to d on the x-axis.

Hints:

a. Remember that the differential of area in polar coordinates is $r \, d\theta \, dr$.

b. You will need the obscure integral

$$\int_{-\infty}^{\infty} e^{-p^2 x^2 \pm qx} \, dx = e^{q^2/4p^2} \frac{\sqrt{\pi}}{p},\tag{13.60}$$

which is Equation 3.323.2 in Gradshteyn & Ryzhik (1980).

c. You will also need a table of the fractional area in the tail of a Gaussian distribution.

This illustrates an important point in statistics which can easily lead to confusion if one is not careful: projected confidence limits of multivariate distributions do not have a simple relationship with distributions where the other variable(s) take on their mean values.

13.4 Derive Equation 13.51 from Equation 13.50.

13.5 Derive Equation 13.52 from Equation 13.50.

14

Neutrino detectors

14.1 Neutrinos

Leptons are elementary particles which are subject to the weak interaction but not the strong interaction. There are 12 known leptons: the electron (e^-), the muon (μ^-), and the tau lepton (τ^-), their charge conjugates (e^+, μ^+, τ^+), and neutrinos (ν_e, $\bar{\nu}_e$, ν_μ, $\bar{\nu}_\mu$, ν_τ, $\bar{\nu}_\tau$) associated with each of the three lepton flavors. Neutrinos have small masses. They were once thought to be massless, but neutrino oscillation experiments have now shown that at least two neutrinos must have finite mass. Neutrinos interact with matter via only the weak and gravitational interactions, enabling them to easily penetrate matter. The weakness of their interaction is, in fact, the primary obstacle to their detection.

As a starting point we adopt the Standard Model of particle physics, in which the weak interaction is mediated by the exchange of $W^\pm$ bosons, in what are known as charged current (CC) interactions, or by Z bosons, in neutral current (NC) interactions. The weak interaction is observed to be maximally parity violating. Neutrinos are observed to have negative helicity and antineutrinos to have positive helicity. The vector minus axial vector (V−A) theory of weak interactions requires the helicities described above and is inherently maximally parity violating. We assume that neutrinos are Dirac particles,[1] that there are no "sterile" neutrinos, and that there are only three neutrino flavors.

For the most part we will pass by a number of interesting aspects of neutrino theory and experimental questions being addressed primarily in the field of high energy (reactor and accelerator) particle physics, concentrating instead on issues most directly related to observational astrophysics. We will not discuss grand unified theories or supersymmetry. Thorough coverage of all aspects of neutrino physics is given by Zuber (2004) and Giunti & Kim (2007).

---

[1] The distinction between Dirac and Majorana neutrinos is that in the Majorana case neutrinos and antineutrinos are identical. For most of this chapter we will simply be ignoring the Majorana case and assume that Dirac neutrinos have left-handed chirality.

A key question being addressed by astronomical detectors is that of neutrino mass. The existence of finite neutrino mass requires physics beyond the Standard Model. The nature of any extension beyond the Standard Model is a longstanding, open, and central question in particle physics.

14.2 Solar neutrino production

In the core of the Sun, the thermonuclear synthesis of each ^{4}He nucleus is accompanied by the production of two electron neutrinos and the release of 26.731 MeV of energy, with an effective net reaction of

$$4p + 2e^- \rightarrow {}^4\text{He} + 2\nu_e. \tag{14.1}$$

This energy is released as a combination of kinetic energy and gamma rays, and during subsequent positron annihilation. Some energy is carried away by the neutrinos themselves, although that amount is small for the most common reaction pathways (0.53 MeV on average). The remaining energy is rapidly thermalized.

The proton–proton chain dominates the energy production in the solar interior. Proton–proton fusion proceeds predominantly through the reaction

$$p + p \rightarrow d + e^+ + \nu_e, \tag{14.2}$$

where the neutrinos, dubbed *pp* neutrinos, have a continuous energy distribution up to a maximum of 420 keV. A minor contribution to the synthesis of deuterium is made by the reaction

$$p + e^- + p \rightarrow d + \nu_e, \tag{14.3}$$

with mono-energetic neutrinos at 1.442 MeV, dubbed *pep* neutrinos. The main branch of the reaction network through ^{3}He and the reaction of two ^{3}He to produce ^{4}He generates no further neutrinos. Since two ^{3}He nuclei are required and a neutrino was released in the generation of each, two ν_e are produced for each ^{4}He. A rare reaction takes a single ^{3}He nucleus directly to ^{4}He,

$$^3\text{He} + p \rightarrow {}^4\text{He} + e^+ + \nu_e, \tag{14.4}$$

with a continuous neutrino spectrum up to a maximum energy of 18.773 MeV (so called *hep* neutrinos). The remainder of the network proceeds from ^{3}He to ^{7}Be and then either

$$^7\text{Be} + e^- \rightarrow {}^7\text{Li} + \nu_e, \tag{14.5}$$

with mono-energetic neutrinos at either 862 or 384 keV (the ^{7}Be neutrinos), or from ^{7}Be to ^{8}B and then

$$^8\text{B} \rightarrow {}^8\text{Be}^* + e^+ + \nu_e, \tag{14.6}$$

with a continuous spectrum of neutrinos up to a maximum energy of about 15 MeV. No further neutrinos are produced as $^8Be^*$ spontaneously decays into two alpha particles or as ^7Li reacts with a proton to produce two alpha particles.

The CNO cycle plays a minor role in solar energy production, with neutrino-producing reactions

$$^{13}N \rightarrow {}^{13}C + e^+ + \nu_e \quad (E_\nu \leq 1.199 \text{ MeV}), \tag{14.7}$$

$$^{15}O \rightarrow {}^{15}N + e^+ + \nu_e \quad (E_\nu \leq 1.732 \text{ MeV}), \tag{14.8}$$

$$^{17}F \rightarrow {}^{17}O + e^+ + \nu_e \quad (E_\nu \leq 1.739 \text{ MeV}). \tag{14.9}$$

The rates of these reactions are sensitive to temperature and density, which vary through the solar core. So an understanding of solar structure and composition is necessary to predict the emergent neutrino fluxes. According to the well-confirmed Standard Solar Model or SSM (e.g. Bahcall *et al.*, 2001), the predicted solar neutrino spectrum is as shown in Figure 14.1. This level of detail is necessary in

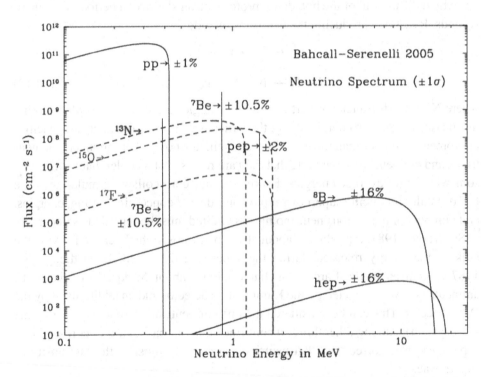

Figure 14.1 Neutrino flux predictions at 1 AU for the Standard Solar Model (Bahcall *et al.*, 2005). For continuum radiation the correct units for the vertical scale are $\text{cm}^{-2}\,\text{s}^{-1}\,\text{MeV}^{-1}$. Contributions from the CNO cycle are shown in blue (in electronic version). Energy thresholds for radiochemical detectors are 233 keV for gallium and 814 keV for chlorine. Čerenkov detectors have thresholds of around 5 MeV.

understanding the spectrum because the efficiencies of neutrino detectors can be very energy dependent. Initial results indicated a detection rate well below that set by the SSM. The nature of these observations and the resolution of their discrepancy will be discussed in later sections.

14.3 Supernova production

Unlike solar nucleosynthesis, core-collapse supernovae are capable of producing all three flavors of neutrinos, although not in equal numbers. We will use the notation ν_x to indicate the possibility of any neutrino flavor ($x = e$, μ, or τ). It had been thought that the beginning of core collapse was accompanied by the *prompt* release of electron neutrinos originating from inverse beta decay reactions such as

$$e^- + p \rightarrow n + \nu_e. \tag{14.10}$$

This was followed by thermal production of all flavors of neutrinos and antineutrinos which diffuse out of the hot, dense proto-neutron star over a period of about 10 seconds. Reactions producing the neutrinos include

$$e^+ + e^- \rightarrow \nu_x + \bar{\nu}_x, \tag{14.11}$$

$$N + N \rightarrow N + N + \nu_x + \bar{\nu}_x, \tag{14.12}$$

where N stands for nucleon. Particle kinetic energies are of order 20 MeV during this phase, so neutrino energies are of this order as well. Electron neutrino opacities are somewhat higher than those of electron antineutrinos and of muon and tau neutrinos and antineutrinos, meaning that they are released at a cooler, later stage with somewhat lower average energies. However, since core-collapse simulations have had difficulty producing supernova explosions, details concerning timing, energies, and numbers of the various neutrino flavors emitted must be regarded as uncertain.

Supernova 1987A produced about 10^{58} neutrinos, which carried away the bulk of the energy released during the supernova explosion. Even though SN 1987A occurred in the Large Magellanic Cloud, about 50 kpc from Earth, 12 antineutrinos were detected by the Kamiokande detector and an additional 8 by the IMB detector. This can be considered the birth of neutrino astronomy (beyond the solar system). Among other things it confirmed that our basic view of core-collapse supernovae was correct and that neutrinos carry away most of the luminosity of supernovae.

14.4 Atmospheric neutrinos

A substantial neutrino flux is produced in Earth's atmosphere by energetic cosmic rays. These can rightly be considered as a problematic foreground for observing extraterrestrial neutrinos. Or, more constructively, they can be viewed as one of several tools for observing cosmic rays. The methodologies of cosmic ray observations are discussed in Chapter 15. It turns out that observations of atmospheric neutrinos are important in understanding the properties of neutrinos themselves.

When energetic cosmic rays hit nuclei in Earth's atmosphere, they produce hadron showers containing pions and some kaons and baryons.[2] Charged pions decay into positive or negative muons plus muon neutrinos (or antineutrinos). Relativistic muons are penetrating particles which eventually decay into electrons or positrons plus muon neutrinos and electron antineutrinos (or vice versa), as shown in Figure 14.2. Typically, therefore, cosmic ray showers at GeV energies may contain about twice as many muon neutrinos as electron neutrinos.

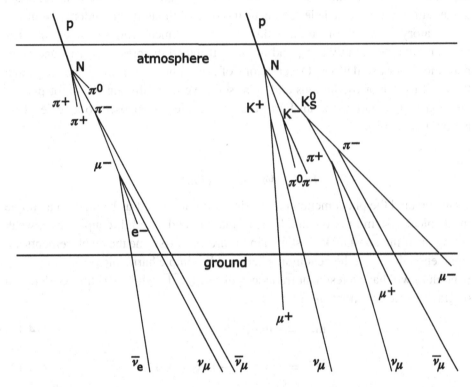

Figure 14.2 A cosmic ray of a few GeV (left) interacts with a nucleus in Earth's atmosphere. Such cosmic ray showers typically produce twice as many ν_μ (and $\bar\nu_\mu$) as ν_e (and $\bar\nu_e$). Above 100 GeV (right) kaon decays dominate and the muons are more energetic and longer lived, producing a different mixture of neutrino flavors.

---

[2] Gamma rays in the TeV range may also produce extensive air showers and cannot be ruled out as a source of energetic muons. However, this is of secondary importance.

Results from Super-Kamiokande revealed a deficit of ν_μ (relative to ν_e) primarily for *upward* traveling muons. The interpretation is flavor oscillations with a portion of the upward ν_μ having converted to ν_τ during their passage through Earth. The nature of flavor oscillations is discussed in the following section.

14.5 Neutrino oscillations

The first instrument designed to detect solar neutrinos was the chlorine radio-chemical detector built by Ray Davis in the 1960s at the Homestake gold mine in South Dakota. It quickly became clear that the Homestake detector saw only about one-third of the solar electron neutrino flux expected from the SSM. Later the Kamiokande Čerenkov detector and the GNO gallium detector both recorded about half of the expected electron neutrino flux. Ultimately the Sudbury Neutrino Observatory (SNO) resolved these discrepancies by measuring separately the flux in ν_e and that due jointly to ν_μ and ν_τ, and attributed the presence of ν_μ and ν_τ to neutrino flavor oscillations. Observations of atmospheric neutrinos also suggested the need for flavor oscillations and were sensitive to a different region of parameter space (fluctuations on an Earth diameter scale, as opposed to a scale of an astronomical unit).

14.5.1 Vacuum oscillations

To understand the phenomenon of neutrino oscillations, let us begin with a simple mechanical system of two pendulums of mass m and length L coupled by a weak spring of spring constant k, as shown in Figure 14.3. Assume the displacements are small enough so that the pendulums behave like simple harmonic oscillators. Each pendulum will have a resonant frequency $\omega_0 = \sqrt{g/L}$, where g is the acceleration of gravity. The equations of motion are

$$m\ddot{x}_1^2 = -m\omega_0^2 x_1 + k(x_2 - x_1), \qquad (14.13)$$

$$m\ddot{x}_2^2 = -m\omega_0^2 x_2 - k(x_2 - x_1). \qquad (14.14)$$

Solving simultaneously we find that the eigenmodes of the system occur at frequencies $\omega = \omega_0$ and $\omega = \sqrt{\omega_0^2 + 2k/m}$ and correspond to the two pendulums moving together ($x_1 = x_2$) and in opposite directions ($x_1 = -x_2$) with equal amplitudes. An initial state with just one pendulum moving is a superposition of these eigenstates and will evolve through states with both oscillating, then just the other pendulum oscillating, then both, and then back to the original state, with the whole sequence repeating.

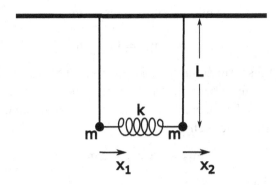

Figure 14.3 System of two coupled pendulums.

An analogous evolution occurs in the neutral kaon system K^0 and $\bar{K}^0$ whose eigenstates K_S^0 and K_L^0 have a small mass difference. Ignoring decays, an initially pure K^0 beam will evolve into a mixture of K^0 and $\bar{K}^0$. The particle is said to oscillate between the states K^0 and $\bar{K}^0$.

The neutrino system is essentially different only in that there are three flavors of neutrinos. But the mass hierarchy is not known, and it is not known *a priori* whether 1, 2, or all 3 of these neutrinos have finite mass. The masses are known to be small enough that, at the energies of interest, neutrinos are ultrarelativistic.

Neutrinos are produced in flavor states (ν_e, ν_μ, ν_τ). For simplicity, let us assume that there are only two neutrino flavors, ν_e and ν_μ, and that these have eigenstates ν_1 and ν_2 such that

$$\nu_e = \nu_1 \cos\theta + \nu_2 \sin\theta, \tag{14.15}$$

$$\nu_\mu = -\nu_1 \sin\theta + \nu_2 \cos\theta. \tag{14.16}$$

In this way, a beam which was originally pure ν_e will, over time, develop a ν_μ component, depending on θ and the mass difference Δm^2. The transition probability from ν_e to ν_μ is

$$P(\nu_e \to \nu_\mu) = \sin^2 2\theta \, \sin^2 \frac{\Delta m^2 \, L \, c^3}{4 \, \hbar \, E}, \tag{14.17}$$

where L is the path length.[3] So the oscillations occur most rapidly for low energy neutrinos.

Notation conventions vary. We adopt the convention that all mass differences are positive:

$$\Delta m_{12}^2 = \Delta m_{21}^2 = |m_1^2 - m_2^2|, \tag{14.18}$$

---

[3] This equation is most often written in a system of natural units where $c = \hbar = 1$.

$$\Delta m_{23}^2 = \Delta m_{32}^2 = |m_2^2 - m_3^2|, \tag{14.19}$$

$$\Delta m_{13}^2 = \Delta m_{31}^2 = |m_1^2 - m_3^2|, \tag{14.20}$$

that Δm_{12}^2 is the mass term acting in solar neutrino studies, and that Δm_{23}^2 is the mass term acting in atmospheric neutrino studies. The mixing angles θ_{12}, θ_{23}, and θ_{13} are, in principle, independent. The mass differences are coupled, in a fashion determined by the mass hierarchy,

$$\Delta m_{13}^2 = |\Delta m_{23}^2 \pm \Delta m_{12}^2|, \tag{14.21}$$

in the notation used here.

14.5.2 Matter oscillations

When neutrinos pass through matter, the natural vacuum oscillation can be enhanced by the Mikheyev–Smirnov–Wolfenstein (MSW) effect. The matter in question might be the solar interior, in the case of solar neutrinos, or Earth's interior, for upward traveling neutrinos. In passing through normal matter, all neutrino flavors undergo NC interactions, whereas only electron neutrinos undergo CC interactions. This can be thought of as producing a different effective mass for the electron neutrino, which under the right conditions can enhance any vacuum oscillations. The MSW effect depends in detail on the electron density in the matter, the gradient in electron density, and the neutrino energies. As with vacuum oscillations, the conversion of ν_e to ν_μ and ν_τ is more effective at low energy.

14.5.3 Conclusions

Solar neutrinos are of rather low energy, of order 1 MeV. To explain the solar ν_e deficit in terms of vacuum oscillations would require $\Delta m_{12}^2 \approx 10^{-10}$ eV^2. If this were the case, one might hope to observe an annual variation due to the ellipticity of Earth's orbit. No such effect has been seen, other than the expected $1/r^2$ variation in flux. An explanation using the MSW effect within the solar interior, on the other hand, yields $\Delta m_{12}^2 \approx 10^{-4}$ eV^2. The details of the conversion are dependent on solar properties such as the variations in electron density along the propagation path, and on neutrino energies. There would also be the possibility of day/night variations due to the MSW effect as the neutrinos pass through Earth's interior (ν_e regeneration). No such effect has been seen, further limiting available parameter space.

The bulk of the evidence, primarily from Super-Kamiokande and SNO, is that for solar neutrinos the so-called large mixing angle (LMA) model is correct, with $\Delta m_{12}^2 \approx 8 \times 10^{-5}$ eV^2 and a "large" mixing angle θ_{12}. Further information from the nuclear reactor experiment KamLAND pins down $\theta_{12} \approx 32°$, as shown

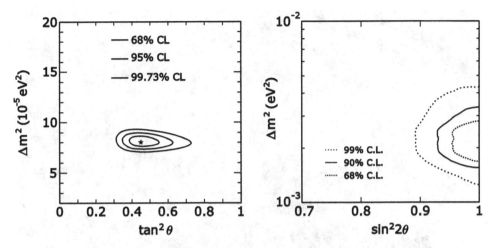

Figure 14.4 (Left) Allowed solar neutrino mixing parameters Δm_{12}^2 and $\tan^2 \theta_{12}$ (Aharmim *et al.*, 2005). (Right) Allowed atmospheric neutrino mixing parameters Δm_{23}^2 and $\sin^2 2\theta_{23}$ (Ashie *et al.*, 2005).

in Figure 14.4. These measurements demonstrate the existence of some physics beyond the Standard Model of particle physics, giving rise to finite neutrino masses and MSW oscillations of the electron neutrinos generated in the solar core. There is no longer a solar neutrino "problem," and the SSM is shown to be basically correct. For atmospheric neutrinos with typical energies of 1 GeV, the evidence favors $\Delta m_{23}^2 \approx 2.4 \times 10^{-3}$ eV^2 and $\theta_{23} \approx 45°$. Little is known yet about the 1–3 channel; the mixing angle θ_{13} appears to be small but may be non-zero.

14.6 Radiochemical (transmutational) detectors

Radiochemical detectors record neutrino interactions with atomic nuclei which transmute the nuclei to a different atomic number, i.e. a different chemical element. One limitation of such detectors is that they are sensitive only to ν_e. On the other hand, solar neutrinos are expected to be produced purely as ν_e, so this seems like a good match. Other limitations are that no information is obtained about neutrino energies (other than being above some threshold), about neutrino directions (presumably known for solar neutrinos), or about timing on any scales shorter than a few months. These make radiochemical detectors generally poor choices for studying supernova neutrinos.

14.6.1 Chlorine

The experiment designed by Ray Davis and set up in the Homestake gold mine in South Dakota was the pioneering instrument for solar neutrino detection

Figure 14.5 Ray Davis during construction of his historic neutrino detector. Courtesy of Brookhaven National Laboratory.

(Figure 14.5). Neutrino detectors need to be deep underground so they are shielded from atmospheric muons. The target nuclide was ^{37}Cl in a tank containing 615 metric tons[4] of C_2Cl_4 (perchloroethylene). The most abundant isotope of chlorine is ^{35}Cl with about three times the abundance of ^{37}Cl, so only about one out of four chlorine atoms was ^{37}Cl. The reaction

$$\nu_e + {}^{37}\text{Cl} \rightarrow {}^{37}\text{Ar} + e^- \tag{14.22}$$

produces a radioactive isotope of argon. The energy threshold of this reaction is 814 keV, precluding detection of any *pp* neutrinos. But chlorine detectors are sensitive to ^{8}B neutrinos. A smaller contribution to the detectable neutrino flux comes from the higher energy (862 keV) ^{7}Be neutrinos with small contributions from the *pep* neutrinos and those from the CNO cycle. The *hep* neutrinos can be neglected.

After typically two months of operation a near steady state concentration of argon atoms was achieved. During this time some 10^{23} solar neutrinos passed

---

[4] A metric ton = 10^3 kg.

through the detector, generating on average only 29 atoms of ^{37}Ar. In those two months some of these decayed, leaving 16 ± 4 argon atoms in the tank. The problem of detecting these extremely weakly interacting particles had been transformed into the proverbial problem of finding the needle in the haystack (16 atoms among more than 2×10^{30} molecules)! With difficulty the ^{37}Ar was extracted, concentrated, and placed in gas proportional counters, which eventually recorded the Auger electrons released in the decay of the argon back to ^{37}Cl. This experiment operated nearly continuously over about 25 years to achieve good statistics (Cleveland *et al.*, 1998). The rate of solar neutrino detection was only 34% of that predicted by the Standard Solar Model, a discrepancy eventually explained as being due to neutrino flavor oscillations.

14.6.2 Gallium

The GALLEX detector, located in the Gran Sasso underground laboratory in Italy, was in operation from 1991 to 1997 (Hampel *et al.*, 1999). The detector contained 30 metric tons of gallium, a significant fraction of the world's supply.[5] The gallium was used as a target for the inverse beta decay reaction

$$v_e + {}^{71}\text{Ga} \rightarrow {}^{71}\text{Ge} + e^-. \tag{14.23}$$

The neutrino energy threshold for this reaction is low (233 keV), making the detector sensitive to the numerous *pp* neutrinos released from the solar core. The gallium was present in the form of a highly purified solution of $GaCl_3$ in water. The task was to detect the individual atoms of ^{71}Ge. The isotope ^{71}Ge is radioactive, with a half life of 11.4 days.[6] The decay was by electron capture, which produced x-rays of energy 1.2 keV (L shell capture) or 10.4 keV (K shell capture).

There were formidable obstacles to this technique. The Ge atoms were allowed to accumulate over a period equal to several half lives of ^{71}Ge until a steady state concentration of ^{71}Ge was reached. Then the ^{71}Ge had to be extracted from the massive target in a time short compared to the radioactive half life (within a few days) and with high efficiency. This was achieved by a chemical desorption of $GeCl_4$. All of this had to take place in an environment well shielded from cosmic ray muons (i.e. deep underground). And all of the equipment was made of high purity materials to minimize background from naturally occurring radioactive isotopes such as ^{40}K, ^{238}U, ^{232}Th, ^{222}Rn, ^{85}Kr, and ^{39}Ar.

The germanium in the detector was ultimately converted to gaseous GeH_4 (germane). The low energy x-rays from its decay were then detected with highly

---

[5] Current worldwide production of gallium is of order 100 metric tons per year, and demand is exceeding the rate of production.

[6] Equivalent to a 1/e decay time of 16 days.

specialized gas proportional counters with pulse-shape discrimination against background beta particles or Compton scattered electrons. Solar electron neutrinos were detected at a rate of about 60% of that expected from the SSM. Neutrino oscillations accounted for the apparent discrepancy.

GALLEX was upgraded and renamed GNO (Gallium Neutrino Observatory). It also used 30 metric tons of gallium. Results for the period 1998 to 2002 included improvements in both the statistical and systematic uncertainties in the GALLEX result. There was interest in attempting to monitor both seasonal variations and variations with the solar activity cycle. Plans to increase the gallium mass to 60 or 100 metric tons did not materialize. GNO was terminated in 2004 after a safety incident at the Gran Sasso laboratory involving another experiment. The GNO result for solar neutrinos was about 50% of that predicted by the SSM.

SAGE was another radiochemical Ga $\rightarrow$ Ge detector, located in the Caucasus as a joint effort between the USA and Russia operating from 1990 through 2001. Its target contained, on average, about 50 metric tons of liquid metallic gallium. The target mass varied, in part, due to theft of about 2 tons of gallium.[7] The technology of extracting germanium from metallic gallium was different from the extraction technology of GALLEX. SAGE detected about 55% of the flux expected from the SSM, also pointing to neutrino flavor oscillations (Abdurashitov *et al.*, 1994). The discrepancy between the SSM and the gallium results is different than that for the chlorine experiment due to the different neutrino energies probed by the two techniques.

14.6.3 Other targets

Other radiochemical targets have been proposed. Foremost among these appears to be ^{127}I, which has an energy cutoff somewhat below that of ^{37}Cl. It would detect the same types of neutrinos as a chlorine detector, but with greater sensitivity. A prototype ^{127}I detector was constructed in the Homestake mine. After 2001 the future useability of the Homestake mine was in some doubt. In 2003 the dewatering pumps were shut off, and the mine was allowed to flood. But the US National Science Foundation and others are funding continuing use of that site under the names Sanford Underground Laboratory and Deep Underground Science and Engineering Laboratory (DUSEL, Figure 14.6). Construction at the level 4850 feet (1478 m) below surface, the level of the original Davis experiment, was proceeding during 2008–2010. Plans are to have an additional facility at the 7400 foot (2255 m) level, which would be the deepest neutrino detector site. A prototype liquid ^{40}Ar tracking/scintillation detector called ICARUS has been built at Gran

---

[7] Worth about US $1M at current prices.

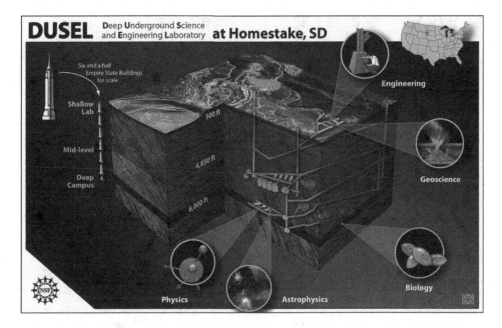

Figure 14.6 DUSEL laboratory. Credit: Zina Deretsky, National Science Foundation.

Sasso. The tracking nature of the ICARUS detector makes it rather different than other radiochemical detectors and not subject to many of their limitations. The nuclide ^{7}Li is another possible radiochemical target, along with a long list of other proposed nuclides.

14.7 Čerenkov detectors

Čerenkov detectors are sensitive to neutrinos scattering off of electrons in water via either charged current (CC) elastic scattering,

$$v_e + e^- \rightarrow v_e + e^-, \tag{14.24}$$

or neutral current (NC) elastic scattering,

$$v_x + e^- \rightarrow v_x + e^-, \tag{14.25}$$

as shown in Figure 14.7. The scattered electrons are traveling at nearly c, whereas the speed of light in water is reduced by the index of refraction n $\approx$ 1.33. The electrons will therefore emit Čerenkov radiation in a cone with geometry determined by the equation

$$\cos \theta = \frac{1}{\beta n}, \tag{14.26}$$

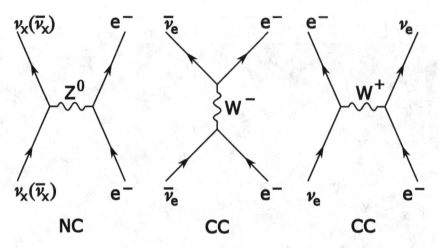

Figure 14.7 Feynman diagrams of neutrino–electron elastic scattering. (Left) NC scattering of all neutrino and antineutrino flavors. (Center) CC scattering of electron antineutrinos. (Right) CC scattering of electron neutrinos.

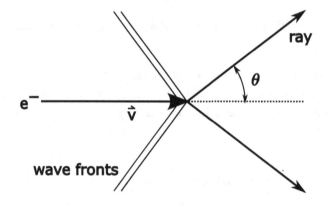

Figure 14.8 Geometry of Čerenkov radiation.

giving an opening angle of about $42°$ as shown in Figure 14.8. The visible and ultraviolet light is detected by an array of photomultiplier tubes surrounding the target. The center of the circle of photomultipliers which see a signal gives information on the direction of arrival for the neutrino, and the precise geometry of the event helps discriminate against certain types of background events. The angular resolution of this scheme can be of order a few degrees, with energy resolution of order 20%, and timing down to about 1 ns. Čerenkov detectors are sensitive to neutrinos of energy greater than about 5 MeV, since the neutrino energy is transferred to the scattered electron and must be sufficient to make it relativistic. Unlike radiochemical detectors, Čerenkov detectors are also capable of seeing ν_μ and ν_τ, but only by neutral current scattering. To determine the neutrino flavor, one must look carefully at the event geometry. An electron neutrino event will produce an energetic electron which will rapidly lose energy in an electromagnetic shower. The energy will be distributed in a shower of scattered electrons moving in different directions, producing a diffuse directionality to the Čerenkov light. Muons, on the other hand, are penetrating particles losing energy slowly and producing a sharp, well-defined Čerenkov cone. If a tau neutrino interacts inside the detector, there will be an initial shower from the scattered electron, and then a short time and distance later a second shower when the tauon decays. Čerenkov detectors can also see hadronic interactions, which will be discussed in the following sections.

Čerenkov detectors have the significant advantages of operating in real time and in retaining information about the energies and directions of the incoming neutrinos. Also they are sensitive to all neutrino flavors. A disadvantage is that they have energy thresholds higher than those of radiochemical detectors. Čerenkov detectors require many expensive photomultiplier tubes, but the signal processing is relatively straightforward compared to the elaborate chemical processing of radiochemical detectors.

14.7.1 Kamiokande and Super-Kamiokande

IMB was the first water Čerenkov detector, but we begin our discussion with the detector located in one of the Kamioke Mining Company's mines on the island of Honshu in Japan. Kamiokande was a pioneering Čerenkov neutrino detector[8] containing 3000 metric tons of purified water surrounded by about 1000 photomultiplier tubes. As a solar neutrino detector, it was able to provide directional information from electron scattering events, proving that the neutrinos that it detected were coming from the Sun. It was also online in 1987 and provided crucial timing, directional, and flux information on the neutrinos from SN 1987A (as did the IMB detector).

Its successor, Super-Kamiokande, is a Čerenkov detector containing 50 000 metric tons (5×10^7 kg) of highly purified water viewed by over 11 000 photomultiplier tubes. The water is divided between an inner portion (inner detector) which is the primary target, and an outer portion which serves as an anti-coincidence shield to discriminate against atmospheric muons. The inner portion of Super-Kamiokande is shown in Figure 14.9 and part of the outer portion is shown in Figure 14.10. A simulated event is shown in Figure 14.11. For solar neutrinos, the flux measured by Super-Kamiokande is about 45% of that predicted by the Standard Solar Model (Fukuda *et al.*, 2001).

---

[8] The original purpose of the Kamiokande detector was the search for proton decay.

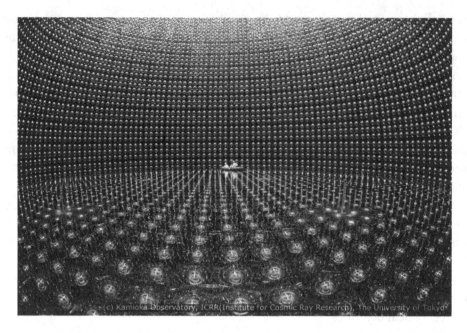

Figure 14.9 The inner portion of the Super-K detector on April 23, 2006, as it begins to be filled with water. Note the boat at the far side. For color version of figure, see plate section.

Figure 14.10 The outer portion of the Super-K detector with outward-looking PMTs. For color version of figure, see plate section.

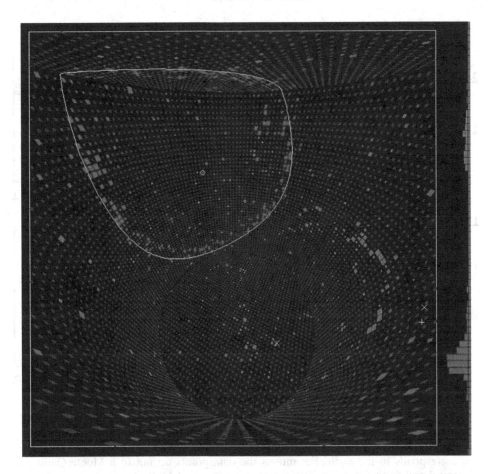

Figure 14.11 Event topology at Super-K. In this event a simulated 481 MeV muon neutrino produces a relativistic muon which emits Čerenkov light, coded red. The muon decays into an electron with its own Čerenkov light, coded yellow/green. Color coding of arrival times is shown at right, along with a histogram of photomultiplier hits. Credit: The Super-Kamiokande Collaboration. For color version of figure, see plate section.

Super-Kamiokande provided dramatic evidence of oscillation for atmospheric neutrinos as shown in Figure 14.12. Atmospheric electron neutrinos were seen in the expected quantity, for all directions. Atmospheric muon neutrinos, however, displayed a clear deficit for those coming from below the horizon. The interpretation is that there were matter-enhanced (MSW) oscillations turning about half of the ν_μ into ν_τ during passage through Earth (Fukuda *et al.*, 1998).

Hadronic cross sections in water become large above a few MeV, the dominant interaction being

$$\bar{\nu}_e + p \to n + e^+, \tag{14.27}$$

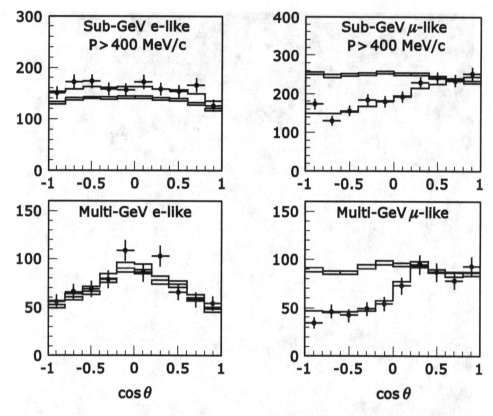

Figure 14.12 Number of electron-like and muon-like events seen at Super-Kamiokande as a function of the cosine of the zenith angle; $\cos\theta = -1$ corresponds to the nadir. For muons the data points do not fit a Monte Carlo simulation in the absence of mixing, but do fit a model with large mixing angle. Adapted from Ashie *et al.* (2005).

which is very important at typical supernova neutrino energies. Other hadronic reactions of importance include charged and neutral current cross sections on oxygen nuclei. The elastic CC and NC electron scattering cross sections continue to rise with increasing neutrino energy, but hadronic cross sections are dominant above 30 MeV, as shown in Figure 14.13.

In 2001 over half of the photomultiplier tubes were destroyed in a chain reaction by shock waves, the first of which appears to have been created by an implosion of one of the PMTs near the bottom of the tank. Full operation was restored in 2006 under the name Super-Kamiokande-III. Neutrons are readily captured by gadolinium. A proposed modification of Super-Kamiokande adding $GdCl_3$ or $Gd_2(SO_4)_3$ would greatly improve the sensitivity to $\bar{\nu}_e$ by allowing detection of the neutrons generated in Equation 14.27.

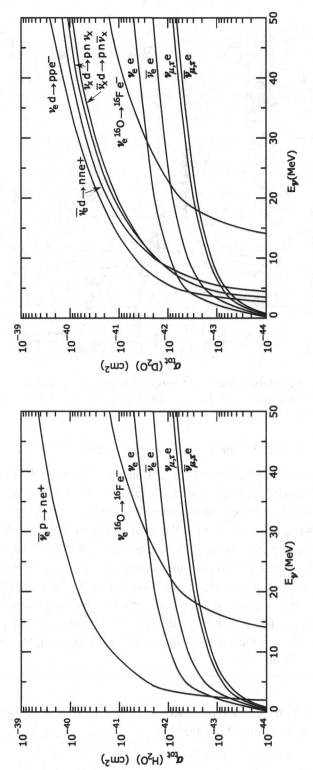

Figure 14.13 (Left) Hadronic (red in electronic version) and leptonic (black) neutrino cross sections per molecule in light-water Čerenkov detectors. (Right) Hadronic (red and blue) and leptonic (black) neutrino and antineutrino cross sections per molecule in heavy-water Čerenkov detectors. Adapted from Raffelt (1996), Virtue (2001), and Giunti & Kim (2007).

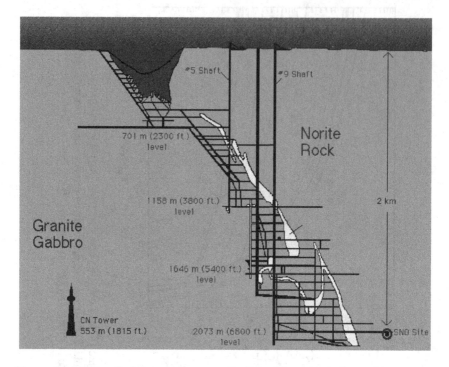

Figure 14.14 The Creighton mine near Sudbury, Ontario, is located at a seam between norite rock and granite gabbro. The SNO site is located well away from other mine activities. Courtesy of SNO.

14.7.2 *Sudbury Neutrino Observatory*

The Sudbury Neutrino Observatory (SNO) in Canada is located 2073 m below ground in an active nickel mine, as shown in Figure 14.14. At this location the atmospheric muon flux is reduced by an order of magnitude or more compared to previously used sites, other than the Kolar gold mines in India. At the center of the detector is a sphere containing 1000 metric tons (10^6 kg) of heavy water (D_2O), surrounded by about 10 000 photomultipliers, as shown in Figures 14.15 and 14.16. This in turn is surrounded by a 7000 ton light water (H_2O) shield. SNO can see all of the electron scattering reactions seen by Super-Kamiokande, and shares Super-K's greater sensitivity to ν_e scattering. In addition, the presence of deuterium allows the hadronic reactions shown in Figure 14.13 (right), including the charged current reaction

$$\nu_e + d \rightarrow p + p + e^- \tag{14.28}$$

(Ahmad *et al.*, 2001), with the relativistic electron producing Čerenkov light. There is also the hadronic neutral current reaction

$$\nu_x + d \rightarrow p + n + \nu_x, \tag{14.29}$$

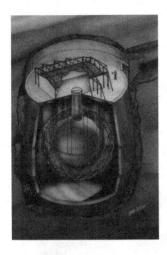

Figure 14.15 Artist's conception of the SNO detector. Courtesy of SNO.

Figure 14.16 SNO detector during construction, showing the inner acrylic vessel and the outer cage partially populated at the top with photomultipliers. Courtesy of SNO. For color version of figure, see plate section.

where x can stand for either e, μ, or τ (see Figure 14.13). Because of the latter reaction, the heavy water in SNO has significantly larger cross sections for muon and tau neutrinos than light water. The detection of the neutron requires an additional step. Neutron capture on deuterium is possible, but inefficient. During the second phase of SNO operation (Ahmed *et al.*, 2004; Aharmim *et al.*, 2005), the more efficient neutron capture on ^{35}Cl (from added NaCl) was used,

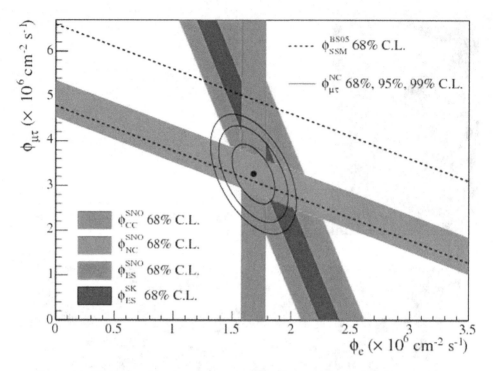

Figure 14.17 Measured solar electron neutrino flux versus neutrino flux for other flavors (Aharmim *et al.*, 2005). For color version of figure, see plate section.

$$n + {}^{35}Cl \rightarrow {}^{36}Cl + \gamma. \tag{14.30}$$

In both cases the resulting gamma ray Compton scatters off of electrons producing isotropic Čerenkov radiation. In the third phase of SNO operation the salt was removed and thermalized neutrons were captured on ^{3}He inside one of many proportional counters suspended on a grid in the D$_2$O (Prior, 2009). The total, flavor-independent flux of solar neutrinos detected by SNO is in good agreement with the SSM (Figure 14.17), indicating that the shortfalls in the chlorine and gallium experiments were due to neutrino flavor oscillations.

For supernova detection, the hadronic CC and NC reactions given above have cross sections larger than electron elastic scattering cross sections at the energies expected for supernova neutrinos. In addition, supernova bursts are expected to contain antineutrinos, in which case the hadronic CC reaction

$$\bar{\nu}_e + d \rightarrow n + n + e^+ \tag{14.31}$$

and the hadronic NC reaction

$$\bar{\nu}_x + d \rightarrow p + n + \bar{\nu}_x \tag{14.32}$$

would also become important. There were no nearby supernovae between 1999, when SNO started taking data, and 2006, when it was shut down. SNO was thought to be sensitive to supernovae out to a distance of 30 kpc in real time and much further *a posteriori*. In the future, experiments such as SNO+ or HALO may exist at the SNO site, which will operate under the umbrella name of SNOLAB.

14.7.3 IceCube

The extreme smallness of neutrino scattering cross sections suggest using the largest possible target mass. If Čerenkov light is to be detected, the target must be transparent. The largest naturally occurring transparent masses on Earth, other than the atmosphere, are in the form of water or ice.

IceCube uses as a target one cubic kilometer of antarctic ice (Achterberg *et al.*, 2006). Surface ice tends to have air bubbles which create a short light scattering length. But the antarctic ice sheet is several kilometers thick, and at sufficient depth the ice has a sufficiently long scattering length. A prototype detector was made, named AMANDA, by drilling vertical holes into the ice using hot water jets. Before the ice could refreeze, long strings of photomultiplier tubes were lowered into the holes, as shown in Figure 14.18.

The IceCube collaborators are primarily interested in exploring the highest energy neutrinos, those above 100 GeV. Such neutrinos would be produced by the

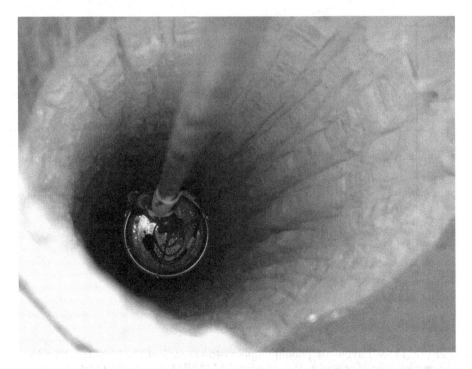

Figure 14.18 Photomultiplier being lowered down into a 2 km hole in the antarctic ice. Courtesy of IceCube/NSF. For color version of figure, see plate section.

same sorts of processes that produce high energy cosmic rays and TeV gamma rays. These include supernova shocks, active galactic nuclei (AGN), and gamma ray bursts (GRB). But energetic cosmic rays and gamma rays have limited propagation lengths, unlike neutrinos which may provide the least biassed views of such events. Both electron neutrinos and muon neutrinos are likely to be produced, but not tau neutrinos. However, ν_τ can be produced by flavor oscillations.

There are 80 strings of photomultipliers in IceCube, each with 60 *downward facing* photomultipliers, laid out as shown in Figure 14.19. The desired events are upward traveling neutrinos, which may be ν_e, ν_μ, or ν_τ. The pattern of energy distribution in the ice is sufficient to differentiate the different neutrino flavors and in some cases allows good measurement of the direction the neutrino was traveling. Event characterization and background discrimination require precise timing ($\sim$5 ns). Use of timing information is shown in Figure 14.20.

Detection of an upward traveling ν_τ would require that the neutrino undergo a deep inelastic scattering with a nucleus within the detector volume, producing a τ lepton and a hadronic shower. The hadronic shower would be detected through Čerenkov radiation. The lifetime of a τ lepton, even relativistically enhanced, is

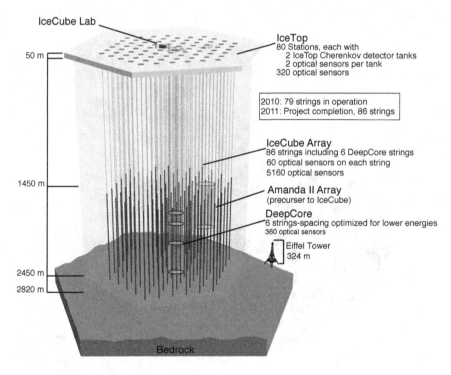

Figure 14.19 Layout of photomultiplier strings in IceCube. Shaded cylinder shows location of prototype system named AMANDA. Courtesy of IceCube/NSF. For color version of figure, see plate section.

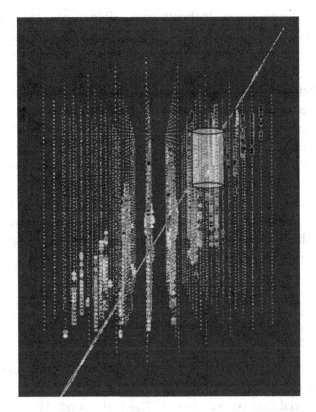

Figure 14.20 Simulation of an upward travelling 6 PeV muon in IceCube (Ahrens *et al.*, 2001). Photomultiplier timing is color coded with red indicating the earliest signals. For color version of figure, see plate section.

rather short so the τ lepton would travel a short distance before decaying, most likely, in another hadronic shower. But if the original ν_τ were energetic enough, it may be possible to spatially separate the two showers. The event topology and timing would be unmistakable, although the rate of such events is not likely to be large.

Neutrinos from galactic supernovae have lower energy, $\sim$10 MeV, than those of primary interest, $\sim$100 GeV, so the event signature would be different. However, they would be observable by IceCube through

$$\bar{\nu}_e + p \rightarrow e^+ + n. \tag{14.33}$$

The positron would travel a relatively short length and produce Čerenkov light in a localized burst.

For muon neutrinos, the characteristic signature will be upward traveling muons. Muons are more highly penetrating than electrons, and will emit Čerenkov light along kilometer-length tracks. It should be possible to get very good directional information from these muon tracks. The dominant background is caused by cosmic ray air showers above the detector, which produce downward traveling muons. However, there is also a background produced by cosmic ray showers in the northern hemisphere, which produce neutrinos that travel through Earth and then appear as upward traveling neutrinos in IceCube. Such events have been seen by Ice Cube (Achterberg *et al.*, 2007) at TeV to PeV energies and demonstrate that the instrument is well calibrated and that the event rate is understood.

Variations of this technique have been tried in both seawater and fresh water. An instrument in Lake Baikal (Russia) pioneered underwater neutrino detection. Projects ANTARES and NESTOR are prototyping deep sea detectors in the Mediterranean. They face backgrounds produced by bioluminescence and ^{40}K, but seem to be viable technologies. A future cubic kilometer version of a deep sea detector has been dubbed KM3NeT.

14.8 Scintillation detectors: Borexino

Borexino looks superficially like a Čerenkov detector, but is quite different. Its target is a liquid scintillator made of 300 metric tons of the aromatic solvent pseudocumene (an isomer of trimethyl benzene), into which small amounts of fluorescent material have been dissolved. This is surrounded by a non-fluorescing 900 metric ton buffer of pseudocumene. All of that is contained in a spherical stainless steel tank, which in turn is surrounded by a 2100 metric ton water shield. Photomultiplier tubes are affixed to both sides of the stainless steel sphere (looking both inwards and outwards). The outward looking photomultipliers are used to discriminate against Čerenkov radiation from any cosmic ray muons surviving passage through the rock overburden at Gran Sasso. Extremely low levels of background radioactivity are required for all the materials used in this experiment.

The main goal of the experiment is to detect neutrino elastic scattering off of electrons, particularly for the mono-energetic ^{7}Be solar neutrinos, made possible by the low energy threshold of 250 keV. The inward looking photomultiplier tubes record the amount of scintillation light, which is proportional to the energy transferred in the scattering event. Directional information is lost.

Since the Borexino target is pseudocumene, there will be hadronic interactions on protons (hydrogen) as in light-water detectors like Super-Kamiokande. But the other hadronic target will be carbon instead of oxygen, with the potential interaction

$$\nu_e + {}^{12}\text{C} \rightarrow {}^{12}\text{N} + e^-. \tag{14.34}$$

14.9 Cosmological implications

A consequence of the standard Big Bang cosmology, in addition to the 2.73 K Cosmic Microwave Background (CMB) radiation, is the existence of numerous *relic*[9] or *primordial* neutrinos at a temperature of 1.95 K. The current number density of neutrinos plus antineutrinos is expected to be 111.9 cm^{-3} per generation (Giunti & Kim, 2007). Purely from an observational point of view, detecting such neutrinos is extremely challenging. As we have seen, neutrino cross sections are small enough at energies of order 1 MeV. But they drop off dramatically at lower energies ($\sim 10^{-4}$ eV), as indicated in Figure 14.13.

Since we now know that at least two neutrinos have mass, the question arises as to whether neutrinos may be responsible for much of the *missing mass* or *dark matter* in the universe. Relic neutrinos would have been relativistic at decoupling, so they would contribute a *hot* dark matter component. Our present understanding of large-scale structure formation requires substantial *cold* dark matter (CDM) and can be made consistent only with a limited amount of hot dark matter. Hot dark matter suppresses the growth of small-scale structure, contrary to evidence which indicates a *bottom-up* scenario with small-scale structures forming before large-scale structures. From these considerations we can set $\Omega_\nu \lesssim 0.02$ corresponding to

$$\sum_j m_j \lesssim 1 \text{ eV}. \tag{14.35}$$

Fortunately this is consistent with the values of Δm_{12}^2 and Δm_{23}^2 determined for solar and atmospheric neutrino measurements. Since those data were only sensitive to differences in the squares of the masses, this provides an additional constraint on the absolute mass scale.

14.10 Background of supernova neutrinos

There is also a *relic* background of supernova neutrinos and antineutrinos, often called the diffuse supernova neutrino background (DSNB). These presumably were emitted at energies typical of current supernova neutrinos (~ 20 MeV), but may appear significantly redshifted in current detectors. The difficult task is to identify a particular event as emanating from a supernova without the key signal of multiple, nearly concurrent events. The Sun produces neutrinos, not antineutrinos, and flavor oscillations do not convert any to antineutrinos. Supernovae, on the other hand, produce neutrinos and antineutrinos in equal numbers. The detection of antineutrinos of an appropriate energy therefore is a signature of core-collapse

---

[9] The term *relic* unfortunately is also used for neutrinos due to supernovae throughout the universe back to the earliest stars.

supernovae. Hopkins & Beacom (2006) and Horiuchi *et al.* (2009) have used limits set by Super-Kamiokande as a constraint on global star-formation history. The proposed addition of gadolinium to Super-Kamiokande would likely allow direct measurement of the DSNB.

Exercises

14.1 Calculate the distance that a 1 GeV muon will travel before decay. (Look up the lifetime of a muon and calculate its Lorentz factor.)

14.2 Calculate the solar neutrino flux at 1 AU, (a) from the solar luminosity, (b) from Figure 14.1.

14.3 Estimate the present day number density of relic Big Bang neutrinos.

14.4 Kamiokande-II and IMB detected, respectively, 12 and 8 antineutrinos from SN 1987A. How consistent are these results, assuming a fiducial mass of 2140 metric tons for Kamiokande-II and 5000 metric tons for IMB?

14.5 Derive Equation 14.17. Note that the neutrinos are ultrarelativistic. This is important in determining the time evolution of the states.

14.6 Calculate the neutrino luminosity of SN 1987A from the event rates observed with Kamiokande and IMB.

15

Cosmic ray detectors

The detection on Earth of ionizing radiation whose strength increased with altitude was the first evidence for the existence of what, today, are known as cosmic rays. In the discussion which follows we will, for the most part, bypass the early history of controversies in this field over whether the radiation consisted of particles or gamma rays and over whether or not the radiation was of extraterrestrial origin. Instead, we will begin with our modern understanding that cosmic rays are indeed of *cosmic* origin and consist of energetic particles, most of which are charged.[1] The focus thus will be on the measurable properties of such particles and the best ways to make such measurements.

15.1 Properties of cosmic rays

The most readily measurable properties of energetic particles are charge, mass, and energy. Charged particle trajectories can also be well determined locally. Such trajectories can be used within a detector to measure a particle charge to mass ratio from the curvature of a track in a magnetic field or to relate multiple secondary particle tracks back to a common point of interaction. However, except possibly for the very highest energy cosmic rays, a primary cosmic ray trajectory does not lead back to the location of the astrophysical source of the cosmic ray.

Observationally, cosmic rays at GeV energies are found to consist mostly of protons, with about a 10% contribution of helium nuclei, 1% of heavier nuclei, and an approximately 1% contribution of electrons and positrons. The proton differential energy spectrum follows an approximate power law

$$\frac{dN}{dE} \propto E^{-\gamma}, \tag{15.1}$$

---

[1] Although the term *cosmic rays* is sometimes used to include gamma rays and neutrinos, these are not included in this chapter but are discussed separately in Chapters 11 and 14, respectively.

285

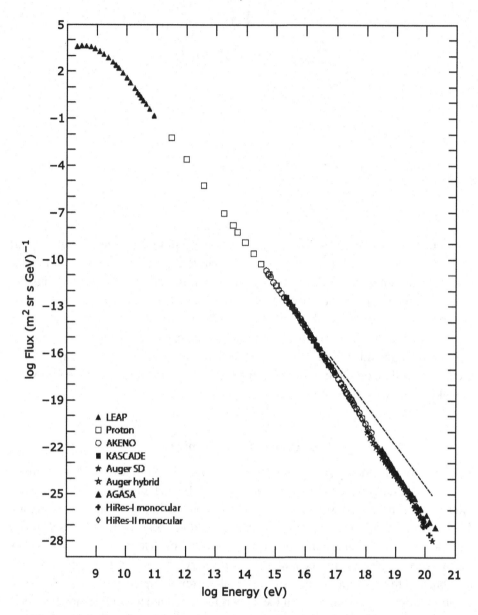

Figure 15.1 Cosmic rays are observed over 12 orders of magnitude in energy and nearly 32 orders of magnitude in flux (Beatty & Westerhoff, 2009). Note the "knee" around 3×10^{15} eV and the "ankle" around 3×10^{18} eV.

with $\gamma \approx 2.7$ over many decades in energy, as shown in Figure 15.1. Small apparent deviations from power law behavior include a "knee" at about 3×10^{15} eV and an "ankle" around 3×10^{18} eV. The "ankle" is usually interpreted as a cross-over from galactic to extragalactic cosmic rays. A chief problem in cosmic ray physics is providing a plausible explanation for the origin of these ultrahigh energy cosmic rays (UHECRs) above 3×10^{18} eV.

15.2 Intervening regions

In Chapter 1 we looked at various ways in which regions between a source and an observer affect the flow of electromagnetic radiation. Intervening regions are also significant for the flow of energetic particles, but the physical processes involved are quite different. As with electromagnetic radiation, we can either consider the intervening regions as obstacles blocking the flow of information from distant astrophysical sources, or we can consider the intervening regions as targets of interest being probed by cosmic rays.

15.2.1 Magnetic fields

The Larmor radius (gyroradius) for a charged particle of mass m and charge q moving in a uniform magnetic field B is given by

$$r_L = \frac{m \, v_\perp}{q \, B}, \tag{15.2}$$

where $v_\perp$ is the velocity component perpendicular to the field. For ultrarelativistic nuclei of atomic number Z and energy E moving perpendicular to the magnetic field this is

$$r_L \approx \frac{E/c}{Z \, e \, B}. \tag{15.3}$$

It is customary to define the magnetic *rigidity*, $\mathcal{R}$, as the relativistic momentum divided by the charge,

$$\mathcal{R} = \frac{p_\perp}{q} = r_L \, B. \tag{15.4}$$

Table 15.1 gives characteristic magnetic fields and corresponding characteristic gyroradii for 10 GeV protons in various regions that cosmic rays traverse. Clearly the geomagnetic field will have a strong influence on the direction of propagation of low energy cosmic rays, with gyroradii of the same order as Earth's radius. Since the majority of cosmic rays are positively charged, this results in an excess of particles moving from west to east, as shown in Figure 15.2. The presence of this effect can be taken as proof that the majority of cosmic rays have positive charge. There is also a latitude effect, with the flux of lower energy cosmic rays being larger at higher geomagnetic latitudes. Just as solar wind particles are concentrated in the polar regions, producing the aurora, cosmic rays are similarly funneled into the polar regions.

Table 15.1. *Characteristic magnetic fields and gyroradii*

Region	B (gauss)	r_L (meters)	
Earth magnetosphere	0.3	1.1×10^6	$\approx 0.17\,R_\oplus$
Interplanetary medium	50×10^{-6}	6.7×10^9	≈ 0.045 AU
Interstellar medium	3×10^{-6}	1.1×10^{11}	$\approx 3.6 \times 10^{-6}$ pc
Intergalactic medium	10^{-9}	3.3×10^{14}	$\approx 1.1 \times 10^{-2}$ pc

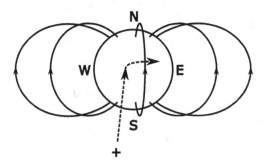

Figure 15.2 A positively charged cosmic ray incident from the zenith near the geomagnetic equator is deflected from west to east.

The interplanetary magnetic field is neither dipolar nor static. The lowest energy cosmic rays are deflected on scales of order an astronomical unit. They reach Earth's orbit only to the extent that they are able to diffuse against the outward advection of the solar wind, which in turn is dependent on the solar cycle. The interstellar medium causes cosmic rays of low and medium energy to gyrate with radii which are fractions of a parsec. Consequently, cosmic rays from galactic sources such as supernova remnants, which may be of order kiloparsecs distant, reach the solar system only by circuitous routes. Galactic cosmic rays have an observed isotropic distribution, which is why determining observationally the direction of propagation of individual cosmic rays is relatively unimportant. The propagation direction retains little information about the cosmic ray point of origin. The galactic field determines whether or not cosmic rays of different rigidities are able to escape from the galactic plane (Cesarsky, 1980). The very highest energy cosmic rays are thought to be extragalactic and may have notable anisotropy. The strength of the intergalactic magnetic field is not well known and may, in places, be much higher than the value of 10^{-9} gauss given in Table 15.1.

Energetic neutrons are not produced directly in plausible cosmic ray acceleration scenarios, but energetic protons may be converted into neutrons as discussed below in Section 15.2.8. Neutron trajectories are not affected by interstellar magnetic fields. Free neutrons at rest have mean lifetimes of 887 seconds and beta decay as

$$n \rightarrow p + e^- + \bar{\nu}_e. \tag{15.5}$$

But at Lorentz factors of $\gamma \approx 10^6-10^9$ ($10^{15}-10^{18}$ eV) neutron lifetimes are of order hundreds or thousands of years, allowing them to travel significant galactic distances.

15.2.2 Spallation reactions

Cosmic rays consist mostly of protons, but contain atomic nuclei out through U. The light nuclei Li, Be, and B are not abundant products of either Big Bang nucleosynthesis or stellar nucleosynthesis, yet they are present in cosmic rays, enhanced by of order 10^6 over cosmic abundances. The rare light elements seen in cosmic rays are presumed to have been created from cosmic ray protons or alpha particles in *spallation* reactions. Some typical spallation reactions and their shorthand notations are

$$\begin{array}{lll}
p + {}^{12}\text{C} \rightarrow {}^{11}\text{B} + p + p & {}^{12}\text{C}(p, \quad 2p)^{11}\text{B}, & (15.6) \\
p + {}^{12}\text{C} \rightarrow {}^{9}\text{Be} + p + p + p + n & {}^{12}\text{C}(p, \quad 3p\,n)^{9}\text{Be}, & (15.7) \\
p + {}^{16}\text{O} \rightarrow {}^{6}\text{Li} + \alpha + \alpha + p + p + n & {}^{16}\text{O}(p, 2\alpha\,2p\,n)^{6}\text{Li}, & (15.8) \\
p + {}^{16}\text{O} \rightarrow {}^{10}\text{B} + \alpha + p + p + n & {}^{16}\text{O}(p, \quad \alpha\,2p\,n)^{10}\text{B}, & (15.9) \\
\alpha + {}^{16}\text{O} \rightarrow {}^{6}\text{Li} + \alpha + \alpha + \alpha + p + n & {}^{16}\text{O}(\alpha, \quad 3\alpha\,p\,n)^{6}\text{Li}. & (15.10)
\end{array}$$

The transition elements Sc through Mn ($Z = 21$–25), which are spallation products of Fe, are enhanced in abundance in cosmic rays by factors of 10 to 1000. The odd-Z elements F, P, Cl, and K (and to a lesser extent Na and Al), whose nuclei are less tightly bound and which are produced in lower abundance in stellar nucleosynthesis than adjacent even-Z nuclei, are also enhanced in cosmic rays.

In a similar vein, cosmic ray spallation is a source of replenishment of short-lived isotopes such as ^{14}C. Cosmic rays produce secondary neutrons in Earth's atmosphere, which then react with nitrogen to produce ^{14}C,

$$n + {}^{14}\text{N} \rightarrow {}^{14}\text{C} + p. \tag{15.11}$$

Other radioactive isotopes thought to be cosmic ray secondaries, produced in the interstellar medium and in Earth's atmosphere, are listed in Table 15.2. An overview of the differential energy spectra of some of the most abundant components of cosmic rays is shown in Figure 15.3.

Table 15.2. *Radioactive isotopes*
replenished from cosmic rays

Isotope[a]	Half-life (years)
^{10}Be	$1.51 \pm 0.06 \times 10^6$
^{14}C	5700 ± 30
^{26}Al	$7.17 \pm 0.24 \times 10^5$
^{36}Cl	$3.01 \pm 0.03 \times 10^5$
^{41}Ca	$1.02 \pm 0.07 \times 10^5$
^{129}I	$1.61 \pm 0.07 \times 10^7$

[a] IAEA, The Live Chart of Nuclides,
ENSDF 2009; www-nds.iaea.org.

15.2.3 Interstellar ionization losses

For galactic sources a direct line from a source to the Earth might be of order 1 kpc long ($\sim 3 \times 10^{21}$ cm). However, the paths of cosmic rays will not be direct, due to interstellar magnetic fields, and the actual path lengths may be of order 100 or 1000 times as long. For an average interstellar density of 3×10^{-24} g cm^{-3} (about 1–2 hydrogen atoms per cubic centimeter), this corresponds to a column density of order 1–10 g cm^{-2}. The theory of ionization losses is discussed, below, in Section 15.3.1. But for now, we will note that this will result in significant loss of particle flux for protons of energy 100 MeV and below and for electrons below 10 MeV. In other words, the lower end of the cosmic ray energy spectrum will be significantly modified by interstellar ionization losses. These are also energies where cosmic ray fluxes are heavily modified by the solar wind.

The low energy end of the electron spectrum will be replenished, in part, by what are known as knock-on electrons (sometimes called delta rays). These are cosmic ray secondaries, produced by the collisions of higher energy protons with the interstellar medium.

15.2.4 Bremsstrahlung

Cosmic ray electrons also lose energy through bremsstrahlung (braking radiation). When an electron encounters the Coulomb field of an atomic nucleus of charge Ze, its path will be deflected, and it will emit radiation whose instantaneous power is given by the relativistic generalization of the Larmor formula,

$$P = \frac{1}{4\pi\epsilon_0} \frac{2\,e^2}{3\,c^3} \left(\gamma^4 a_\perp^2 + \gamma^6 a_\parallel^2 \right). \tag{15.12}$$

We will assume that the component of acceleration $a_\perp(t)$ perpendicular to the velocity is relevant and that $a_\parallel = 0$. A low mass particle, such as an electron, will

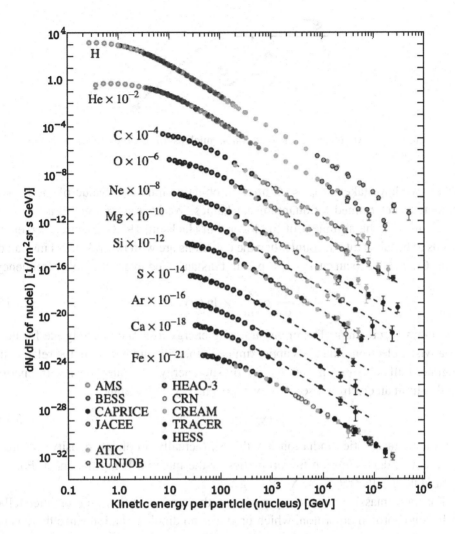

Figure 15.3 Differential fluxes of different nuclear components of cosmic rays (Gaisser & Stanev, 2008). Credit: P. J. Boyle & D. Muller.

experience a large acceleration and radiate. A classical picture of a typical interaction is shown in Figure 15.4. The ultrarelativistic expression for the power radiated at low frequencies ($\hbar\omega \ll \gamma m_e c^2$) by a steady flux of electrons is

$$\frac{dP}{d\omega} \propto \frac{Z^2}{m_e^2} \ln\left(\frac{\gamma^2 c}{\omega b_{min}}\right) \tag{15.13}$$

$$= \frac{Z^2}{m_e^2} \ln\left(\frac{\gamma^2 m_e c^2}{\hbar\omega}\right), \tag{15.14}$$

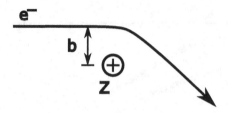

Figure 15.4 An electron incident on a nucleus of charge Ze at an impact parameter b.

where we have taken $b_{min} \approx \hbar/m_e c$, an effective minimum value of the impact parameter determined by quantum mechanics. We have also ignored a factor of order unity in the argument of the logarithm (Jackson, 1998). Electron screening needs to be taken into account when the collisions are with neutral atoms instead of ions. Converting to an energy loss per unit distance and integrating over frequency,

$$\frac{dE}{dx} \propto \frac{Z^2}{m_e} \gamma \, \ln(\gamma), \qquad (15.15)$$

for minimal screening. The bremsstrahlung energy loss rate is negligible for non-relativistic electrons and rises approximately linearly with energy in the relativistic regime. Full screening makes the relativistic energy loss rate linear with energy. In this limit an electron loses energy exponentially with distance,

$$E(x) = E_0 \, e^{-x\rho/X_0}, \qquad (15.16)$$

with a characteristic "radiation length" X_0 (actually a column density) of order 10 g cm^{-2}, depending on the properties of the material being traversed. For the neutral interstellar medium, $X_0 \approx 65 \text{ g cm}^{-2}$.

For more massive cosmic ray particles, the most common type of interstellar collision is proton on proton, which produces no dipole radiation since there is no changing dipole moment. Other nuclear collisions such as protons on ^{4}He nuclei radiate small amounts of power due to unfavorable kinematic factors, which result in much smaller accelerations than for electrons.

Generally speaking both ionization and bremsstrahlung losses are small in the low density interstellar medium. Bremsstrahlung losses become relevant for relativistic electrons. However, at the highest energies, synchrotron and inverse Compton losses are the largest sources of energy loss.

15.2.5 Synchrotron losses

A relativistic electron moving with velocity v perpendicular to a magnetic field B radiates energy at a rate

$$P = 2\sigma_T \, c \, u_{mag} \, \beta^2 \gamma^2 = \sigma_T \, c \, \frac{B^2}{\mu_0} \beta^2 \gamma^2, \tag{15.17}$$

where u_{mag} is the magnetic field energy density, $\beta = v/c$, and the Lorentz factor $\gamma = E/m_e c^2$. This power must be reduced by a factor of 2/3 if one wishes to average over an isotropic distribution of possible orientations of the velocity with respect to the magnetic field. The Thomson cross section is

$$\sigma_T = \frac{8\pi}{3} \left(\frac{e^2}{4\pi\epsilon_0 m_e c^2} \right)^2 = \frac{8\pi}{3} \left(\frac{\alpha\hbar}{m_e c} \right)^2 = 6.65 \times 10^{-25} \text{ cm}^2. \tag{15.18}$$

This rate of energy loss is significant for an electron above a few GeV. The time scale for radiating most of an electron's energy is

$$t \approx \frac{1}{\sigma_T} \frac{m_e c^2}{E} m_e c \frac{1}{\beta^2} \frac{\mu_0}{B^2}. \tag{15.19}$$

For a 10 GeV electron in a $3\,\mu\text{G} = 3 \times 10^{-10}$ T magnetic field, this corresponds to about 10^8 years.

For protons the synchrotron power radiated is reduced by the square of the ratio of the electron mass to the proton mass, a factor of order 3×10^{-7}. For heavy nuclei the rate of energy loss is further multiplied by the ratio $(Z/A)^4$ where Z is the atomic number and A is the atomic mass number. Synchrotron radiation from protons and heavy nuclei is generally small, except for particles of very high energy in strong magnetic fields.

15.2.6 *Inverse Compton losses*

The inverse Compton effect can be written as

$$e^- + \text{photon} \rightarrow e^- + \text{higher energy photon.} \tag{15.20}$$

Inverse Compton losses follow the same loss formula as for synchrotron radiation but with the magnetic energy density $u_{mag} = B^2/2\mu_0$ replaced by the photon energy density. If the energy density of photons (starlight, CMB, etc.) is comparable to the energy density in magnetic fields, then inverse Compton losses will be comparable to synchrotron losses. Since these energy densities are indeed comparable in the interstellar medium, so will be the inverse Compton and synchrotron losses. Together they may be referred to as radiative losses, following the relation

$$P_{rad} = \frac{4}{3} \sigma_T \, c \left(u_{mag} + u_{photon} \right) \beta^2 \gamma^2. \tag{15.21}$$

Both mechanisms will be significant for electrons above a few GeV.

15.2.7 Pair production

Pair production by energetic protons interacting with photons from the CMB can be represented by

$$\gamma_{CMB} + p \rightarrow p + e^+ + e^- \tag{15.22}$$

above a proton energy threshold of about 5×10^{17} eV. Each interaction produces only a small amount of energy loss since the bulk of the momentum and energy in the rest frame of the CMB remains with the cosmic ray proton. Nevertheless, this effect may be important since it is the main source of energy loss before the onset of the GZK effect, discussed in the next section. Energetic electrons and positrons are also produced by high energy photons in the field of the CMB,

$$\gamma_{CMB} + \gamma \rightarrow e^+ + e^-. \tag{15.23}$$

15.2.8 GZK effect

The space through which cosmic rays propagate is filled abundantly with photons from the CMB. For highly relativistic cosmic ray particles, these photons appear as gamma rays in the rest frames of the particles. Protons above a threshold of 5×10^{19} eV have sufficient energy to produce pions, in what is known as the Greisen–Zatsepin–Kuzmin (GZK) effect,

$$\gamma_{CMB} + p \rightarrow p + \pi^0, \tag{15.24}$$

$$\gamma_{CMB} + p \rightarrow n + \pi^+, \tag{15.25}$$

$$\gamma_{CMB} + n \rightarrow p + \pi^-, \tag{15.26}$$

$$\gamma_{CMB} + n \rightarrow n + \pi^0. \tag{15.27}$$

Multiple pion production dominates at higher energies. Heavy nuclei usually photodisintegrate rather than produce pions. For example, an Fe nucleus near the GZK cutoff can undergo photodisintegration reactions with either CMB or infrared photons.

For protons of energy 10^{20} eV, the GZK mean free path is about 10 Mpc. For iron nuclei, the photodisintegration mean free path is of similar magnitude. After traversing several mean free paths, essentially no cosmic rays should remain above the GZK cutoff. Observations of such high energy cosmic rays, therefore, imply a source of production which, although extragalactic, is relatively nearby on cosmological scales.

15.2.9 Decays

Electrons, protons, and most atomic nuclei are stable components of cosmic rays. But the GZK mechanism produces neutrons and neutral and charged pions. Neutrons decay according to Equation 15.5, and pions decay primarily as

$$\pi^\pm \rightarrow \mu^\pm + \nu_\mu(\bar{\nu}_\mu), \tag{15.28}$$

$$\pi^0 \rightarrow \gamma + \gamma. \tag{15.29}$$

Muons decay into electrons (positrons) and muon and electron neutrinos (antineutrinos). Decays therefore must be considered as possible sources of cosmic ray protons and electrons.

15.2.10 Atmospheric interactions

Earth's atmosphere is opaque to primary cosmic rays. By the time vertically incident cosmic ray particles reach a residual pressure of 7 millibar (35 km altitude), they will have traversed a column density of order 7 g cm^{-2}. This is of order 1/10 of the mean free path of an energetic proton, so a fraction of the energetic primary cosmic rays will have undergone an interaction and created secondary particles. Primary cosmic rays may be studied at this altitude if sufficient attention is paid to corrections for secondary particle fluxes. The production of secondaries peaks around a column density of 55 g cm^{-2} (20 km altitude).

At a cosmic ray particle energy above a few GeV, a hadronic interaction with an atmospheric nucleus leads to an *extensive air shower* containing both a hadronic component (mesons and baryons) and an electromagnetic component (leptons and photons). The concept of an air shower was introduced in the previous chapter (Figure 14.2). The topic of extensive air showers is covered, below, in Section 15.5.

A penetrating component of secondary cosmic rays exists in the form of muons. The mean muon lifetime at rest is 2.2×10^{-6} s, so relativistic muons can easily survive to reach Earth's surface. Muons do not participate in the strong interaction, so they also survive passage through substantial amounts of matter and are detectable kilometers underground. The other penetrating component of secondary cosmic rays is neutrinos, discussed in Chapter 14 (where they are called *atmospheric neutrinos*).

15.3 Detectors

Cosmic ray detectors are based on one of severals means by which energetic charged particles interact with matter, as will be discussed in the following sections. Many of these interactions produce either high energy photons (ultraviolet, x-ray, or gamma ray) or byproducts such as ionization trails, which have been discussed in Chapter 10. Some of the technologies of charged particle detection, therefore, are similar to technologies employed in the detection of high energy photons.

In the discussion that follows we assume that cosmic ray nuclei remain fully stripped of their electrons. We also ignore nuclear reactions that may lead to neutron stripping. However, fragmentation of heavy nuclei is an important process when considering the development of extensive air showers.

15.3.1 Ionization detectors

One type of particle interaction is ionization. A charged particle of charge Z and mass M ($M \gg m_e$) passing through neutral material will experience Coulomb interactions with the electrons it encounters. Some of these interactions will lead to ionizations while others will lead to electronic excitations. Let I be the logarithmic mean excitation potential (ionization potential) of the electrons (Ahlen, 1980). The rate of energy loss by ionization is well approximated by the relativistic Bethe formula,

$$\frac{dE}{dx} = -\frac{4\pi}{m_e c^2} \frac{N Z^2}{\beta^2} \left(\frac{e^2}{4\pi \epsilon_0}\right)^2 \left[\ln \frac{2 m_e c^2 \beta^2}{(1 - \beta^2) I} - \beta^2\right], \qquad (15.30)$$

where $\beta = v/c$ and N is the number density of electrons in the material.

The rate of energy loss is high at low energies, decreasing to a minimum around a few times Mc^2, as shown for protons in Figure 15.5. More slowly moving particles remain near the electrons longer and are able to deliver larger kicks. The rate of energy loss then increases again for highly relativistic energies. For relativistic particles near the minimum in energy loss, the rate of energy loss is a sensitive function of the particle charge, Z. The ionizations are relatively easy to detect, and a trail of ionizations is visible in a variety of types of detectors. The direction of the trail indicates the propagation direction of the particle. Detectors sensitive to the rate of ionization are often called "dE/dx detectors." Ionization loss for electrons follows a somewhat different relationship due to, among other things, the different reduced mass of the collision. Broadly speaking, minimum ionization for electrons occurs around 1 MeV with a generally 1/E behavior at low energies. At high energies electrons lose energy primarily by bremsstrahlung.

Ionization may be detected by any of the classical methods such as cloud chambers, bubble chambers, and spark chambers, but most frequently scintillators and solid state detectors are used (see Chapter 11). Although scintillators are dE/dx detectors, their energy response is non-linear and may be different for particles of different mass and charge. Careful calibration is needed, for example, with an accelerator producing beams of known energy. Solid state detectors, in contrast,

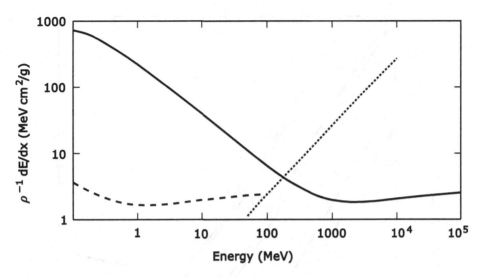

Figure 15.5 Ionization energy loss rates in air for protons (solid line) and electrons (dashed line) and bremsstrahlung energy loss rate in air for electrons (dotted line). Based on the NIST ESTAR and PSTAR databases.

have good linearity. As can be seen from Figure 15.6, accurate measurement of the rate of energy loss at minimum ionization is sufficient to determine atomic number (charge), and monitoring ionization over some distance as the particle loses energy is sufficient to determine the isotope and the energy. Mass resolution can be a small fraction of an a.m.u., as illustrated in Figure 15.7.

Ionizing particles leave permanent damage tracks in various solids, and the damage tracks can be revealed by chemical etching. Exploration of fossil tracks in meteorite grains has been used to study the history of cosmic rays throughout the life of the solar system. Modern detectors specifically designed for exploring cosmic ray irradiance are typically stacks of transparent dielectrics such as sheets of Lexan polycarbonate (Price & Fleisher, 1971). Polymers of this sort are particularly sensitive to radiation damage. Figure 15.8 schematically shows a detector stack and the nature of the etched regions. Empirically, the damage rate is found to be dependent on the ratio Z/β.

15.3.2 Bremsstrahlung

As we have seen, bremsstrahlung (braking radiation) is the result of interactions of an energetic charged particle with the Coulomb fields of nuclei it encounters. For an acceleration (deceleration) a, along the direction of particle motion, the power radiated by a particle of charge q is given by

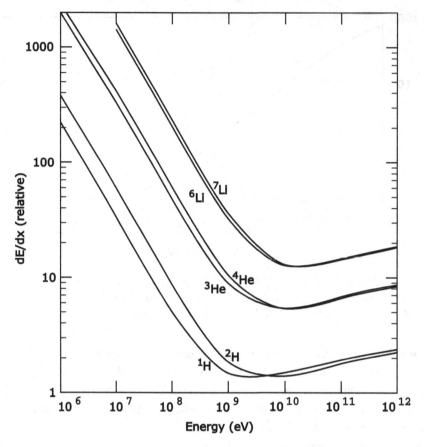

Figure 15.6 Relative rates of energy loss as a function of particle energy for ^{1}H, ^{2}H, ^{3}He, ^{4}He, ^{6}Li, and ^{7}Li nuclei estimated from the Bethe formula.

$$P = \frac{q^2 a^2 \gamma^6}{6\pi \epsilon_0 c^3}. \tag{15.31}$$

At fixed energy, electrons have higher Lorentz factors than protons by the ratio m_p/m_e. Since the dependence on the Lorentz factor is γ^6, it is clear that electrons will produce much greater losses due to bremsstrahlung than protons (for acceleration perpendicular to the direction of motion, the loss goes as γ^4). Bremsstrahlung is a relatively important process for electrons, as was shown in Figure 15.5. Heavier charged particle such as protons exhibit significant bremsstrahlung only at high energies.[2]

---

[2] Relativistic heavy charged particles lose energy to e^+e^- pair production at roughly the same rate as to bremsstrahlung.

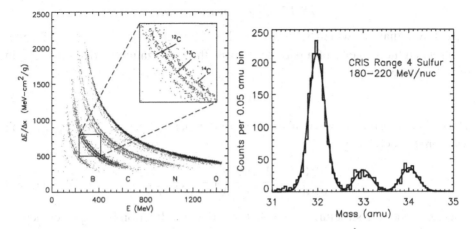

Figure 15.7 (Left) dE/dx versus E calibration data for the Solar Isotope Spectrometer (Stone *et al.*, 1998). (Right) Isotopic mass resolution for CRIS (Ogliore, 2007).

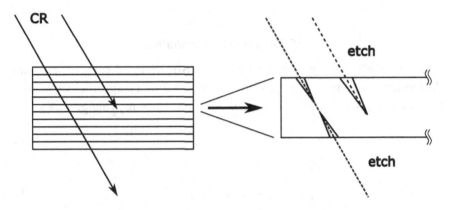

Figure 15.8 Ionization tracks in transparent polymer sheets are evident through the roughly conical cavities produced by etching.

To use bremsstrahlung as a cosmic ray diagnostic, one must detect the broadband radiation which is produced. For electrons of energies where bremsstrahlung is the dominant source of energy loss, the radiation is composed primarily of x-rays and gamma rays. Since x-rays and gamma rays passing through matter generally initiate electromagnetic cascades, the principal use of bremsstrahlung is in calorimetric detectors measuring the total rate of energy deposition. Bremsstrahlung is important in the development of the electromagnetic component of extensive air showers, described below in Section 15.5.

15.3.3 Čerenkov radiation

Čerenkov radiation, which was discussed in Chapter 14 (see Figure 14.7), will occur when the velocity of a particle exceeds the propagation velocity of light, in other words, when

$$\beta n > 1. \tag{15.32}$$

The classical result for the number of photons generated per unit distance per unit wavelength is given by

$$\frac{d^2N}{d\lambda\, dx} = \frac{2\pi\alpha Z^2}{\lambda^2}\left(1 - \frac{1}{n^2\beta^2}\right), \tag{15.33}$$

where α is the fine structure constant. The index of refraction is frequency dependent, and radiation is emitted only at frequencies for which the condition $n > \beta^{-1}$ is satisfied. Typically this will be a region somewhat below the plasma frequency for the material, and the light emitted will be blue and near-ultraviolet. Čerenkov light may be detected by photomultiplier tubes. The amount of radiation is proportional to Z^2. Energy loss to Čerenkov radiation is small.

15.3.4 Transition radiation

Transition radiation is generated when relativistic charged particles encounter a dielectric inhomogeneity. For the simplest case, that of a single dielectric interface between materials 1 and 2, the dielectric constants for x-rays are given by

$$\epsilon_1 = 1 - \frac{\omega_1^2}{\omega^2}, \tag{15.34}$$

$$\epsilon_2 = 1 - \frac{\omega_2^2}{\omega^2}, \tag{15.35}$$

where ω_1 and ω_2 are the plasma frequencies of the two materials. The total x-ray power radiated by a particle of charge Ze is given by

$$S = \frac{\alpha\hbar}{3} Z^2 \frac{(\omega_1 - \omega_2)^2}{\omega_1 + \omega_2}\, \gamma, \tag{15.36}$$

and is strongly peaked in the forward direction (Cherry *et al.*, 1974). If material 2 is vacuum, this simplifies to

$$S = \frac{\alpha\hbar}{3} Z^2 \omega_1 \gamma = \frac{Z^2 e^2}{3c}\omega_1\gamma. \tag{15.37}$$

An intuitive picture of the origin of transition radiation is that of the moving charge and its image charge on the other side of the dielectric boundary (Ginzburg &

Tsytovich, 1979). Transition radiation is very weak. For a particle of unit electric charge and $\gamma \approx 10^4$, this corresponds to a fraction of an x-ray photon. So multiple dielectric interfaces are required to give a significant signal. The usefulness of transition radiation devices is enhanced by the Z^2 dependence of their response to atomic nuclei of charge Ze (Wakely, 2002). The energy loss for charged particles with large Lorentz factors is negligible.

15.4 Balloon-borne and spacecraft missions

Cosmic ray experiments may be either balloon-borne or satellite based. Those carried by balloons suffer from the effects of the residual atmosphere, such as ionization loss. Satellites, however, tend to be in near-equatorial orbits and therefore experience stronger geomagnetic shielding than balloons at high geomagnetic latitudes. Both balloon and satellite missions are appropriate primarily for low to moderate energy cosmic rays, where fluxes are sufficient for detector areas of, at most, a few square meters.

15.4.1 1990s and early 2000s

The Advanced Composition Explorer satellite (ACE) was launched in 1997. Among other things, it carried the Cosmic Ray Isotope Spectrometer (CRIS), which contained silicon energy loss detectors followed by scintillating tracking detectors[3] and measured isotopic composition out to $Z = 30$ at energies of around 100 MeV/nucleon with an étendue of 0.25 m^2 sr.

Other missions from this period included the Balloon-borne Experiment with a Superconducting Spectrometer (BESS) and the Advanced Thin Ionization Calorimeter (ATIC). BESS used a superconducting solenoid with a magnetic field of 0.8 T and drift chambers to unambiguously measure the sign of particle charges. One focus of BESS was the study of low energy antiprotons. ATIC had silicon energy loss detectors, scintillating tracking detectors, and a BGO (BiGeO) calorimeter. It was used at energies of 10^{10}–10^{14} eV/nucleon.

15.4.2 TRACER and CREAM

The Transition Radiation Array for Cosmic Energetic Radiation (TRACER) and Cosmic Ray Energetics And Mass (CREAM) projects are long duration balloon-borne experiments designed to study charges and energies of cosmic ray nuclei around 10^{12}–10^{15} eV. The low flux of heavy nuclei at these energies requires large detector cross sections and long integration times. Long duration balloon flights are

---

[3] In cosmic ray work, particle tracking devices of this type are often called *hodoscopes*.

typically made at altitudes of 110 000 to 125 000 feet (about 35 km) and can last from weeks up to a few months. The residual atmosphere at such altitudes corresponds to mass column densities of order 4 g cm^{-2}. Such flights are typically made in the arctic or antarctic to take advantage of the low geomagnetic cutoffs and the mild circumpolar winds. Environmental factors can be severe for these latitudes, altitudes, and flight durations. Geopolitical factors make long antarctic flights more practical than long arctic flights, although the expense associated with working in the antarctic is greater. The US Columbia Scientific Balloon Facility (formerly the National Scientific Balloon Facility) supports flights from several locations worldwide, including McMurdo Station (Antarctica, 77.5° S), Esrange (Sweden, 68° N), and previously Fairbanks (Alaska, 65° N) and Lynn Lake (Manitoba, Canada, 57° N).

The TRACER instrument layout is shown in Figure 15.9. The rate of ionization loss in scintillators at the top and bottom of the instrument is proportional to Z^2, providing good determination of Z as well as providing a trigger for the instrument and a coincidence check. Proportional tube arrays consist of crossed grids of single wire proportional tubes over an area of 2 m × 2 m, with the overall instrument having an étendue of 5 m^2 sr. In all there are 16 layers of proportional tubes. The first four pairs measure the track position, the ionization loss rates, and give an approximate measure of the particle energy. The remaining four pairs are used to

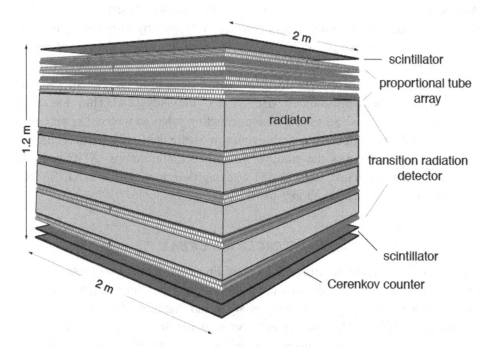

Figure 15.9 The TRACER instrument (Gahbauer *et al.*, 2004). The sections labelled "radiator" are battings of 2–5 μm polyolefin or 17 μm olefin fibers.

detect x-ray transition radiation. Transition radiation provides a determination of the Lorentz factor in the range $500 < \gamma < 10^4$. Transition radiation intensity scales as Z^2. A Čerenkov counter is used to reject non-relativistic particles.

CREAM (Ahn *et al.*, 2007) combines a transition radiation detector similar to that of TRACER with a calorimeter to allow observations of particles down to $Z = 1$ (that is, to include protons and helium nuclei). A disadvantage of the calorimeter is that it employs high-Z elements whose large cross sections create significant backscatter into the transition radiation detector. In part, the effects of this are minimized by providing a precise timing cutoff to exclude backscatter events. CREAM has a smaller étendue than TRACER. In addition to measuring nuclei down to hydrogen, it provides highly redundant measurements of the particle charge Z.

15.4.3 PAMELA

The Payload for Antimatter Matter Exploration and Light-nuclei Astrophysics (PAMELA) was launched in 2006 and is still in orbit aboard the Russian Resurs-DK1 satellite. Like BESS, one of the main aims of PAMELA is the study of antimatter. The instrument contains a magnetic spectrometer with silicon tracker, a tungsten/silicon imaging calorimeter, and a set of time of flight scintillators, with an étendue of 0.002 m² sr. It has a sensitivity of order 3×10^{-8} for the anti-helium/helium ratio and detects an antiproton/proton ratio of order 10^{-4} (energy dependent). For studying cosmic ray spectra, it is sensitive out to energies of about 3×10^{11} eV.

15.4.4 Alpha Magnetic Spectrometer

In May 2011 the Alpha Magnetic Spectrometer (AMS-02) was carried into orbit by the space shuttle Endeavour and installed on the International Space Station, where it will collect data for at least three years. Like BESS, AMS-02 has a magnet for antiparticle identification, but it has additional capabilities such as a transition radiation detector. AMS-02 is designed to measure the antihelium/helium flux ratio to a sensitivity of one part in 10^9. It also has capability of detecting certain types of exotic matter such as strangelets.

15.5 Extensive air showers

At high energies the flux of primary cosmic rays is low, of order 10^{-21} m^{-2} s^{-1} sr^{-1}GeV^{-1} at 10^{18} eV. This implies that a detector with an area of 1 m², a solid angle of 1 sr, and a bandwidth of 10^{18} eV will see a flux of much less than one event per year. This is clearly impractical for the types of detectors discussed so far. Detectors with areas of order kilometers squared are needed to get appreciable flux, necessitating use of the atmosphere as the primary detector medium. Fortunately, interactions of high energy primary cosmic rays with Earth's atmosphere leave several observable signatures.

When an energetic cosmic ray interacts in the atmosphere, it initiates a shower containing a baryonic component and pions. Neutral pions decay by emitting gamma rays, initiating an electromagnetic cascade of photons, electrons, and positrons. Gamma rays interact with the atmosphere (with a mean free path of order 50 g cm^{-2}) and produce e^+e^- pairs. The electrons and positrons (with radiation lengths $X_0 \approx 37$ g cm^{-2} in air) emit photons via bremsstrahlung to continue the cascade. Charged pions interact with the atmosphere producing more pions, muons, and electrons. The baryonic component has further collisions, creating more pions. The baryonic components and the charged pions are often called the hadronic shower. Eventually a large fraction of the initial energy is deposited in the electromagnetic cascade.

One signature of an energetic cosmic ray is the existence of a hadronic shower. If the energy came entirely from an initial high energy gamma ray, the shower produced would be almost entirely electromagnetic. Muons are major products of a hadronic shower. The ratio of muons to electrons is one measure of the strength of the hadronic shower and an indication of the composition of the cosmic ray primary. At fixed energy, heavier nuclei produce more muons, and protons produce fewer muons.

The slant depth of maximum charged particle density in a cosmic ray induced shower, illustrated in Figure 15.10, is referred to as X_{max}, a column density measured in units of g cm^{-2}. A shower reaches its X_{max} when the initial particle energy has been subdivided sufficiently by the cascade that the remaining particles lack the energy to continue the cascade process. The depth into the atmosphere of the shower maximum is a measure of the primary particle energy. For a particular chemical species (e.g. protons) the rate of change of X_{max} with particle energy is known as the elongation rate,

$$ER = \frac{dX_{max}}{d(\log E)}. \tag{15.38}$$

Lighter elements penetrate deeper than heavier elements of the same energy. Lighter elements also exhibit larger fluctuations in X_{max} from event to event, at fixed energy, since heavier element showers are the result of averaging over the nucleonic components.

Arrival times at various stations of a ground array can be used to determine the curvature of the shower front, thereby indicating the altitude at which the

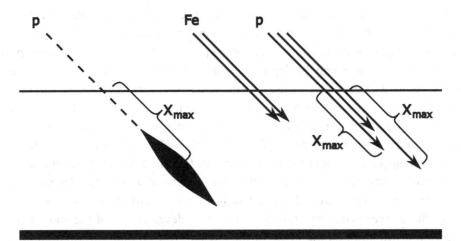

Figure 15.10 X_{max} is the slant depth into the atmosphere at which an extensive air shower has the maximum number of charged particles. The width shown indicates the number of particles, not the lateral spread of the shower. Protons penetrate deeper into the atmosphere and show a wider variation of X_{max} than iron nuclei.

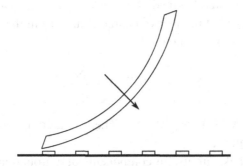

Figure 15.11 The arrival time of air shower particles at ground array stations can be used to determine direction and curvature (and therefore height and X_{max}).

shower originated, as shown in Figure 15.11. Direction of propagation can also be determined. The lateral spread of the shower is a measure of the primary particle energy.

The signature of a heavy nucleus (instead of a proton of the same energy) is a cascade beginning higher in the atmosphere and developing more rapidly. For a surface array these characteristics can be seen as a shower front with a large radius of curvature and a relatively small thickness. The number density of electrons some distance from the shower core gives an indication of the energy of the primary since the more distant shower particles were generated early in the shower development.

Energy deposited in the atmosphere will excite nitrogen molecules, giving rise
to fluorescence. On average about five photons are emitted in the strong 2P(0, 0)
band at 337 nm for every MeV of energy deposition. Since most of the energy of
the primary will ultimately be deposited in this form, measurement of fluorescence
is essentially a calorimetric measurement. The fluorescent light is primarily in the
300–400 nm wavelength range, and detectors are filtered accordingly. The fluorescence yield over this range is about 18 photons per MeV at sea level, increasing
rapidly with altitude (Arqueros *et al.*, 2008). Scattered Čerenkov radiation provides
a small competing signal at these wavelengths, for which the fluorescent light signal must be corrected. Fluorescence detectors consist of wide field telescopes with
cameras containing arrays of photomultiplier tubes. With these, one can reconstruct the shower geometry and the longitudinal development of the shower from
the light distribution on the plane of the sky and from timing information on the
shower development. Stereoscopic viewing from separate ground stations allows
even tighter measures of the characteristics of the shower.

Secondary particles produced by the interaction of an energetic primary will
be relativistic, producing Čerenkov radiation even at low atmospheric densities.[4]
Showers of secondary particles will contain many particles which will survive and
reach the ground. The lateral spread of secondary particles is an important shower
diagnostic. On average the number of charged particles in the shower is related to
the energy of the primary cosmic ray by

$$E \approx 4 \times 10^{15} \, eV \left(\frac{N}{10^6} \right)^{0.9} \tag{15.39}$$

for vertical showers with $E \approx 10^{15}$ eV. There is a complicated dependence on
particle energy and altitude.

Water Čerenkov tanks are useful components of a ground array to measure the
distribution and flux of muons. Muons are penetrating particles which will not be
significantly attenuated by the residual atmosphere above the detectors. Muons are
broadly distributed in air showers and will form the dominant particle flux away
from the shower core. Shielding is necessary to block the electrons and positrons
in the electromagnetic cascade without blocking the penetrating muon component.
Unshielded plastic scintillators are sensitive to electrons and positrons; they also
detect photons, with lower sensitivity.

Ground-based calorimeters are used to detect energetic hadrons. Absorption
lengths of order 1000 g cm^{-2} are required to reliably absorb the particle energy.
The simplest method of doing this is with large area scintillators, although any

---

[4] A distinction must be made between air Čerenkov detection (discussed here) and water Čerenkov detectors
(which may form part of a ground array).

dE/dx ionization detector can be used. Shielding must be used to block the electromagnetic part of the air shower.

To be an effective particle detector, an air shower experiment must be able to discriminate against showers initiated by high energy gamma rays. Such showers are almost entirely electromagnetic, and the shower energy is divided fairly evenly between the multiple electrons and positrons produced. These relativistic particles produce Čerenkov light, and the smooth distribution of Čerenkov light is a signature of a gamma ray event. Hadronic showers, in contrast, maintain a large amount of energy in individual hadrons, giving a Čerenkov image dominated by distinct sub-showers. Methods for discriminating between hadronic and gamma ray showers are discussed by Hillas (1996). Above 10^{19} eV photon interactions are suppressed by the Landau–Pomeranchuk–Migdal (LPM) effect, and photons may penetrate deeply into the atmosphere before developing showers.

Gamma rays above 10^{19} eV also have the possibility of pair production through interaction with Earth's magnetic field. In that case an e^+e^- pair is created well above the atmosphere, where the particles then emit synchrotron radiation. The initial energy is subdivided between a multitude of photons, electrons, and positrons even before the shower reaches the atmosphere. Such events would appear with a highly anisotropic distribution.

In principle, ultrahigh energy neutrinos are capable of initiating extensive air showers. Limits on the rate of neutrino-induced events can be set by searching for shallow, upward-going tracks originating from electron neutrino interactions, as described by Abbasi *et al.* (2008b).

All extensive air shower detectors are complicated instruments requiring extensive calibration and Monte Carlo simulations of shower development and detector characteristics. This is particularly true of ground array detectors.

15.5.1 High Resolution Fly's Eye

The High Resolution Fly's Eye array (HiRes) is an air fluorescence detector in Utah, which operated from 1999 through 2006 as a successor to the original Fly's Eye. There were 64 light collecting elements divided between two sites 12.6 km apart, allowing for stereoscopic (binocular) views of air shower events. One site had 22 telescopes in a single ring, observing 320° in azimuth at low elevation (3–17°) in order to optimize sensitivity for the highest energy events. The other site had 42 telescopes in two rings, one for low elevation and one for elevations 17–31°. The telescopes had effective collecting areas of 3.75 m^2 and hexagonally packed focal planes each with 256 photomultipliers.[5] Each photomultiplier covered about

---

[5] The assemblies of photomultipliers have an appearance like that of the compound eyes found in insects, hence the name Fly's Eye.

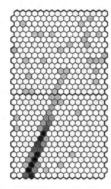

Figure 15.12 The fluorescence light from a nearby high energy HiRes air shower event extending over the focal planes of two adjacent telescopes (Boyer *et al.*, 2002).

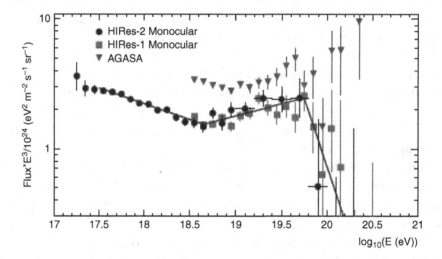

Figure 15.13 Confirmation of the GZK cutoff by HiRes (Abbasi *et al.*, 2008a).

1 square degree of sky. An example of an event with an estimated energy above 10^{20} eV is shown in Figure 15.12.

The HiRes results confirmed the existence of the GZK cutoff at an energy of about 6×10^{19} eV (Abbasi *et al.*, 2008a), as shown in Figure 15.13, contrary to results from the AGASA experiment (Akeno Giant Air Shower Array). Earlier indications of possible cosmic ray anisotropy associated with active galactic nuclei (AGNs) are not confirmed by the HiRes data (Matthews, 2010). The most recent analysis by Abbasi *et al.* (2010b) indicates that the distribution of UHECRs with HiRes is consistent with isotropy, unlike the claimed correlation with BL Lac objects by Gorbunov *et al.* (2004). A statistical analysis of HiRes results appears to show that the highest energy cosmic ray events result from protons (Abbasi *et al.*, 2005, 2010a; Matthews, 2010), contrary to the Pierre Auger Observatory results discussed below.

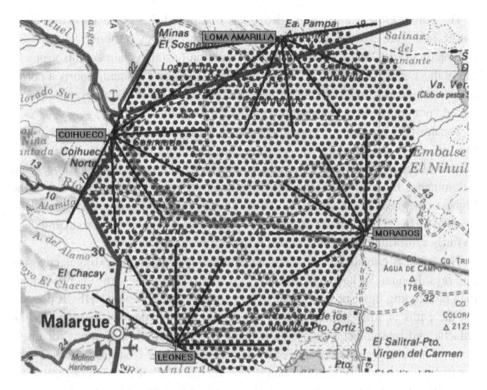

Figure 15.14 Map of the Pierre Auger Observatory in Argentina. Dots (red)
represent surface array detectors. Lines (green) delimit the fields of view of
the fluorescence detectors. Courtesy of Auger Observatory. For color version of
figure, see plate section.

15.5.2 Pierre Auger Observatory

The surface detector of the Pierre Auger Observatory in Argentina consists of 1640
water Čerenkov tanks spread out over 3000 km^2. A map is shown in Figure 15.14.
The fluorescence detector consists of six inward-looking Schmidt telescopes at
each of four sites around the perimeter of the Čerenkov field. The telescopes cover
$1.5-30°$ of elevation. By itself the surface detector can estimate the primary particle trajectory and X$_{max}$ from the arrival times at different detectors. The distribution
of shower particles can be used to estimate the composition and energy of the
primary cosmic ray. Whereas the surface array operates nearly full time, the fluorescence detectors are limited to use under ideal lighting conditions (nighttime, not
near full moon, clear skies). The best information is obtained in "hybrid" mode,
when both the surface array and the fluorescence detectors are triggered. Abraham
et al. (2010) discuss data based on 3754 events above 10^{18} eV which are detected
in hybrid mode and survive a variety of cuts relating to data quality and event
geometry.

The Auger Observatory, like HiRes, has confirmed the existence of the GZK cutoff, and their team of researchers report a cutoff energy of about 4×10^{19} eV (Abraham *et al.*, 2008a). The Auger collaboration has also claimed anisotropy in the highest energy cosmic rays at a confidence level of 99% and reports a correlation with nearby AGNs (Abraham *et al.*, 2008b). As discussed above, the HiRes data do not appear to be consistent with the Auger claim.

The most recent Auger results suggest that the highest energy cosmic rays, above the "ankle" in the cosmic ray spectrum, may be from heavy nuclei such as Fe instead of protons. In other words, the composition of cosmic rays may vary from a dominance by protons at lower energies to a dominance by heavy nuclei at the highest energies (Abraham *et al.*, 2010). This is directly opposite the conclusion reached by the HiRes collaboration. The Auger Observatory is still producing data, so further information on this issue may be forthcoming.

The ankle is considered to be the transition from galactic to extragalactic cosmic rays. Iron nuclei have lower rigidities than protons of the same energy. Rather than proceeding more or less directly from source to observer as UHECR protons do, the paths of fully ionized iron nuclei would be considerably distorted by the galactic and intergalactic fields. So it makes a considerable difference for questions of anisotropy whether the UHECRs are protons or iron-group nuclei. The situation regarding survivability would also change. Rather than being limited by the GZK effect, the range of iron UHECRs would be limited by photofragmentation. It is possible that an upper energy cutoff of UHECRs is due to a limitation on the acceleration mechanism instead of limits on propagation. It is unclear whether a mechanism for accelerating Fe nuclei to the highest energies would be able to do so while leaving the nuclei intact. As discussed above, the HiRes data are interpreted as favoring protons for the highest energy events.

The Auger Observatory has two low energy enhancement projects to allow better exploration of the ankle region. The HEAT (High Elevation Auger Telescopes) project adds high elevation capability ($30-60°$) to what was essentially a low elevation fluorescence system. The AMIGA (Auger Muons and Infill for Ground Array) project, as the name suggests, would fill in a part of the ground array with 66 water Čerenkov tanks and buried muon scintillation counters at a combination of 750 m and 433 m separations. The construction of a larger, northern hemisphere version of the Auger Observatory will be a path toward improved high energy performance as well as full sky coverage. The proposed Auger North would have a coverage of $20\,000$ km^2, nearly an order of magnitude increase over Auger South.

15.5.3 *Telescope Array (TA) project*

The Telescope Array is an air shower detector near Hinckley, Utah. It consists of a ground array of 576 scintillation detectors with 1.2 km spacing and three fluorescence sites set in a triangle around the ground array. The fluorescence sites have on average about 12 telescopes each, covering $3-33°$ in elevation and about $108°$ in azimuth.

The TALE (Telescope Array Low Energy) extension of the Telescope Array contains additional "infill" scintillator stations for the ground array since lower energy cosmic rays produce narrower showers. The spacing of these stations is 500 m. There are also towers with fluorescence telescopes covering nearly the entire range from zenith to horizon.

15.5.4 *Atmospheric Čerenkov Telescope Array*

A priority of the 2010 Astronomy Decadal Survey report is ACTA (Atmospheric Čerenkov Telescope Array). It is anticipated that this would involve US scientists joining the European CTA (Čerenkov Telescope Array) project. ACTA is viewed primarily as a TeV (10^{15} eV) gamma ray telescope in place of the proposed US Advanced Gamma-ray Imaging System (AGIS). Current instruments of this sort include the High Energy Stereoscopic System (HESS), Major Atmospheric Gamma Imaging Čerenkov telescope (MAGIC), the Very Energetic Radiation Imaging Telescope Array System (VERITAS), and the Collaboration of Australia and Nippon for a GAmma-Ray Observatory in the Outback (CANGAROO).

A key requirement to use a Čerenkov array for detecting gamma rays is the ability to reject cosmic ray showers based on the existence of hadronic fragments and sub-showers generated away from the shower core (Hillas, 1996). If the cosmic ray primary is a heavy nucleus, one may also hope to detect the Z^2-enhanced Čerenkov light from the primary cosmic ray before its first hadronic interaction in the atmosphere. Atmospheric Čerenkov arrays are not directly useful for studying showers induced by rare energetic cosmic rays since the shower cores would need to arrive close to the positions of the Čerenkov telescopes. Although the science goals are different, both the gamma ray and cosmic ray efforts will benefit from similar technologies and improved understanding of the development of air showers.

15.5.5 *JEM-EUSO*

The Japanese Experiment Module–Extreme Universe Space Observatory (JEM-EUSO) is a project to monitor the isotropic fluorescence light from cosmic rays from above, on the International Space Station. From this vantage point it will be possible to monitor extensive air showers over a larger atmospheric volume than existing and proposed ground-based systems. This would improve statistical accuracy for the rarest, highest energy cosmic ray events. S-EUSO (Super-Extreme

Universe Space Observatory) is a proposed, enlarged, free-flying version of EUSO.

15.6 Particle acceleration

Since the method of producing UHECRs is a central question in cosmic ray physics, it is worth examining the theoretical possibilities for accelerating charged particles to ultrahigh energies. One possibility is the first order Fermi mechanism, in which particles gain energy by repeatedly crossing a shock front. This mechanism is not rapid, except in highly relativistic shocks, since it is diffusive in nature and requires multiple shock crossings. Millions of years are required for shocks traveling at velocities of 0.3c to accelerate protons to 10^{20} eV in magnetic fields of order 10 μG. Not only must the shocks persist for this time scale, but the shock regions must be large enough so that the particles do not diffuse away before they are sufficiently accelerated. This requires regions of size

$$L > \frac{2\,E}{B\,\beta\,Z}\,\frac{pc\,\mu G}{10^{15}\,eV} \tag{15.40}$$

(Hillas, 1984). The requirement is somewhat milder for heavy nuclei than for protons, due to the dependence on the nuclear charge Z, but it is still difficult to find galactic sources satisfying this requirement for E = 10^{20} eV. The "knee" feature in the cosmic ray spectrum is interpreted as an indication that this limit is being reached at energies of only 3×10^{15} eV.

Shocks are present in galactic supernova remnants, but they fail to meet the above size requirement by two or three orders of magnitude. Other possibilities exist such as unipolar induction by the rotating magnetic dipole fields of pulsars, which have strong magnetic fields but are much smaller. Any energy imparting mechanism must also compete with energy loss mechanisms, which for pulsars would include synchrotron radiation. Neutron stars may be able to accelerate protons to energies of about 10^{13} eV, but not much higher. Somewhat more plausible are AGNs, gamma ray bursts (GRBs), and radio galaxies. This is consistent with the interpretation of the "ankle" in the spectrum as representing a transition to extragalactic origins for cosmic rays.

It is also possible that there are some yet unknown ultramassive particles whose decays produce energetic cosmic rays. This is usually known as the "top-down" scenario (as opposed to "bottom-up" scenarios involving successive stages of acceleration). It has been hypothesized that Grand Unification will be associated with particles at masses around 10^{24} eV, whose decay would lead to UHECRs. Another possibility would be generating UHECRs from ultrahigh energy neutrinos, although it is unclear how such neutrinos could be generated. Chang *et al.* (2009) have recently proposed a novel "bottom-up" acceleration scenario.

Exercises

15.1 For a cosmic ray incident at $60°$ from the zenith, what is X_0 for an air shower which reaches maximum at an elevation of 10 km? Assume an isothermal atmosphere with a scale height of 7 km and a sea level pressure of 1033 g cm^{-2}. What changes if the atmosphere has a lapse rate of 10 K km^{-1}?

15.2 Give a simple, classical, non-relativistic explanation for the isotope shift of the dE/dx vs. E ionization energy loss curves of Figure 15.6. Concentrate on comparing ^{1}H and ^{2}H and make your explanation both qualitative and quantitative.

15.3 Approximate Earth's magnetic field by a pure dipole at Earth's center. Consider charged particles orbiting Earth in the geomagnetic equatorial plane. Positive particles can circulate clockwise (east to west) in circular orbits slightly larger than Earth if they have rigidities slightly greater than $R_\oplus B_\oplus$. Such orbits are closed, therefore they *cannot* correspond to trajectories of cosmic rays (they do not connect to infinity). Show that particles arriving horizontally from the east can correspond to cosmic rays if they have rigidities larger than the above value but not if they have lower rigidities. Regions of disallowed cosmic ray trajectories are known as Störmer cones.

16

Gravitational waves

Few physicists today doubt the reality of gravitational radiation. The existence of gravitational waves is a firm prediction of the theory of general relativity, and gravitational radiation is observed indirectly through the energy loss and decreasing orbital period of the Hulse–Taylor binary pulsar system PSR J1915+1606 (Hulse & Taylor, 1975). Observing gravitational radiation directly would certainly be a further confirmation of general relativity. But more importantly, observations of gravitational radiation will enable us to determine properties of regions and of times for which we otherwise have little information.

The pioneering work of Joseph Weber using resonant bar detectors (Weber, 1966) is of great importance in this field and has inspired much of the observational work that followed. Most books and reviews on gravitational wave detection begin with a discussion of Weber bars. In this chapter we concentrate on interferometric detectors, which seem at this time to hold the greatest promise for detecting gravitational waves.

16.1 Characteristics of gravitational radiation

Gravitation is described in the field equations of general relativity as curvature of space-time.[1] In quantum field theory the mediator of the gravitational interaction is thought to be the massless, spin-2 graviton. In the classical limit these formulations are equivalent. Both predict the existence of gravitational radiation which is quadrupolar in nature.

In a linearized theory, gravitational waves in free space are described as small perturbations $h_{\mu\nu}$ on the Minkowski flat space-time metric.[2] The perturbation

---

[1] Space limitations force us to gloss over many of the subtleties of general relativity.

[2] In constructing 4-vectors we let $x^0 = ct$ and employ the Einstein summation convention. Greek indices run from 0 to 3 and Latin indices from 1 to 3. Most intermediate or advanced books on electrodynamics or relativity contain a description of 4-vector notation.

tensor is symmetric, and in the transverse traceless (TT) gauge,[3] may be expressed as

$$\mathbf{h} = \begin{pmatrix} 0 & 0 & 0 & 0 \\ 0 & h_{xx} & h_{xy} & h_{xz} \\ 0 & h_{yx} & h_{yy} & h_{yz} \\ 0 & h_{zx} & h_{zy} & h_{zz} \end{pmatrix}, \tag{16.1}$$

where the trace

$$h_{ii} = h_{xx} + h_{yy} + h_{zz} = 0, \tag{16.2}$$

and

$$h_{ji} = h_{ij}. \tag{16.3}$$

Such perturbations satisfy the wave equation

$$\Box^2 h_{\mu\nu} = 0, \tag{16.4}$$

where the d'Alembertian[4] is defined as

$$\Box^2 = -\frac{1}{c^2}\frac{\partial^2}{\partial t^2} + \nabla^2 = \frac{\partial}{\partial x_\nu}\frac{\partial}{\partial x^\nu}. \tag{16.5}$$

For propagation in the z-direction, there are two transverse, linearly polarized, plane wave solutions to the wave equation, corresponding to the "+" and "×" polarization states shown in Figure 16.1. These waves propagate at the speed of light and are of the form

$$h_{xx} = -h_{yy} = h_+ = A^+ e^{i(kz - \omega t)}, \tag{16.6}$$

$$h_{xy} = h_{yx} = h_\times = A^\times e^{i(kz - \omega t)}, \tag{16.7}$$

where A^+ and $A^\times$ are complex amplitudes of the two polarization states (Misner *et al.*, 1973; Blair, 1991). The quantity h is dimensionless and is interpreted as a mechanical strain, a change in distance divided by distance. Indeed, a pair of free masses separated by a distance x along the x-axis will experience a variation in separation δx in response to a gravitational wave traveling along the z-axis, where

$$\frac{\delta x}{x} = \frac{1}{2}h_+ \tag{16.8}$$

or more generally

$$\delta x = \frac{1}{2}h_+ x + \frac{1}{2}h_\times y, \tag{16.9}$$

[3.] In electromagnetism the choice of the Lorentz gauge simplifies certain equations involving the 4-vector potential A_μ. Similarly in gravitational theory the choice of the TT gauge simplifies certain equations involving $h_{\mu\nu}$.
[4] In general relativity the d'Alembertian is most often written as $\Box$ instead of $\Box^2$.

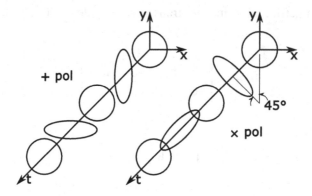

Figure 16.1 The "+" and "×" linear polarization states of gravitational waves traveling along the z-axis are illustrated by their effects on a ring of low mass test particles in the linearized weak field limit. Displacements have been exaggerated.

$$\delta y = -\frac{1}{2} h_+ y + \frac{1}{2} h_\times x. \tag{16.10}$$

The energy density in gravitational waves is

$$U_G = \frac{1}{32\pi} \frac{c^2}{G} \omega^2 \left(h_+^2 + h_\times^2 \right), \tag{16.11}$$

where G is the gravitational constant and ω is the angular frequency of the radiation.

16.2 Sources of gravitational waves

Including the source term, the wave equation for the gravitational wave amplitude is

$$\Box^2 h_{\mu\nu} = -\frac{16\pi G}{c^4} T_{\mu\nu}, \tag{16.12}$$

where $T_{\mu\nu}$ is the stress–energy tensor. This is difficult to solve for the most general case, so initially we will consider only compact sources in the weak field limit.

Gravitational waves are emitted by systems with changing mass quadrupole moment tensors. Dipole gravitational radiation is prohibited by the conservation of linear and angular momentum. The quadrupole moment[5] of a mass density distribution ρ is

$$Q_{ij} = \int \rho \left(x_i x_j - \frac{r^2}{3} \delta_{ij} \right) d^3 x. \tag{16.13}$$

[5] This definition of a quadrupole moment is 1/3 of the definition normally used in electromagnetism.

In the weak field, low velocity limit, a system radiates gravitational waves with a luminosity

$$L_G = \frac{1}{5} \frac{G}{c^5} \sum_{ij} \left\langle \left(\dddot{Q}_{ij} \right)^2 \right\rangle, \tag{16.14}$$

where a dot implies a time derivative. Many sources of interest, those generating strong gravitational waves, will not be in this weak field limit, and fully relativistic luminosity calculations are needed.

Let us consider as an example a binary star system with masses $m_1 = m_2 = m$ and separation a. Taking circular orbits and assuming the Keplerian value for the angular frequency

$$\omega_0 = \frac{2\pi}{P} = \sqrt{\frac{G(m_1 + m_2)}{a^3}}, \tag{16.15}$$

the mass quadrupole moment would be

$$Q_{ij} = \begin{pmatrix} \cos^2 \omega_0 t - 1/3 & \sin \omega_0 t \cos \omega_0 t - 1/3 & 0 \\ \sin \omega_0 t \cos \omega_0 t - 1/3 & \sin^2 \omega_0 t - 1/3 & 0 \\ 0 & 0 & -1/3 \end{pmatrix} \frac{m}{2} a^2. \tag{16.16}$$

Differentiating the mass quadrupole moment thrice with respect to time gives

$$\dddot{Q}_{ij} = \begin{pmatrix} 4 \sin 2\omega_0 t & -4 \cos 2\omega_0 t & 0 \\ -4 \cos 2\omega_0 t & -4 \sin 2\omega_0 t & 0 \\ 0 & 0 & 0 \end{pmatrix} \left(\frac{2Gm}{a^3} \right)^{3/2} \frac{m}{2} a^2. \tag{16.17}$$

The frequency of the radiation will be at $2\omega_0$, twice the orbital frequency. This system's luminosity in gravitational waves is

$$L_G = \frac{64}{5} \frac{G^4 m^5}{c^5 a^5}. \tag{16.18}$$

A more complete derivation of this result including polarization and angular dependences requires the use of tensor spherical harmonics (Mathews, 1962; Peters & Mathews, 1963). For two 1 $M_\odot$ stars at a separation of 1 AU this corresponds to a luminosity of about 4.3×10^{13} W.[6] That may sound like a lot, but consider that it is only a small fraction of a solar luminosity, indeed even a small fraction of the luminosity of the faintest known white dwarfs. It is comparable to the internal heat loss from Earth (which is mostly from radioactive decay). So gravitational radiation is weak. The radiation pattern in the above example is different for the two

---

[6] In contrast, the binary pulsar PSR J1915+1606 radiates about 7×10^{24} W. At a distance of 6400 pc, this implies a total strain of $h \approx 10^{-22}$ at a frequency of 7×10^{-5} Hz and higher harmonics.

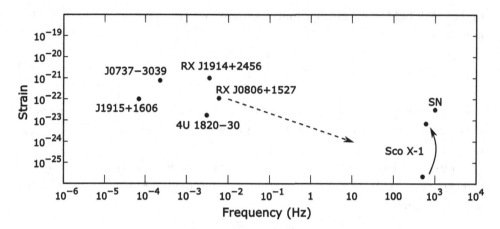

Figure 16.2 Estimated frequencies and instantaneous strains for possible sources of gravitational radiation. Sources shown include the binary neutron star systems J1915+1606 and J0737–3039, the white dwarf binaries RX J0806+1527 and RX J1914+2456, and the neutron star–white dwarf system 4U 1820–30. The dashed line shows the evolution of binary systems due to energy loss by gravitational radiation. Sco X-1 is a low mass x-ray binary with a deformed spinning neutron star (data from Bildsten (1998) and implied sensitivity gain for 20 days integration by Cutler & Thorne (2002)). The supernova estimate is for an energy conversion efficiency of 10^{-6} and a distance of 10 Mpc (Thorne, 1987). Authors of diagrams of this sort often plot an "effective detectable strain," making allowance for such things as integration times and geometrical effects. Such diagrams are easily misinterpreted.

polarizations, but in both cases is greatest along the axis of rotation.[7] The strain from such a source at a distance of 1 kpc would be of order h $\approx 10^{-24}$.

Some likely astrophysical sources of gravitational radiation are shown in Figure 16.2. Sources can be characterized by the nature of the expected waveform into four categories: chirp,[8] periodic, burst, and stochastic. Observationally it is necessary to understand the characteristic frequencies of gravitational waves, the expected strains produced at Earth, and the rates at which typical events occur.

An example of a chirp waveform is shown in Figure 16.3. Such a waveform would be expected from a coalescing binary system of two compact objects, such as a neutron star–neutron star binary. The waveform would initially be nearly periodic. As energy is lost from the system through gravitational radiation, the orbit becomes tighter and the frequency increases. The amplitude also increases. Typical

---

[7] Sometimes it is more convenient to calculate the strain using the formula $h_{ij}(t) = 2G\ddot{Q}_{ij}(t - r/c)/rc^4$, where r is the distance to the source and the quadrupole moment has been projected onto the transverse traceless gauge of the outgoing radiation.

[8] Chirp waveforms are quasi-periodic but with frequencies which evolve with time. Such signals arise in non-linear optics, in radio frequency signal processing, and in bird calls.

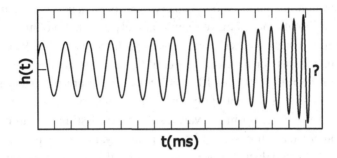

Figure 16.3 Temporal evolution of the observable strain from a neutron star–neutron star coalescence.

frequencies are of order 100 Hz and durations of order 10 s. Eventually the neutron stars would coalesce, possibly into a black hole. The details of the final coalescence and ringdown may be complicated. The final stages also will take place under conditions of strong gravitational fields and high velocities, requiring calculations beyond the linearized theory presented above. Events producing strains at Earth of order 10^{-22}–10^{-21} may occur at a rate of a few per year from sources as distant as 200 Mpc. The inspiral of compact objects onto massive black holes can produce rich waveforms due to frame dragging (Hughes, 2001).

Supernovae, the prototype of burst sources, certainly have enough energy to create strong gravitational waves. But any resulting gravitational waves will depend on the amount of asymmetry present. Absent any initial rotation, isolated type II core-collapse supernovae may have a high degree of spherical symmetry. On the other hand, evidence from double neutron star systems and isolated neutron stars indicates that kick velocities of several hundred kilometers per second were imparted during some supernova explosions, implying high asymmetry (Fryer & Kalogera, 1997). Convective instability or other instabilities in proto-neutron stars may provide significant asymmetry. Type Ia supernovae, in contrast, are likely candidates for gravitational wave generation since they start from a stage with significant quadrupole asymmetry (Falta *et al.*, 2010). The scenarios for generating gravitational waves are varied, as are the predicted waveforms. In all cases the gravitational signal lasts only a fraction of a second. If type II supernovae are sufficiently asymmetric, events producing strains at Earth of order 10^{-22} may be observable once per year out to the Virgo cluster. Supernova signals are likely to be in the kilohertz range.

Nearly periodic signals could be produced by non-axisymmetric neutron stars. Neutron stars rotate. Since pulsars radiate due to non-axisymmetric magnetic fields, some degree of departure from axial symmetry is certainly present. Rapidly spinning (millisecond) pulsars with ellipticities of order $\epsilon > 10^{-7}$ could be detectable throughout much of our Galaxy. The persistence and nearly periodic nature of these signals should make them relatively easy to recognize. Frequencies may be of order 10 Hz – 1 kHz and strains as large as 10^{-25}. Binary star systems with white dwarfs or neutron stars could produce strains as large as 10^{-21} at frequencies in the range $10^{-4} - 10^{-1}$ Hz. Since these are persistent signals, the concept of event rates does not apply. Instead, the question is how many of these sources exist at the present time at various distances. Estimates can be made, but the population of such sources is not known precisely. However, it seems likely that below about 10^{-3} Hz, galactic binary systems are sufficiently numerous to produce a confusion-limited background of gravitational waves.

The Big Bang is one probable source of stochastic gravitational radiation. It is very hard to predict the characteristics of such radiation since much of it may have been produced in the very early states, about which little is known. An energy density in gravitational waves of order 10^{-8} of closure density or greater would probably be necessary for detectability. The multitude of galactic binary star systems will also produce a stochastic background of weak, overlapping, low frequency signals. Estimates are that signals will be present in the 10^{-6}–10^{-3} Hz range at strains of order 10^{-20}.

16.3 Ground-based interferometric detectors

The conceptual design of most current ground-based gravitational wave detectors begins with the Michelson interferometer, introduced in Chapter 9. A list of interferometers and their locations is given in Table 16.1. Aerial views of some of

Table 16.1. *Gravitational wave interferometers*

Name	Arm length[a] (meters)	Latitude[b]	Longitude[b]	Az1[c] (deg)	Az2[c] (deg)
LIGO (Hanford)	4000 2000	46°27′19″ N	119°24′27″ W	234.00	324.00
LIGO (Livingston)	4000	30°33′46″ N	90°46′27″ W	162.28	252.28
Virgo	3000	43°37′53″ N	10°30′17″ E	19.4	289.4
GEO 600	600	52°14′42″ N	9°48′26″ E	68.388[d]	334.057[d]
TAMA 300	300	35°40′36″ N	139°32′10″ E	180	270
CLIO	100	36°25′28″ N	137°18′30″ E	180	270
AIGO ?	4000	31°21′28″ S	115°42′50″ E	90	180

[a] Nominal.
[b] Coordinates of apex (beamsplitter).
[c] Azimuth angle of vector from apex to end mirror, measured east of north.
[d] Unique in not being a nominal 90° apart.

these interferometers are shown in Figure 16.4. At the current stage of development there is a high degree of cooperation between these various projects, with ideas and designs being shared worldwide. We will concentrate on instruments with kilometer or longer arms (LIGO and Virgo). The LIGO (Laser Interferometer Gravitational-wave Observatory) experiment consists of three interferometers. Interferometers with 4 km arms are located in Livingston parish, Louisiana, and at the Hanford Nuclear Reservation in Washington. The vertices of the Hanford and Livingston sites are separated by 3002 km (on a straight line path through the Earth). The southwest arms at the two LIGO sites are roughly parallel and the northwest arm at Hanford and the southeast arm at Livingston are roughly antiparallel, making these interferometers sensitive, more or less, to the same regions in the sky and the same polarizations. The Hanford site also houses a 2 km interferometer oriented parallel to the 4 km instrument. Virgo is a French, Italian, and Dutch instrument with 3 km long arms located near Cascina, Italy. Geo 600 is an interferometer with 600 meter arms built by Germany and the UK and located near Sarstedt, Germany. TAMA 300 is an interferometer with 300 meter long arms located in Tokyo. CLIO (Cryogenic Laser Interferometer Observatory), which is located 1 km underground in the Kamioka mine, is a prototype for a future Large-scale Cryogenic Gravitational Telescope (LCGT) with 3 km arms. If built, LCGT will also be located in the Kamioka mine with a proposed apex about 1.5 km SSW of CLIO at an elevation of 400 m with a minimum overburden of 200 m (Kuroda *et al.*, 2010). The arms will be oriented at azimuths of approximately 60° and 330°.

In all designs the interferometer beamsplitter and mirrors act as free masses (in the horizontal plane). For strains of order 10^{-23} and arms 1 km in length, the required positional precision is of order 10^{-20} m or 10^{-5} fm. Measuring the position of a macroscopic object with sub-nuclear precision is a challenging activity! Yet several instruments currently are approaching the required sensitivity.

The existence of several gravitational wave detectors at different locations on Earth's surface is necessary to fully recover the propagation direction and polarization of any detected wave (Abbott *et al.*, 2005; Abadie *et al.*, 2010a,b). Coincidence detections from widely separated instruments are also good insurance against spurious signals. The worldwide distribution of interferometers is shown in Figure 16.5. For all of these interferometers, maximum sensitivity occurs perpendicular to the plane of the interferometer, at the zenith and at the nadir. Omnidirectional coverage with good sensitivity requires instruments widely distributed across Earth's surface, as shown in the figure. Determining the direction of propagation, however, requires measuring precise timing differences as the wave passes the Earth. To obtain good timing and positional information it is critical to have actual instruments in the southern hemisphere (the nadir points of northern hemisphere instruments do not enter into consideration). The Australians hope to

Figure 16.4 Satellite images of (a) LIGO Hanford, WA, (b) LIGO Livingston, LA (Credit: USDA Farm Service Agency); (c) Virgo, (d) GEO 600 (Credit: Google Earth). For color version of figure, see plate section.

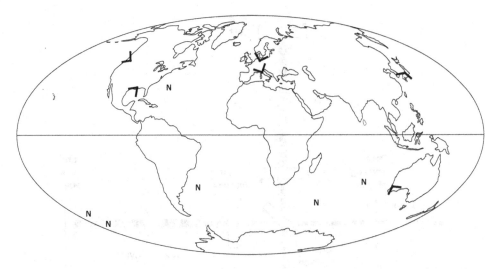

Figure 16.5 Hammer–Aitoff (equal area) projection of Earth's surface showing locations of existing and planned interferometric gravitational wave detectors and orientations of their arms. Nadir directions for the instruments shown are marked "N".

build a major interferometer at the Australian International Gravitational Observatory (AIGO) located near Gingin in Western Australia, as shown. It appears that this may be done in collaboration with LIGO, creating LIGO-South. If so, one of the Advanced LIGO detectors planned for the Hanford site would be installed in Australia instead, leaving only one Advanced LIGO interferometer at Hanford.

16.3.1 Fabry–Perot

For a Michelson interferometer with arms of length L, light of wavelength λ experiences a relative phase shift of

$$\delta\phi = \frac{4\pi L}{\lambda} \, h,$$ (16.19)

when illuminated by a gravitational wave of strain h at optimal orientation (for example, arms oriented in the x- and y-directions for "+" polarization and propagation in the z-direction). For L = 1 km, λ = 1.064 μm, and h = 10^{-23}, the phase difference between the arms is of order 10^{-13} radians, which is impractically small to be measured directly.

A way to increase the phase shift is to make each arm a Fabry–Perot[9] cavity as shown in Figure 16.6. Such a cavity can be made using a highly reflective concave

---

[9] Fabry–Perot interferometers were introduced in Chapter 9.

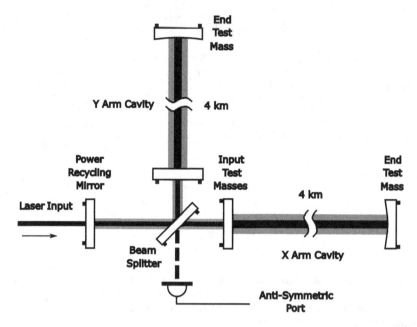

Figure 16.6 Simplified optical layout of LIGO (Abadie *et al.*, 2010c). Note the x and y Fabry–Perot cavities and the power recycling mirror.

mirror at the far end of the arm and a high reflectivity mirror with a small transmittance at the near end of the arm. One way of looking at a Fabry–Perot cavity is that it forces the light to make several round trips through the cavity before returning to the beamsplitter, effectively making the cavity longer by a factor $2\mathcal{F}/\pi$, where $\mathcal{F}$ is the finesse of the Fabry–Perot. Then

$$\delta\phi = \frac{8\mathcal{F}L}{\lambda}\,h. \tag{16.20}$$

For a finesse of 100 the phase shift in an arm is increased by a factor of $2\mathcal{F}/\pi \approx 64$. The power stored in the optical cavity at resonance is increased by a factor of $2\mathcal{F}/\pi \approx 64$ also, the relevance of which we will see in a moment. In LIGO the Fabry–Perot arm finesse has been of order 220. Plans for Advanced LIGO have called for a finesse as high as 1250, although 450 is a more recently proposed value. During its initial science run, Virgo had a Fabry–Perot finesse of 50 (Accadia *et al.*, 2010). Some plans for Advanced Virgo called for the finesse to increase to as much as 600. An increased effective length of the arms is useful only up to the point where the multi-pass light transit time in an arm equals half the period of the gravitational wave. Beyond that, some of the light will be subject to an out of phase signal, decreasing the sensitivity.

16.3.2 Recycling interferometers

The LIGO and Virgo interferometers are designed to operate around nulls in the recombined beams. For small displacements the observed signal at the detector is linear in the strain and linear in the stored laser power. So if the system is limited by signal to noise at the detector, the sensitivity can be increased by increasing the stored power. The Fabry–Perots already increase the stored power. A further increase can be achieved by taking the light off the other beamsplitter path, which otherwise would propagate backwards towards the laser, and "recycling" it with a high reflectivity mirror placed so that the light adds coherently. An implementation of this technique is shown in Figure 16.6. This configuration puts strict limits on allowable light loss throughout the optical system (typically much less than 10^{-4} per surface). A variety of recycling schemes are possible, including "signal recycling," which improves performance at certain resonant frequencies but decreases the bandwidth. A detailed discussion of each of these is beyond the scope of this book. Interested readers can consult Meers (1988) or Blair (1991).

16.3.3 Lasers

The lasers employed in these systems are Nd:YAG lasers with output powers of order 10 W at an infrared wavelength of 1.064 μm. This choice was determined in part by the power available from various laser systems. The infrared wavelength also permitted very small scattering losses, much smaller than is possible at optical wavelengths, a requirement for recycling. Higher power lasers would help to keep shot noise low without the need for as much recycling.

Ideally the laser output should be entirely in the fundamental TEM_{00} Gaussian mode at fixed frequency, fixed power, and fixed polarization. For example, frequency stability is necessary for the light to remain in resonance with the Fabry–Perot cavities, requiring feedback to adjust the cavity lengths, the laser frequency, or both. And although sensitivity to power fluctuations is reduced by using null detection, power stability is also important.

This ideal stability will not generally be the case, so we would like to stabilize these various laser characteristics (Blair, 1991). One way to do this is with a ring cavity mode cleaner (Goßler *et al.*, 2003). A ring cavity is a resonant cavity like a Fabry–Perot except that the laser circulates around a triangular path set by three mirrors. At least one of the mirrors should be weakly concave to compensate for diffraction. Preferably this mirror should be at near normal incidence to avoid introducing excessive astigmatism. A typical geometry is shown in Figure 16.7. With mirror reflectivities exceeding 99.8% it is possible to achieve a cavity finesse of order $\mathcal{F} \approx 2000$. For a cavity of path length L = 5 m, frequency and power

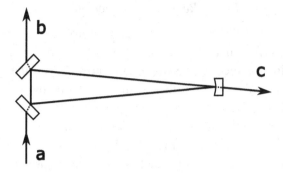

Figure 16.7 Mode cleaner with input port a and possible output ports b and c.

stability are improved by averaging on time scales of $2\mathcal{F}L/\pi c \approx 21$ μs, a characteristic dwell time for light in the cavity. The phase shift upon reflection is different for S- and P-polarizations, so only one linear polarization can be resonant at a given time.

The power density is high within the Fabry–Perot cavities and the mode cleaners, so even small losses can produce significant heating. Care must be taken to minimize the effects of distortion due to surface heating of reflective optics and internal heating of transmitting optics.

16.3.4 Seismic noise

A gravitational wave interferometer must be well isolated from the local environment, and one of the greatest challenges is seismic noise. We use this term to include the effects of distant seismic events, microseismic events, vibration from local human activity (automobiles, trains, logging, etc.), ocean waves, wind loading, earth tides, etc. Seismic noise is omnipresent and broadband, with an approximate ν^{-2} displacement power spectrum at frequencies above 1 Hz.

Vibration isolation is achieved through multiple systems acting as mechanical filters. Take, for example, a simple pendulum of length L consisting of a rigid mass suspended by wires. Neglecting the mass of the wires, the pendulum mode will have a resonant frequency of

$$\nu_0 = \frac{1}{2\pi}\sqrt{\frac{g}{L}}, \tag{16.21}$$

where g is the acceleration of gravity. For 1-meter long wires, this corresponds to a 0.5 Hz resonance. The frequency response of such a system is

$$\frac{(\Delta x)_{out}}{(\Delta x)_{in}} = \frac{1 + i\nu/\nu_0 Q}{1 - \nu^2/\nu_0^2 + i\nu/\nu_0 Q}, \tag{16.22}$$

where Q is the "quality factor," which is mostly determined by the wires. For high Q, the response at frequencies well above resonance goes as ν^{-2}, reducing vibration at high frequencies. Additional seismic isolation may be achieved by using several isolation stages in series. GEO 600 uses a three stage compound pendulum. The Virgo "superattenuator" is a five stage compound pendulum providing seismic isolation of over 10^8 at 4 Hz in both horizontal and vertical directions (Ballardin *et al.*, 2001; Braccini *et al.*, 2009). Advanced LIGO will use a quadruple pendulum, such as that shown in Figure 16.8, which theoretically provides a ν^{-8} rolloff well above the resonant frequencies.

Additional modes of oscillation are present, including the "violin" modes of the support wires. Neglecting stiffness, a wire of length L under tension T has a fundamental frequency of

$$\nu_0 = \frac{1}{2L}\sqrt{\frac{T}{\mu}}, \tag{16.23}$$

where μ is the mass per unit length of the wire. For example, when a mass of 10 kg is supported by two steel wires of length L = 1 m, cross section 0.1 mm^2, and mass per unit length of 1 g/m, the lowest violin mode will be 110 Hz. For LIGO the fundamental violin mode falls at 345 Hz. This mode and its harmonics fall within

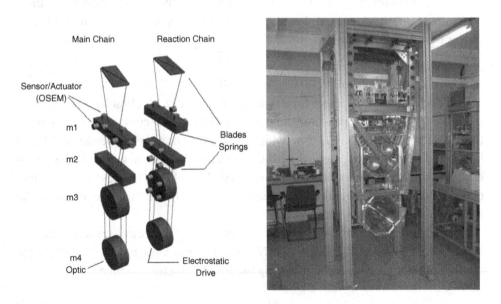

Figure 16.8 Quadruple pendulum prototype at Rutherford Appleton Laboratory in the UK as planned for Advanced LIGO (Copyright Science and Technology Facilities Council and Brett Shapiro/LIGO). For color version of figure, see plate section.

LIGO's target frequency range, but as long as Q is large they will be relatively narrow resonances which can be ignored.

The positioning control systems for the beamsplitter and cavity end mirrors are quite complex since they not only have to provide seismic isolation but also cavity alignment, including angular alignments at the level of 10^{-8} radians (10^{-9} radians for Advanced LIGO) and a cavity length fixed to an accuracy of 10^{-11} m (10^{-14} m for Advanced LIGO).

16.3.5 Quantum limit, shot noise, and radiation pressure fluctuations

The Heisenberg uncertainty principle states that in simultaneously determining conjugate variables such as position and momentum, there is a minimum uncertainty

$$\Delta x \, \Delta p_x \geq \frac{\hbar}{2}. \tag{16.24}$$

In measuring x, one introduces an uncertainty in momentum which, in a measurement time τ, leads to a further uncertainty in x,

$$\Delta x' = \frac{\Delta p_x}{m} \tau. \tag{16.25}$$

These uncertainties Δx and $\Delta x'$ add in quadrature. The combination is smallest when the two terms are equal, leading to an uncertainty in the position of a *single* mirror of mass m:

$$\Delta x'' \geq \sqrt{\frac{\hbar \tau}{m}}. \tag{16.26}$$

The above argument appears, for example, in Caves *et al.* (1980b). For a mirror mass of 10 kg and a measurement time of 0.01 s this corresponds to a positional uncertainty of 3×10^{-19} m, which means that strain measurements of order $h \approx 10^{-21}$ or better are practical with kilometer-scale baselines. This is the "standard quantum limit" described by Caves (1980a, 1981). The way in which this quantum uncertainty is enforced depends on the technique used to measure the mirror position.

Ignore for the moment the Fabry–Perot cavities and the power recycling mirror, and consider just the Michelson interferometer. For a laser power P, the average number of photons entering and leaving the interferometer in a time τ is

$$N = \frac{P\tau}{\hbar\omega}. \tag{16.27}$$

Counting statistics says that there will be shot noise, fluctuations in this number equal to $\sqrt{N}$ or fractional fluctuations of $1/\sqrt{N}$. For a perfect detector looking at the interference signal from the beamsplitter, greatest sensitivity is achieved near a "dark fringe," and the resulting uncertainty in differential arm length is

$$\Delta x_{shot} = \frac{c}{2\omega} \frac{1}{\sqrt{N}}. \tag{16.28}$$

For a power of 10 W, a measurement time of 0.01 s, and a laser wavelength of 1.064 μm, this corresponds to an uncertainty in length of 1.2×10^{-16} m. Photon counting error can be a major limitation, although it can be reduced by increasing the laser input power or using power recycling. The laser power at the beamsplitter is the relevant quantity in this calculation. One effect of the Fabry–Perot cavities is to increase the storage time of the light and hence the value of τ which should be chosen. Another is to increase the effective arm lengths against which this path difference is to be compared. The power recycling increases the number of photons N, also decreasing the shot noise.

The laser light built up inside the Fabry–Perot cavities exerts pressure which pushes the mirrors apart. If the light intensity is constant this presents no problem. But maintaining a steady pressure requires intensity stabilized lasers and Fabry–Perot cavities with stable finesse (hence stable alignment). The averaging effects of mode cleaners and of the Fabry–Perot cavities themselves help improve short term stability.

Beyond this there are more fundamental fluctuations related to radiation pressure. The same shot noise described above contributes a fluctuating force on the end mirrors. The random photon arrivals constitute impulsive events, each imparting a momentum of $2\hbar\omega/c$, where the factor of 2 corresponds to the fact that the photons are reflected. This fluctuating momentum leads to an uncertainty in differential arm length

$$\Delta x_{rad.press.} = \frac{\hbar\omega\tau}{mc} \sqrt{N}. \tag{16.29}$$

The relevant power in this case is that built up inside each Fabry–Perot cavity. In this case the fluctuations increase with increased power. So minimum uncertainty in the quadrature combination of shot noise and radiation pressure effects is achieved at an optimal laser power

$$P = \frac{mc^2}{\omega\tau^2}. \tag{16.30}$$

And at this optimal power the uncertainty reduces to the standard quantum limit of $\sqrt{\tau\hbar/m}$. Too little power corresponds to excess shot noise. Too much power leads to large radiation pressure fluctuations.

16.3.6 Thermal noise

At finite temperature all components of the system introduce thermal noise. Consider damped mechanical systems, which can be represented by mass m, resonant frequency ν_0, and damping factor Q. According to the fluctuation–dissipation theorem (Callen & Welton, 1951), the presence of dissipation means there will be thermal fluctuations in position

$$\langle \Delta x \rangle^2 = \frac{kT}{2\pi^3\,Q\,\nu_0^3\,m} \left[\left(1 - \left(\frac{\nu}{\nu_0}\right)^2\right)^2 + \frac{1}{Q^2}\frac{\nu^2}{\nu_0^2}\right]^{-1}. \tag{16.31}$$

Now the mass of one of the Fabry–Perot cavity mirrors may be 10 kg. Its normal modes of vibration are at frequencies of order 30 kHz, but the resonant wings extend down into the passband of the interferometer. A high Q value for the mirror substrates is necessary to minimize the impact of this noise. The mirror coatings are made of lower-Q materials and the thermal noise of the present coatings can be a dominant limiting effect on interferometer performance (Agresti, 2008).

The violin modes of the pendulum wire suspensions are also thermally excited. The origin of the noise is primarily due to friction at the attachment points. As we have seen, these modes lie within the desired sensitivity range of the interferometer, and their effects must be minimized. Steel suspension wires generally give Q of order 10^4–10^5 (Gillespie & Raab, 1993). Most interferometer teams plan to use higher-Q materials such as fused silica wires in their ultimate designs.

Most gravitational wave interferometer groups have given some thought to the possibility of future cryogenic systems to reduce thermal noise. As mentioned above, the CLIO interferometer is a prototype for a Large Cryogenic Gravitational-wave Telescope (LCGT). Factors making cryogenic cooling difficult include the laser heating of the mirrors and the need to maintain seismic and acoustic isolation.

16.3.7 Other factors

There are a number of reasons why the interferometer beams require high vacuum environments in which to propagate. For one, evacuation of the beam paths prevents acoustic waves from perturbing the optics. Residual gas would also cause scattering of light and result in multipath interference. In addition, residual gas would produce fluctuations in the refractive index, causing optical path length variations. The last two effects would result in excess phase noise. The vacuum quality also affects the cleanliness and useful life of the mirror coatings. Most facilities have adopted vacuum requirements of order 10^{-9} torr or better (much less for some gases).

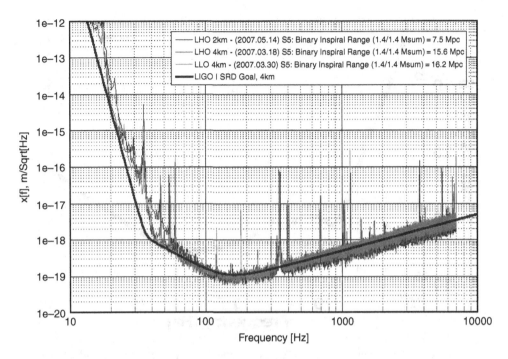

Figure 16.9 Displacement sensitivity of LIGO interferometers as of 2007. Divide vertical scale by 4000 m to obtain strain sensitivity. Credit: LIGO Laboratory/NSF. For color version of figure, see plate section.

16.3.8 Performance

The performance of LIGO as of 2007 is illustrated in Figure 16.9. At frequencies below 40 Hz the performance is limited by seismic noise. Between 40 and 150 Hz the limiting factor is thermal noise. And above 150 Hz, shot noise is the main limitation. To broaden and improve the sensitivity, improvements in all three factors are needed. Several other factors remain important just below the current sensitivity limit, including radiation pressure fluctuations at low frequencies and residual gas effects at all frequencies.

Plans for an improved, Advanced LIGO call for it to begin operation in about 2014. The test masses will be increased from 11 kg to 40 kg, which will decrease radiation pressure fluctuations (and the quantum noise limit) while also lowering the thermal noise of the mirrors. The laser power will be increased to 100–200 W, leading to about 6 kW within the power recycling cavity and nearly 1 MW within the Fabry–Perot cavities. The mirror suspension wires will be made of fused silica, which has much higher Q than steel, reducing the thermal noise contributions. The seismic cutoff frequency will be reduced from 40 Hz to about 10 Hz. The electronic

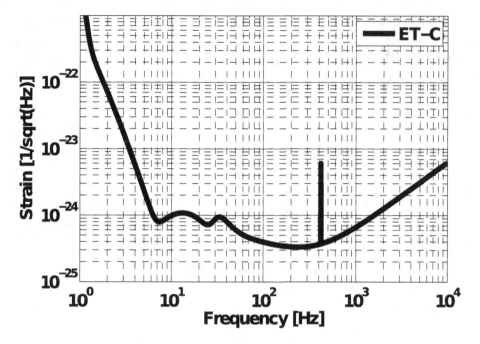

Figure 16.10 Design sensitivity goals for the Einstein Telescope (Hild *et al.*, 2010).

and mechanical control and detection systems will be extensively modified. In all there will be about a factor of 10 improvement in strain sensitivity.

The Einstein Telescope (ET) is a proposed large-scale improvement over the Advanced LIGO generation instruments, by which researchers hope to improve sensitivity by another order of magnitude over the $1-10^4$ Hz band, as shown in Fig 16.10. Details are very much in flux at present. It appears that the ET will be built underground to take advantage of the better seismic and environmental conditions and that it will utilize cryogenic cooling. Both of these ideas are already incorporated in the Japanese LCGT concept. The arm lengths for ET may be increased to of order 10 km. The current plan is to use a triangular configuration (three overlapping 60° interferometers).

16.3.9 Squeezed states

The standard quantum limit may be bypassed by the technique of squeezed states (Caves, 1981). This allows the uncertainty in one conjugate variable to be reduced at the expense of increased uncertainty in the other conjugate variable. This does not violate the principles of quantum mechanics since the uncertainty principle limits only the product of two uncertainties. A demonstration of the plausibility of this technique at audio frequencies for gravitational wave interferometers was given by McKenzie *et al.* (2004). The implementation of such a scheme for the GEO 600 interferometer is discussed by Vahlbruch *et al.* (2010). See also Corbitt *et al.* (2006) and Goda *et al.* (2008).

16.4 Space-based interferometric detectors

The Laser Interferometer Space Antenna (LISA) is a joint project of the ESA and NASA.[10] In the 2010 US National Research Council decadal report on astronomy and astrophysics, LISA was given third priority among large-scale space missions. The project concept is of three spacecraft in heliocentric orbits in a triangular configuration, as shown in Figure 16.11, with nominal separations of 5×10^6 km. The corresponding light crossing time for each arm of the triangle is 16.7 seconds. Orbital eccentricities are likely to be somewhat less than that of Earth. The separations are not static; the orbital dynamics give the spacecraft relative velocities of up to 15 m/s. Arm lengths of this order make LISA primarily sensitive to low frequency gravitational waves, in the range 10^{-4}–10^{-1} Hz, at a strain sensitivity of h $\approx 10^{-23}$ after one year of observation, as shown in Figure 16.12. Ground-based gravitational interferometers cannot operate at these frequencies due to seismic disturbances. Compared to ground-based, two-arm, right-angle interferometers, the

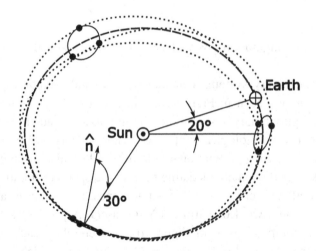

Figure 16.11 LISA configuration viewed from 60° above the ecliptic. The filled circles represent the three satellites in heliocentric orbits, trailing Earth by about 20° and separated by 5×10^6 km. The separation of the satellites has been exaggerated in the drawing.

---

[10] The description here is of the mission concept circa 2009 (ESA/NASA, 2009a). Some aspects may change.

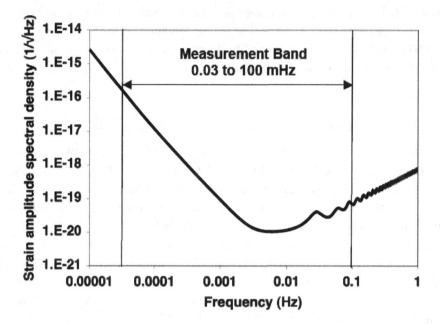

Figure 16.12 Estimated sensitivity of LISA (ESA/NASA, 2009b). Sensitivity is limited at low frequencies by residual acceleration noise on the test masses, at mid-range by shot noise, and at high frequencies by the responsivity drop at wavelengths shorter than the satellite separations.

LISA triangular configuration gives added redundancy and additional polarization information.

Many galactic binary systems generate gravitational radiation in LISA's frequency range, as was seen in Figure 16.2. In addition, LISA is uniquely suited to detecting merging binary systems containing massive black holes in the range 10^4–10^7 $M_\odot$ out to cosmological distances. These events will probe the behavior of matter in the strong gravitational field limit. LISA also has the potential for detecting background fluctuations dating back to the epoch of inflation.

Each spacecraft will carry two 40 cm telescopes, one pointing at each of the other spacecraft, and two 1.064 μm Nd:YAG lasers. A possible schematic view of a LISA spacecraft is given in Figure 16.13. Central to each spacecraft are two 46 mm cubic AuPt alloy test masses (1.96 kg each) which define the ends of each leg of the interferometer. In order for the test masses to follow geodesic variations, they must be protected against external forces. The spacecraft do this, shielding them from drag, interplanetary dust, the solar wind, solar radiation pressure, Poynting–Robertson drag, etc. The spacecraft themselves must not affect the test masses by touching them. So the spacecraft are servo-controlled to maintain

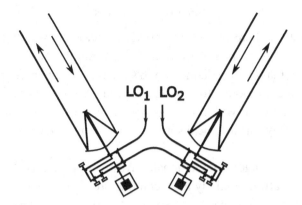

Figure 16.13 Schematic view of a possible optical configuration of one of the LISA spacecraft. The telescope optics are off-axis. Laser light paths are shown (in red in electronic version). All diagonal elements shown are beamsplitters; steering mirrors are not shown. The proof masses are shown in yellow.

precise orientation and location with respect to the masses. Thrusters capable of force precision at the micronewton level keep the spacecraft centered on the test masses (for a 400 kg spacecraft, 1 μN of thrust imparts an acceleration of 2.5 nm s^{-2}).

Each of the six telescopes (two on each of three spacecraft) sends a laser signal to a different spacecraft. Each telescope also receives and interferometrically measures the approximately 100 pW signal from a distant spacecraft and also interferometrically determines the position of the associated local test mass. And at each telescope the phases of the onboard lasers are compared. In this potential configuration there are 18 interferometric measurements which can be used to stabilize and control the lasers and to decipher the gravitational wave signal. It will be possible to remove the effects of laser noise in the analysis process (Tinto & Armstrong, 1999; Armstrong *et al.*, 1999; Estabrook *et al.*, 2000; Tinto & Dhurandhar, 2005).

While the interferometric measurements are ongoing, it is necessary to control to high precision the five remaining degrees of freedom of the test masses: two positions and three orientations. Error signals are derived from capacitance sensors, which can also be used to position the test masses with respect to these remaining degrees of freedom.

LISA Pathfinder is a pilot mission to test some of the technologies required for LISA. It consists of a single spacecraft containing two free-flying test masses separated by 35 cm. The current schedule, based on the Critical Design Review, points towards a launch of LISA Pathfinder in 2012–2013. LISA itself is unlikely to launch before 2020–2025.

16.5 Other systems

Despite the emphasis here on interferometric detectors, it should be emphasized that solid resonant detectors are still in use. For example, there is MiniGRAIL, a spherical 1400 kg mass with a diameter of 68 cm and a resonant frequency of order 3 kHz. With cryogenic cooling to 20 mK, it should have a strain sensitivity of order $h \approx 4 \times 10^{-21}$ Hz$^{-1/2}$. It will have an isotropic response. MiniGRAIL should be very sensitive for any astrophysical sources which emit gravitational waves in this frequency range.

At very low frequencies, gravitational waves can be searched for via Doppler tracking of spacecraft. Essentially the technique probes gravitational waves of frequencies 10^{-6}–0.1 Hz on scales of order 1–10 AU, typical spacings to various spacecraft, such as Cassini (Estabrook & Wahlquist, 1975). Similarly, precise pulsar timing probes gravitational waves from 10^{-9} to 10^{-6} Hz on scales of around 200 pc, a characteristic distance to known pulsars. Such work is being undertaken by the North American Nanohertz Observatory for Gravitational Waves (NANOGrav).

There is also a plan for a DECi-hertz Interferometer Gravitational-wave Observatory (DECIGO), a space-based mission to cover the frequency gap 10^{-2}–10^2 Hz between LISA and the ground-based interferometers and overlapping parts of the frequency range of both, at better sensitivity. Like LISA it would be composed of three spacecraft in a triangular configuration, but with separation reduced to 1000 km. At that separation it would be possible to operate the arms as Fabry–Perot cavities, as with ground-based interferometers, although with only modest finesse.

Another proposed successor to LISA is the Big Bang Observer (BBO), which requires 12 spacecraft. Individual triads would be more compact than LISA, but the separation between triads would in some cases be much larger. The BBO currently faces many technical challenges and unknown costs.

16.6 Data analysis

The variety of waveforms expected presents some special problems in data analysis. Consider, for example, the inspiral problem. What one would like are a variety of templates, representing different masses and spins for the compact objects. Such templates would show the evolution of expected signal amplitudes and frequencies as functions of time. These templates could then be cross correlated with an interferometer waveform to find the physical event most consistent with the received signal. In the time domain, the cross correlation of a signal s(t) with a template k(t) is written as

$$g(\tau) = s \star k = \int_{-\infty}^{\infty} s^*(t)\, k(t + \tau)\, dt, \tag{16.32}$$

or in the Fourier domain as

$$G(f) = S^*(f) \, K(f). \tag{16.33}$$

This would be sufficient for the case of uniform responsivity and white noise of spectral density N. For interferometers of the type discussed here, the dominant noise over much of the frequency band will be shot noise, which is white. But the signal responsivity will fall off roughly as f^{-2} in power (f^{-1} in amplitude) for gravitational wavelengths longer than the effective length of the interferometer arms. In such a case, the approach which optimizes the signal to noise ratio is a "matched filtering," where the template is weighted by the responsivity:

$$G'(f) = S^*(f) \, \frac{K(f)}{f^2}. \tag{16.34}$$

A thorough analysis would need to take into account the various noise sources and their spectral properties and variations in the signal responsivity across the entire frequency band. Ideally the template should include all the stages from inspiral to merger to ringdown, and must keep accurate phasing throughout the encounter. The inspiral phase can begin with a post-Newtonian calculation. The later stages of inspiral through merger to ringdown must be handled by numerical calculations of general relativity.

Exercises

16.1 Consider the Hulse–Taylor binary system which consists of stars with masses 1.4414 $M_\odot$ (pulsar) and 1.3867 $M_\odot$ (companion) with an orbital semi-major axis of 1 949 100 km and an orbital period of 0.322 997 448 930 days (Weisberg & Taylor, 2005). The orbital eccentricity is 0.617 133 8 and it is at a distance of 6400 pc. Note the precision in some of these parameters made possible by high precision pulsar timing. Treat the orbits as Keplerian. The Hulse–Taylor binary radiates much more strongly than the equal mass binary discussed in the text.

a. How much of the difference is due to the greater mass of the Hulse–Taylor binary?

b. How much is due to the smaller separation?

c. How much is due to the eccentricity of the orbit? See Figure 2 of Peters & Mathews (1963) or calculate directly from the formula

$$f(e) = \frac{1 + 73e^2/24 + 37e^4/96}{(1 - e^2)^{7/2}}. \tag{16.35}$$

d. Estimate the eccentricity effect by considering what happens at closest approach.

e. What will be an effect of the higher energy loss at minimum separation?

f. (optional) It may be illuminating to calculate the mass quadrupole and its third derivative for various points along the orbit.

16.2 In polar coordinates the angular distribution of quadrupole gravitational radiation from an equal mass binary system with e $= 0$ in orthogonal linear polarizations is

$$\frac{dP_1}{d\Omega} \propto 1 + 2\cos^2\theta + \cos^4\theta, \tag{16.36}$$

$$\frac{dP_2}{d\Omega} \propto 4\cos^2\theta, \tag{16.37}$$

for a combined distribution of

$$\frac{dP}{d\Omega} \propto 1 + 6\cos^2\theta + \cos^4\theta, \tag{16.38}$$

where θ is measured from the rotational axis.

a. Calculate the solid angles of the radiation patterns P_1, P_2, and P (using the methods of Chapter 11).

b. What fraction of the total power is radiated in each polarization? (Note: The constant of proportionality is the same in the above three equations.)

16.3 An interferometer with 90° arm separation has a sensitivity to signals coming from the direction θ, ϕ (in spherical coordinates with the polar axis perpendicular to the plane of the interferometer) in orthogonal linear polarizations of

$$F^+ = \frac{1}{2}\left(1 + \cos^2\theta\right)\cos 2\phi, \tag{16.39}$$

$$F^\times = \cos\theta \ \sin 2\phi \tag{16.40}$$

(Forward, 1978). Note that these are *amplitude* responses in the sense

$$\frac{\Delta L}{L} = F^+ h^+ + F^\times h^\times. \tag{16.41}$$

Define an antenna pattern solid angle in a way analogous to the *power* patterns of the previous problem,

$$\Omega = \int \frac{|F(\theta, \phi)|}{F_{max}} d\Omega. \tag{16.42}$$

a. What are the solid angles of the + and × polarization patterns?

b. What is the solid angle for unpolarized radiation?

16.4 What is the reception pattern for an interferometer with 60° arm separation like LISA or the proposed Einstein telescope? Consider only two arms. (Hint: Pick a geometry which optimizes the response to one particular polarization.)

16.5 For a binary system consisting of stars of masses m_1 and m_2, there is a parameter known as the "chirp mass"

$$\mathcal{M} = \mu^{3/5} \, M^{2/5}, \tag{16.43}$$

which characterizes the inspiral pattern, where $\mu = m_1 m_2 / (m_1 + m_2)$ is the reduced mass and $M = m_1 + m_2$ is the total mass of the system.

a. What is the chirp mass for an equal mass binary system with $m_1 = m_2 = 1 \, M_\odot$?

b. What is the chirp mass for the Hulse–Taylor binary with $m_1 = 1.4414 \, M_\odot$ and $m_2 = 1.3867 \, M_\odot$?

17

Polarimetry

17.1 Sources of polarized radiation

17.1.1 Synchrotron radiation

Polarization of electromagnetic radiation may occur in a variety of astrophysical circumstances. In some cases the polarization may be considered to be intrinsic to the source (arising simultaneously with the radiation). One example is synchrotron radiation, in which the radiation arises from energetic electrons spiraling along magnetic field lines. The polarization of synchrotron radiation will be in general elliptical and will depend on the angle between the line of sight and the magnetic field. See Figure 17.1 for geometries giving rise to linear, elliptical, and circular polarization. For relativistic electrons, radiation is beamed in the forward direction, making linear polarization more likely to be seen than circular polarization. Linear polarizations approaching 50% are common at centimeter wavelengths in extragalactic radio jets. For a synchrotron spectral index of -1 ($I(\nu) \propto \nu^{-1}$), linear polarization up to 75% is possible (Saikia & Salter, 1988).

17.1.2 Zeeman effect

The Zeeman effect also produces polarized radiation. For a magnetic quantum number M, magnetic fields split energy levels into 2M+1 magnetic sublevels. Radiation will then be linearly or circularly polarized depending on the direction of radiation with respect to the magnetic field and the change in the projected magnetic quantum number, as indicated in Figure 17.2. Relevant magnetic fields vary from 10^{-6} gauss for the interstellar medium, to 10 000 gauss for magnetic stars. Compact objects such as white dwarfs can have even higher fields. The Zeeman effect may be observed from radio wavelengths out to the ultraviolet and from both atoms and molecules. At radio wavelengths one sees the Zeeman effect in neutral hydrogen and in OH in low density regions and in CN in higher density regions.

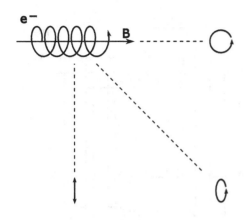

Figure 17.1 Synchrotron radiation is circularly polarized along the magnetic field direction, linearly polarized in the orthogonal direction, and elliptically polarized in between.

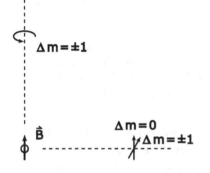

Figure 17.2 The longitudinal Zeeman effect (circularly polarized radiation along the magnetic field direction) and the transverse Zeeman effect (linearly polarized radiation orthogonal to the magnetic field direction).

The neutral atomic hydrogen hyperfine line at 21 cm is split by the Zeeman effect, allowing for determination of magnetic fields in the general interstellar medium. The energy splitting is $\pm\mu_B/h$ where μ_B is the Bohr magneton (9.27 $\times$ 10^{-21} erg/gauss) and h is Planck's constant. The resulting frequency splitting equals ±1.4 Hz/μG for a total of 2.8 Hz/μG. For typical fields of order microgauss this splitting is much less than the linewidth, requiring careful spectropolarimetry to disentangle the two circularly polarized components.

The OH radical is a sensitive probe of magnetic fields. It has a $^2\Pi$ ground state with Λ-doubling[1] of the J states for both $^2\Pi_{1/2}$ and $^2\Pi_{3/2}$. The four Λ-doubling transitions in the lowest energy $^2\Pi_{3/2}$ J = 3/2 state fall at 1612, 1665, 1667, and 1721 MHz, as shown in Figure 17.3. All four are strong maser transitions. The Zeeman energy splittings are given by

$$\Delta E = \pm g_F \, m_F \, \mu_B \, B, \tag{17.1}$$

---

[1] OH in its lower energy states is intermediate between Hund's coupling cases a and b. Λ-doubling arises from an interaction between the orbital momentum of the unpaired electron and the end over end rotation of the molecule.

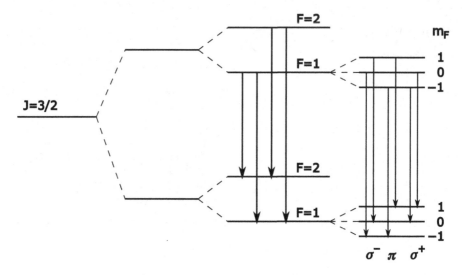

Figure 17.3 The J = 3/2 level of OH in the $^2\Pi_{3/2}$ electronic state is split successively by Λ-doubling, hyperfine structure, and Zeeman splitting. The hyperfine
lines are, from left to right, 1612, 1665, 1667, and 1721 MHz. The 1665 MHz
line is divided into σ^-, π, and σ^+ components. Splittings are not to scale.

where g_F is the Landé g-factor and B is the magnetic field. The g-factor is of order
1.17 for the F = 1 levels and 0.70 for the F = 2 levels. For the 1665 MHz line,
a 1 milligauss magnetic field corresponds to a splitting of $\Delta E/h \approx \pm 1640$ Hz or
± 0.29 km/s and for the 1667 MHz line, $\Delta E/h \approx \pm 980$ Hz or ± 0.18 km/s. These
can be greater than the maser line widths. For 10 milligauss or more the splitting
would exceed even typical thermal line widths. In the geometry corresponding to
the transverse Zeeman effect (Figure 17.2), a triplet would be seen in which each
component is linearly polarized, with the central component along the direction of
the transverse field and the other two components perpendicular to the transverse
field. The size of the splitting measures the strength of the field in the plane of the
sky. In the longitudinal Zeeman effect only two components are visible and they
have opposite senses of circular polarization. The amount of splitting is a measure
of the line of sight magnetic field (magnitude and sign).

Considering just the 1665 MHz line, in emission one would see the pattern
shown in Figure 17.4. The doublet lines would be circularly polarized in opposite
senses. The triplet lines would be linearly polarized, with the π component perpendicular to the magnetic field and the σ components parallel to the field for the
pure geometries shown in Figure 17.2. Otherwise, for the σ components, elliptical
polarization would be present.

In dense regions of the interstellar medium, H I and OH are not abundant and
are not available for measuring magnetic field strengths. CN has proven to be a

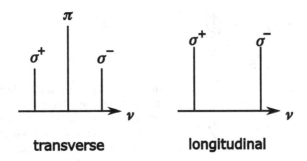

Figure 17.4 Zeeman doublet and triplet showing π and $\sigma^{\pm}$ components.

useful Zeeman probe of molecular clouds, among other things due to the presence of multiple hyperfine components with different Zeeman splittings (Crutcher *et al.*, 1999). Other abundant molecules with unpaired electrons which may prove useful include SO, C_2H, CH, and C_2S.

Line emission from non-magnetic species may be polarized when molecules are subject to anisotropic radiation fields (Goldreich & Kylafis, 1981). The anisotropic radiation unevenly populates the magnetic sublevels (basically the reverse of the process shown in Figure 17.2), leading to emission which is linearly polarized.

17.1.3 Thermal emission

Dust grains are important sources of emission from the infrared out to the submillimeter. In general, dust grains are non-spherical. If their spatial orientations were random, this would not be particularly noteworthy. But if they are aligned they will emit and scatter preferentially in one polarization and absorb preferentially in the same linear polarization.

The Davis and Greenstein (1951) mechanism proposes that paramagnetic relaxation results in prolate grains (for example) preferentially oriented with their major axes perpendicular to the magnetic field direction. Lazarian *et al.* (1997) and Lazarian & Cho (2005) discuss a variety of alternate alignment mechanisms.

17.1.4 Scattering

There are many other astrophysical circumstances in which the polarization process is extrinsic to the emission process, that is, the polarization is imposed or modified in some region external to that which generated the radiation. Processes such as Thomson, Rayleigh, or Mie scattering can create polarized light from initially unpolarized light. This is a purely geometrical effect; single scattering at 90°

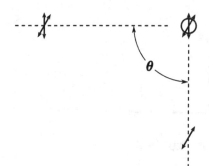

Figure 17.5 Unpolarized light incident from the left polarizes an object in two orthogonal directions. For scattering at an angle of 90°, only one of those has a component orthogonal to the line of sight, so the observer will see linear polarization.

is 100% linearly polarized. For scattering at an angle θ the fractional polarization is given by

$$\pi = \frac{\sin^2 \theta}{1 + \cos^2 \theta}, \qquad (17.2)$$

as shown in Figure 17.5. An example is the existence of *reflection nebulae* whose radiation is highly polarized. In the interstellar medium there is also extinction by non-spherical dust grains. If these grains are aligned by interstellar magnetic fields, light passing through the region will acquire a small degree of linear polarization. Light from supernovae is typically partially polarized (Wang & Wheeler, 2008), indicating that supernovae are not spherically symmetric. Polarimetric studies played a critical role in the development of a unified model of AGNs (Antonucci, 1993). Oppenheimer & Hinkley (2009) discuss the role polarimetry can play in detecting extrasolar planets.

17.1.5 Primordial polarization

Polarization of the cosmic microwave background (CMB) radiation contains potentially important information. As discussed briefly in Chapter 12, current efforts at studying the CMB include the WMAP and Planck satellites, both of which have polarimetric capabilities. E-mode polarization of the CMB has been seen (Chiang *et al.*, 2010; Larson *et al.*, 2011). It can arise from Thomson scattering. B-mode polarization is potentially more interesting as primordial B modes are a result of vector and tensor perturbations. Tensor perturbations are associated with primordial gravitational waves and are diagnostics of the epoch of inflation (Amblard *et al.*, 2007). B-mode polarization on large spatial scales can only be produced by tensor perturbations. Detecting B modes will be very challenging; they are expected to be considerably weaker than the E modes at moderate spatial scales ($l \approx 100$).

17.2 Propagation effects

Faraday rotation can change the direction of linear polarization of radio waves, as mentioned in Chapter 1. This effect is due to the difference in propagation velocity for left- and right-hand circular polarizations in a plasma with a component of magnetic field along the line of sight. The original linear polarization state may be decomposed into left- and right-hand circular components. In passing through the magnetized plasma these components acquire a relative phase shift. When reexpressed in a linear polarization basis, this phase shift leads to a rotation of the angle of linear polarization. At long wavelengths Faraday rotation in Earth's ionosphere, as well as the interstellar medium, must be taken into account.

Faraday rotation can also be a mechanism for depolarization. The fact that the degree of rotation is a function of wavelength,

$$\Delta\theta = \frac{e^3}{2\pi m_e^2 c^2} \frac{1}{\nu^2} \text{RM}, \tag{17.3}$$

means that over a finite bandwidth the degree of rotation can vary significantly, resulting in a loss of linear polarization, especially at long wavelengths. If the magnetized plasma is inhomogeneous and the rotation measure is large, different parts of the beam will sample unresolved regions with different rotation measures, also leading to depolarization. Wave propagation through an ionized medium permeated by a magnetic field can also lead to conversion of linearly polarized flux to circularly polarized.

17.3 Polarization-sensitive devices

The key optical elements in polarimetry systems are polarizers, polarization beam splitters, quarter-wave plates, half-wave plates, and modulators. A quarter-wave plate, suitably oriented, may be used to convert linear polarization into circular polarization, and vice versa. A half-wave plate will invert the sense of circular polarization. Depending on orientation, it will also rotate the direction of linear polarization. True quarter-wave and half-wave plates impart the correct phase shift only at a specific wavelength. It is possible, however, to make broadband (achromatic) designs in ways similar to those described in Chapter 8 for making broadband anti-reflection coatings. Tinbergen (1996) discusses design approaches to achromatic polarizers and retarders. Rotating half-wave plates may be used as modulators, both for linear and circular polarimeters. Pockels cells may be used as fast retardation modulators, for example as electronically switched quarter-wave plates.

An example of a polarization beam splitter is the Wollaston prism, shown in Figure 17.6. A Wollaston prism is made of two pieces of birefringent crystal,

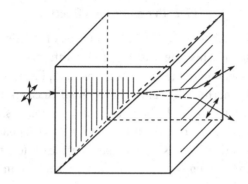

Figure 17.6 A Wollaston prism. Striations indicate the orientation of the optical axis.

oriented so that their optical axes are perpendicular to each other and perpendicular to the direction of the incident beam. At the interface between the two halves, one polarization sees a decrease in index of refraction and the orthogonal polarization sees an increase. Therefore they refract differently and emerge as separate beams. The effect of a Rochon prism is similar to that of a Wollaston prism. A Glan–Thompson prism, however, uses total internal reflection to separate the ordinary and extraordinary rays. At optical wavelengths many detectors, such as CCDs, do not discriminate between polarizations, so polarimeters consist of one or more polarizing elements inserted into the optical path before the detector.

At infrared and radio wavelengths wire grids may be used as polarizers or polarizing beam splitters. The polarization-sensitive bolometers discussed in Chapter 12 are examples in which detectors which are inherently insensitive to polarization may be coupled to specific polarization states. At radio wavelengths the use of linear devices means that detection systems tend to be polarized, often linearly polarized. To measure total intensity one needs to measure orthogonal polarizations, either simultaneously with separate receivers or in alternation. Linearly polarized feeds are frequently converted to circular polarization by placing quarter-wave plates in front of them. Retarders of this sort may be constructed by cutting parallel linear grooves much narrower than a wavelength into dielectric plates.

17.4 Analysis of polarization states

17.4.1 Stokes parameters

The character of polarized radiation may be described by the four Stokes parameters. For a wave propagating in the z-direction (towards the observer), adopt a right-handed coordinate system in which the x-direction is north and the y-direction is east in the plane of the sky. The x and y components of the electric field are given by

$$E_x(t) = a_x(t)\, e^{i(kz - \omega t + \delta_x(t))}, \tag{17.4}$$

$$E_y(t) = a_y(t)\, e^{i(kz - \omega t + \delta_y(t))}. \tag{17.5}$$

Here we follow the conventions and more-or-less the notation of Wolf (2007). The true electric field is, of course, given by the real part of these expressions. The Stokes parameters are then defined by

$$I = S_0 = \langle a_x(t)^2 \rangle + \langle a_y(t)^2 \rangle, \tag{17.6}$$

$$Q = S_1 = \langle a_x(t)^2 \rangle - \langle a_y(t)^2 \rangle, \tag{17.7}$$

$$U = S_2 = 2\langle a_x(t) a_y(t) \cos[\delta_x(t) - \delta_y(t)] \rangle, \tag{17.8}$$

$$V = S_3 = 2\langle a_x(t) a_y(t) \sin[\delta_x(t) - \delta_y(t)] \rangle. \tag{17.9}$$

The total intensity is I. Note that only Stokes I is necessarily non-negative. The parameters Q and U describe the degree and direction of linear polarization. The parameter V describes the degree and sense of circular polarization. Conventions on the sign of V vary. The convention used above is that positive V is called right-hand circular polarization and corresponds to the electric vector rotating clockwise as viewed looking back towards the place from which the light is coming (that is, looking in the negative $\vec{k}$ direction).

Although the above convention for V is widely used in optics, here we must part company with the optical community in order to adhere to the conventions adopted by the IEEE and Commission 40 of the International Astronomical Union (IEEE, 1969; IAU, 1974). So we will redefine Stokes V to be

$$V = S_3 = -2\langle a_x(t) a_y(t) \sin[\delta_x(t) - \delta_y(t)] \rangle. \tag{17.10}$$

Our convention now is that right-hand circular polarization (positive V) corresponds to the electric vector rotating counterclockwise as viewed on the celestial sphere. This convention is illustrated in Figure 17.7.

For polarized light, ignoring the time averaging (assuming coherence), simple algebra gives

$$I^2 = Q^2 + U^2 + V^2. \tag{17.11}$$

For partially polarized light the phases δ_x and δ_y are not definite and

$$I^2 > Q^2 + U^2 + V^2. \tag{17.12}$$

Polarimetry

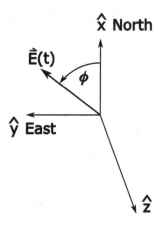

$\hat{x}$ **North**

$\vec{E}(t)$

ϕ

$\hat{y}$ **East**

$\hat{z}$

Figure 17.7 Right circular polarization according to the conventions adopted by the IEEE and Commission 40 of the IAU.

We can define the total degree of polarization

$$\pi = \frac{\sqrt{Q^2 + U^2 + V^2}}{I}, \tag{17.13}$$

or degrees of linear and circular polarization,

$$\pi_L = \frac{\sqrt{Q^2 + U^2}}{I}, \tag{17.14}$$

$$\pi_C = \frac{V}{I}, \tag{17.15}$$

and position angle of the linear component,

$$\theta = \frac{1}{2} \arctan \frac{U}{Q}. \tag{17.16}$$

The Stokes parameters, therefore, are able to describe states of partial polarization. However, absolute phase information is lost. Relative phase information is maintained, as can be seen from the definitions of U and V. But the Stokes parameters are measures of intensities, not electric fields. Therefore there is some loss of information in dealing just with Stokes parameters. We will describe an alternate approach, but first we will treat what can be done with the Stokes representation.

17.4.2 Mueller matrices

Mueller matrices are used to manipulate polarization states represented by Stokes vectors. Each element in an optical system is represented by a Mueller matrix and these matrices are multiplied together and then multiplied by the input Stokes vector to give an output Stokes vector. For example, the Mueller matrix for an ideal linear polarizer oriented in the vertical (x) direction (+Q) is given by

$$M_{\text{Vert}} = \frac{1}{2} \begin{pmatrix} 1 & 1 & 0 & 0 \\ 1 & 1 & 0 & 0 \\ 0 & 0 & 0 & 0 \\ 0 & 0 & 0 & 0 \end{pmatrix}. \tag{17.17}$$

The matrix for rotation by an angle θ is

$$M_\theta = \begin{pmatrix} 1 & 0 & 0 & 0 \\ 0 & \cos 2\theta & \sin 2\theta & 0 \\ 0 & -\sin 2\theta & \cos 2\theta & 0 \\ 0 & 0 & 0 & 1 \end{pmatrix}. \tag{17.18}$$

So the action of a linear polarizer set at an angle θ is given by

$$M_{-\theta}\, M_{\text{Vert}}\, M_\theta = \frac{1}{2} \begin{pmatrix} 1 & \cos 2\theta & \sin 2\theta & 0 \\ \cos 2\theta & \cos^2 2\theta & \sin 2\theta \cos 2\theta & 0 \\ \sin 2\theta & \sin 2\theta \cos 2\theta & \sin^2 2\theta & 0 \\ 0 & 0 & 0 & 0 \end{pmatrix}. \tag{17.19}$$

17.4.3 Jones vectors and matrices

In some cases it is necessary to treat the phase of the radiation correctly, and as we have seen the Stokes parameters and Mueller matrices do not do this. In such cases one can express the complex electric field vector as

$$E = \begin{pmatrix} E_x \\ E_y \end{pmatrix}. \tag{17.20}$$

The operation of optical elements can be represented by 2×2 Jones matrices. Repeating the example used in discussing Mueller matrices, the Jones representation for a linear polarizer is

$$J_{\text{Vert}} = \begin{pmatrix} 1 & 0 \\ 0 & 0 \end{pmatrix}. \tag{17.21}$$

The matrix for rotation by an angle θ is

$$J_\theta = \begin{pmatrix} \cos\theta & \sin\theta \\ -\sin\theta & \cos\theta \end{pmatrix}. \tag{17.22}$$

So the action of a linear polarizer set at an angle θ is given by

$$J_{-\theta}\, J_{\text{Vert}}\, J_\theta = \begin{pmatrix} \cos^2\theta & \sin\theta\cos\theta \\ \sin\theta\cos\theta & \sin^2\theta \end{pmatrix}. \tag{17.23}$$

Jones vectors cannot be used to describe unpolarized or partially polarized light. Instead, the radiation is treated as an incoherent sum of orthogonal fully polarized components. The effects of optical elements on each component may be followed using Jones matrices, with the components being added incoherently at the end.

17.5 Polarization measurement

Operationally, determination of all Stokes parameters may be achieved from intensity measurements from four orientations of linear polarization plus right and left circular polarization:

$$I = I_0 + I_{90}, \tag{17.24}$$

$$Q = I_0 - I_{90}, \tag{17.25}$$

$$U = I_{45} - I_{135}, \tag{17.26}$$

$$V = I_R - I_L. \tag{17.27}$$

In astronomy a position angle $\theta = 0$ generally corresponds to north with positive angles increasing to the east. Positive, right-handed polarization ($V > 0$) is defined as the electric field vector rotating counterclockwise, as viewed by the observer looking back towards the source. In other words, the y-component lags the x-component by 90°. See Hamaker & Bregman (1996) for the conventions underlying the definitions of the Stokes parameters and inconsistencies in previous use in radio astronomy. For elliptical polarization, the major axis of the ellipse is at a position angle

$$\chi = \frac{1}{2} \tan^{-1} \frac{U}{Q}. \tag{17.28}$$

This is also the position angle of the electric field vector for linear polarization, which is a limiting case of elliptical polarization.

Alternatively, Stokes parameters can be determined from correlation characteristics of the electric fields, as in radio interferometry. The appropriate terminology here is to speak of Stokes visibilities. Stokes visibilities, like all visibilities in radio interferometry, are in general complex. For linearly polarized feeds on antennas i and j,

$$V_I = \langle E_{jx}^* E_{ix} \rangle + \langle E_{jy}^* E_{iy} \rangle, \tag{17.29}$$

$$V_Q = \langle E_{jx}^* E_{ix} \rangle - \langle E_{jy}^* E_{iy} \rangle, \tag{17.30}$$

$$V_U = \langle E_{jy}^* E_{ix} \rangle + \langle E_{jx}^* E_{iy} \rangle, \tag{17.31}$$

$$V_V = i \left[\langle E_{jy}^* E_{ix} \rangle - \langle E_{jx}^* E_{iy} \rangle \right]. \tag{17.32}$$

For right and left circularly polarized feeds,

$$V_I = \langle E_{jR}^* E_{iR} \rangle + \langle E_{jL}^* E_{iL} \rangle, \tag{17.33}$$

$$V_Q = \langle E_{jL}^* E_{iR} \rangle + \langle E_{jR}^* E_{iL} \rangle, \tag{17.34}$$

$$V_U = i \left[\langle E_{jL}^* E_{iR} \rangle - \langle E_{jR}^* E_{iL} \rangle \right], \tag{17.35}$$

$$V_V = \langle E_{jR}^* E_{iR} \rangle - \langle E_{jL}^* E_{iL} \rangle. \tag{17.36}$$

17.5.1 Analysis of weak field splittings

When Zeeman splitting is less than the linewidth of the spectral feature, the Zeeman signal is closely related to the shape of the Stokes I spectrum. Consider the longitudinal Zeeman effect where the split components have opposite circular polarizations. For $m_F = 1$, the splitting of Equation 17.1 may be rewritten as

$$\delta\nu = \pm \frac{\mu_B}{h} g_F B, \tag{17.37}$$

$$\delta\lambda = \pm \frac{\mu_B}{hc} \lambda^2 g_F B, \tag{17.38}$$

where $\mu_B/hc = 4.67 \times 10^{-3}$ m^{-1}G^{-1}. Assuming the magnetic field fills the beam, Stokes V is the difference between signals of opposite circular polarizations,

$$V(\lambda) = \frac{1}{2} [I(\lambda + \Delta\lambda) - I(\lambda - \Delta\lambda)]. \tag{17.39}$$

This is a finite difference expression, so for weak fields

$$V(\lambda) = \Delta\lambda \frac{\partial I}{\partial \lambda} + \cdots. \tag{17.40}$$

So Stokes V is proportional to the derivative of Stokes I, and the ratio of Stokes V to Stokes I gives the splitting and thereby a measure of the strength of the line-of-sight component of the magnetic field (Crutcher *et al.*, 1993).

17.6 Optical polarimetry

All elements of an optical system can lead to some degree of depolarization and/or interconversion between polarization states. These will take the form of couplings between Stokes I and V and between Q and U. The polarization characteristics of the telescope itself are important. Generally, one would like to avoid any reflections or transmissions at other than normal incidence. So telescopes should be as close to azimuthally symmetric as possible. Off the optical axis the symmetry is broken and significant polarization of the radiation can occur.

One important application of optical polarimetry is the study of solar magnetic fields on sub-arcsecond scales. Most commonly this is done by observing the Zeeman effect using a pair of Fe I lines at 630.2 nm, for example to measure strong magnetic fields in sunspots. A variety of other low excitation metal lines of, for example, Mg I, Ni I, Sr I, Ti I, Sc I, Ca I, or Ca II, can also be used for this purpose. Some simple molecules such as TiO which can be found in stellar photospheres are also useful. Non-magnetic scattering is a significant contributor to the observed polarization signals. In interpreting the data, it is necessary to take into account the Hanle effect, a magnetic phenomenon which can depolarize linearly polarized signals and rotate the plane of linear polarization. Studies of the coronal magnetic field are done using high ionization states such as Fe XIII or Fe XIV.

17.7 Radio polarimetry and calibration

In a series of papers Hamaker, Bregman, and Sault (1996) lay out the basic procedures underlying radio polarimetry and polarimetric calibration. For an interferometer consisting of telescopes i and j, they define a coherency vector which is the outer product of the Jones vectors at the two telescopes,

$$E = \langle E_i \otimes E_j^* \rangle. \tag{17.41}$$

They show that the operation of Jones matrices J_i and J_j on E_i and E_j can be represented by

$$E_{out} = (J_i \otimes J_j^*) \langle E_{i,in} \otimes E_{j,in}^* \rangle. \tag{17.42}$$

The entire interferometer can be represented by a succession of such operations.

Polarimetric errors can arise from non-ideal behavior of optical elements. For example, a pair of radio feeds on antenna i, instead of being sensitive to individual components may include leakage terms

$$\begin{pmatrix} g_{ix} & 0 \\ 0 & g_{iy} \end{pmatrix} \begin{pmatrix} 1 & D_{ix} \\ D_{iy} & 1 \end{pmatrix}. \tag{17.43}$$

Ideally the gains of the two feeds should be nearly the same ($g_{ix} \approx g_{iy}$) and the leakage terms should be small ($D_{ix}, D_{iy} \ll g_{ix}, g_{iy}$). Calibration is necessary to determine the values of the gains and leakages.

Radio telescope receivers are often placed off axis in order to accommodate feeds for multiple wavelength bands. With off-axis feed systems, azimuthal symmetry is broken, and right-hand and left-hand circular feeds will have different radiation patterns on the sky. This is known as beam squint. In addition, the sidelobes of any radio telescope can be polarized differently than the center of the main lobe, so polarimetric calibration can vary across the field of view.

Sault *et al.* (1996) show that when three or more antennas are present, self calibration of gains and phases (closure) is important, but seven degrees of freedom remain after self calibration. Observing a point source of known polarization allows the number of unknowns to be reduced to three. Typically one would used an unpolarized calibration source. The remaining three degrees of freedom, in the case of parallel linear feeds, correspond to an uncertainty in the angle of linear polarization (an interconversion of Stokes Q and U), a phase error between the two feeds (interconverting U and V), and imaginary values of the leakage terms (interconverting Q and V). Sault *et al.* (1996) gives the corresponding result for circularly polarized feeds. Fortunately none of these contaminate Q, U, or V with I, which could overwhelm a weakly polarized signal. However, incorrect assumptions about the calibrator polarization are able to produce calibration errors that contaminate Q, U, or V with I. A single long observation of a linearly polarized source with altitude/azimuth antennas (so that there is significant parallactic angle rotation) is sufficient to determine these remaining three degrees of freedom as long as the leakage terms remain constant during the integration.

Exercises

17.1 What polarization pattern should be produced by Čerenkov radiation?

17.2 Calculate the Mueller matrices for a horizontal polarizer and for polarizers at $\pm 45°$.

17.3 Calculate the Jones and Mueller matrices for a quarter-wave plate at arbitrary orientation.

17.4 Try to reconcile the definitions of right-circular polarization in Kliger *et al.* (1990), Tinbergen (1996), Hamaker & Bregman (1996), Born & Wolf (1999), and Wolf (2007). Look carefully for ambiguous sentences. Who is (or is not) internally consistent in words, pictures, and complex notation?

Appendix A

Physical constants and units

Table A.1. *Physical constants*

Quantity	mks	cgs
α	$7.297\,352\,538 \times 10^{-3}$	$7.297\,352\,538 \times 10^{-3}$
c	$2.997\,924\,58\ \times 10^{8}\ \mathrm{m\ s^{-1}}$	$2.997\,924\,58\ \times 10^{10}\ \mathrm{cm\ s^{-1}}$
e	$1.602\,176\,487 \times 10^{-19}\ \mathrm{C}$	$4.803\,204\,27\ \times 10^{-10}\ \mathrm{esu}$
ϵ_0	$8.854\,187\,817 \times 10^{-12}\ \mathrm{F\,m^{-1}}$	—
g_n	$9.806\,65 \qquad \mathrm{m\ s^{-2}}$	$9.806\,65 \qquad \times 10^{2}\ \mathrm{cm\ s^{-2}}$
G	$6.674\,28 \qquad \times 10^{-11}\ \mathrm{N\,m^2\,kg^{-2}}$	$6.674\,28 \qquad \times 10^{-8}\ \mathrm{dyne\ cm^2\ g^{-2}}$
h	$6.626\,068\,96\ \times 10^{-34}\ \mathrm{J\,s}$	$6.626\,068\,96\ \times 10^{-27}\ \mathrm{erg\,s}$
k	$1.380\,650\,4 \qquad \times 10^{-23}\ \mathrm{J\,K^{-1}}$	$1.380\,650\,4 \qquad \times 10^{-16}\ \mathrm{erg\,K^{-1}}$
m_e	$9.109\,382\,15\ \times 10^{-31}\ \mathrm{kg}$	$9.109\,382\,15\ \times 10^{-28}\ \mathrm{g}$
m_p	$1.672\,621\,637 \times 10^{-27}\ \mathrm{kg}$	$1.672\,621\,637 \times 10^{-24}\ \mathrm{g}$
μ_B	$9.274\,009\,15\ \times 10^{-24}\ \mathrm{J\,T^{-1}}$	$9.274\,009\,15\ \times 10^{-21}\ \mathrm{erg\,G^{-1}}$
μ_N	$5.050\,783\,24\ \times 10^{-27}\ \mathrm{J\,T^{-1}}$	$5.050\,783\,24\ \times 10^{-24}\ \mathrm{erg\,G^{-1}}$
σ	$5.670\,400 \times 10^{-8}\ \mathrm{J\,m^{-2}\,s^{-1}\,K^{-4}}$	$5.670\,400 \times 10^{-5}\ \mathrm{erg\,cm^{-2}\,s^{-1}\,K^{-4}}$
σ_T	$6.652\,458\,558 \times 10^{-29}\ \mathrm{m^2}$	$6.652\,458\,558 \times 10^{-25}\ \mathrm{cm^2}$

Table A.2. *Other units*

Name	mks	cgs
atm	$1.013\,25 \qquad \times 10^{5}\ \mathrm{Pa}$	$1.013\,25 \qquad \times 10^{6}\ \mathrm{dyne\ cm^{-2}}$
AU	$1.495\,978\,707 \times 10^{11}\ \mathrm{m}$	$1.495\,978\,707 \times 10^{13}\ \mathrm{cm}$
eV	$1.602\,176\,487 \times 10^{-19}\ \mathrm{J}$	$1.602\,176\,487 \times 10^{-12}\ \mathrm{erg}$
Jy	$10^{-26}\ \mathrm{W\,m^{-2}\,Hz^{-1}}$	$10^{-23}\ \mathrm{erg\,s^{-1}\,m^{-2}\,Hz^{-1}}$
$M_\odot$	$1.981\,8 \qquad \times 10^{30}\ \mathrm{kg}$	$1.981\,8 \qquad \times 10^{33}\ \mathrm{g}$
pc	$3.085\,677\,581 \times 10^{16}\ \mathrm{m}$	$3.085\,677\,581 \times 10^{18}\ \mathrm{cm}$

Appendix B

Acronyms

2-DEG	2-Dimensional Electron Gas
AC	AutoCorrelation
ACA	Atacama Compact Array
ACE	Advanced Composition Explorer
ACF	AutoCorrelation Function
ACIS	Advanced CCD Imaging Spectrometer (on Chandra)
ACS	Advanced Camera for Surveys (on HST)
ACT	Atacama Cosmology Telescope
ACTA	Atmospheric Čerenkov Telescope Array
AGASA	Akeno Giant Air Shower Array
AGIS	Advanced Gamma-ray Imaging System
AGN	Active Galactic Nuclei
AIGO	Australian International Gravitational Observatory
ALMA	Atacama Large Millimeter/submillimeter Array
AMANDA	Antarctic Muon And Neutrino Detector Array
AMIGA	Auger Muons and Infill for Ground Array
AMS	Alpha Magnetic Spectrometer
ANTARES	Astronomy with a Neutrino Telescope and Abyss Environmental RESearch
AOS	Acousto-Optical Spectrometer
APEX	Atacama Pathfinder EXperiment
ASCA	Advanced Satellite for Cosmology and Astrophysics
ASIC	Application Specific Integrated Circuit
ASM	All Sky Monitor (on RXTE)
ATIC	Advanced Thin Ionization Calorimeter
AXAF	Advanced X-ray Astrophysics Facility (original name of Chandra)
BAT	Burst Alert Telescope (on SWIFT)
BATSE	Burst And Transient Source Experiment (on CGRO)

BBO	Big Bang Observer
BESS	Balloon-borne Experiment with a Superconducting Spectrometer
BGO	BiGeO
BHI	Black Hole Imager
BIB	Blocked Impurity Band
BIMA	Berkeley–Illinois–Maryland Association (array)
BLIP	Background Limited Infrared Photodetector
CANGAROO	Collaboration of Australia and Nippon for a GAmma-Ray Observatory in the Outback
CARMA	Combined Array for Research in Millimeter-wave Astronomy
CC	Charged Current
CCD	Charge Coupled Device
CDM	Cold Dark Matter
CDMA	Code Division Multiple Access
CGRO	Compton Gamma Ray Observatory
CL	Confidence Limit
CLIO	Cryogenic Laser Interferometer Observatory
CMB	Cosmic Microwave Background
COBE	COsmic Background Explorer
COBRA	Caltech Owens-valley Broadband Reconfigurable Architecture
COMPTEL	imaging COMPton TELescope (on CGRO)
COS	Cosmic Origins Spectrometer
CREAM	Cosmic Ray Energetics And Mass
CRIS	Cosmic Ray Isotope Spectrometer
DC	Direct Current
DECIGO	DECi-hertz Interferometer Gravitational wave Observatory
DM	Dispersion Measure
DMR	Differential Microwave Radiometer (on COBE)
DSNB	Diffuse Supernova Neutrino Background
DUSEL	Deep Underground Science and Engineering Laboratory
EGRET	Energetic Gamma Ray Experiment Telescope (on CGRO)
EPIC	European Photon Imaging Camera
ER	Elongation Rate
ESA	European Space Agency
ET (1)	Ephemeris Time
ET (2)	Einstein Telescope
EUV	Extreme UltraViolet
EUVE	Extreme UltraViolet Explorer
EVLA	Expanded Very Long baseline Array
FET	Field Effect Transistor

FIRAS	Far InfraRed Absolute Spectrophotometer
FIRST	Far InfraRed and Submillimetre Telescope (previous name for Herschel)
FK5	5th Fundamental Catalog
FP	Focal Plane
FPGA	Field Programmable Gate Array
FSR	Free Spectral Range
FUSE	Far Ultraviolet Spectroscopic Explorer
FWHM	Full-Width at Half-Maximum
GAIA	Global Astrometric Interferometer for Astrophysics
GALEX	GALaxy Evolution Explorer
GALLEX	GALLium EXperiment
GAST	Greenwich Apparent Sidereal Time
GBT	Green Bank Telescope
GBM	GLAST Burst Monitor
GIS	Gas Imaging Spectrometer (on ASCA)
GLAST	Gamma-ray Large Area Space Telescope (original name of Fermi)
GMST	Greenwich Mean Sidereal Time
GMT	Greenwich Mean Time
GNO	Gallium Neutrino Observatory
GPS	Global Positioning System
GRB	Gamma Ray Burst
GRP	Gaussian Random Process
GZK	Greisen–Zatsepin–Kuzmin (effect)
HEAT	High Elevation Auger Telescopes
HEMT	High Electron Mobility Transistor
HESS	High Energy Stereoscopic System
HEXTE	High Energy X-ray Timing Experiment (on RXTE)
HFI	High Frequency Instrument (Planck)
HIFI	Heterodyne Instrument for the Far Infrared (Herschel)
HiRes	High Resolution Fly's Eye
HRC	High Resolution Camera (on Chandra)
HRI	High Resolution Imager (on ROSAT)
HRMA	High Resolution Mirror Assembly (on Chandra)
HST	Hubble Space Telescope
HWHM	Half-Width at Half-Maximum
IAU	International Astronomical Union
IBIS	Imager on Board the INTEGRAL Satellite
ICARUS	Imaging Cosmic And Rare Underground Signal

ICRS	International Celestial Reference System
IF	Intermediate Frequency
IGM	InterGalactic Medium
IMB	Irvine–Michigan–Brookhaven (neutrino detector)
INTEGRAL	INTErnational Gamma-Ray Astrophysics Laboratory
IPC	Imaging Proportional Counter
IR	InfraRed
IRAC	InfraRed Array Camera (on Spitzer)
IRAS	InfraRed Astronomical Satellite
IRS	InfraRed Spectrograph (on Spitzer)
ISM	InterStellar Medium
ISO	Infrared Space Observatory
IXO	International X-ray Observatory
JAXA	Japanese Aerospace eXploration Agency
JD	Julian Date
JEM-EUSO	Japanese Experiment Module–Extreme Universe Space Observatory
JEM-X	Joint European X-ray Monitor
JFET	Junction Field Effect Transistor
LAST	Local Apparent Sidereal Time
LAT	Large Area Telescope (on Fermi)
LCGT	Large-scale Cryogenic Gravitational Telescope
LFI	Low Frequency Instrument (on Planck)
LIGO	Laser Interferometer Gravitational-wave Observatory
LISA	Laser Interferometer Space Antenna
LMA	Large Mixing Angle
LMST	Local Mean Sidereal Time
LO	Local Oscillator
LPM	Landau–Pomeranchuk–Migdal (effect)
LSB	Lower SideBand
LSR	Local Standard of Rest
LSST	Large Synoptic Survey Telescope
MAMA	Multi-Anode Microchannel Array
MAXIM	Micro-Arcsecond X-ray Imaging Mission
MCMC	Markov Chain Monte Carlo
MEM	Maximum Entropy Method
MESFET	MEtal Semiconductor Field Effect Transistor
MIPS	Multiband Imaging Photometer (on Spitzer)
MOS	Metal-Oxide-Semiconductor
MOSFET	Metal-Oxide-Semiconductor Field-Effect Transistor

MSW	Mikheyev–Smirnov–Wolfenstein (effect)
MTF	Modulation Transfer Function
NANOGrav	North American Nanohertz Observatory for Gravitational waves
NASA	National Aeronautics and Space Administration
NC	Neutral Current
NCP	North Celestial Pole
NEP (1)	North Ecliptic Pole
NEP (2)	Noise-Equivalent-Power
NESTOR	NEutrinos, from Supernovae and TeV sources, Ocean Range
NGP	North Galactic Pole
NICMOS	Near Infrared Camera and Multi-Object Spectrometer (on HST)
NIR	Near-InfraRed channel (WFC3)
NMR	Nuclear Magnetic Resonance
NTP	Network Time Protocol
OMC	Optical Monitor Camera (on INTEGRAL)
OTF	Optical Transfer Function
OSSE	Oriented Scintillation Spectrometer Experiment (on CGRO)
OVRO	Owens Valley Radio Observatory
PACS	Photodetector Array Camera and Spectrometer (Herschel)
PAMELA	Payload for Antimatter Matter Exploration and Light-nuclei Astrophysics
PCA	Proportional Counter Array (on RXTE)
PDF	Probability Distribution Function
PF	Pupil Function
PPM	Positions and Proper Motions (catalog)
PSD	Power Spectral Density
PSF	Point Spread Function
PSPC	Position Sensitive Proportional Counter (on ROSAT)
PZT	Photographic Zenith Tube
RF	Radio Frequency
RGA	Reflection Grating Array
RM	Rotation Measure
RMS	Root Mean Square
ROSAT	ROentgen SATellite
RXTE	Rossi X-ray Timing Explorer
SAGE	Soviet–American Gallium Experiment
SBC	Solar Blind Channel (ACS)
SC	Sagittal Coma
SCP	South Celestial Pole
SDSS	Sloan Digital Sky Survey

SEP	South Ecliptic Pole
S-EUSO	Super-Extreme Universe Space Observatory
SGP	South Galactic Pole
SI	le Système International (d'Unités)
SIS (1)	Superconductor–Insulator–Superconductor (superconducting tunnel junction)
SIS (2)	Solid-state Imaging Spectrometer (on ASCA)
SIRTF	Space InfraRed Telescope Facility (previous name of Spitzer)
SKA	Square Kilometer Array
SM3B	Servicing Mission 3B (HST)
SM4	Servicing Mission 4 (HST)
SN	SuperNova
SNO	Sudbury Neutrino Observatory
SOFIA	Stratospheric Observatory For Infrared Astronomy
SPI	SPectrometer on INTEGRAL
SPIRE	Spectral and Photometric Imaging REceiver (on Herschel)
SPT	South Pole Telescope
SQUID	Superconducting QUantum Interference Device
SSM	Standard Solar Model
STIS	Space Telescope Imaging Spectrograph
Swift	not an acronym
SZ	Sunyaev–Zel'dovich (effect)
TA	Telescope Array (air shower detector)
TAI	Temps Atomique International
TALE	Telescope Array Low Energy
TAS	Transverse AStigmatism
TC	Transverse Coma
TCB	Barycentric Coordinate Time
TCG	Geocentric Coordinate Time
TDB	Barycentric Dynamical Time (deprecated)
TDT	Terrestrial Dynamical Time (deprecated)
TEM	Transverse ElectroMagnetic (mode)
TES	Transition Edge Sensor
TRACER	Transition Radiation Array for Cosmic Energetic Radiation
TSA	Transverse Spherical Aberration
TT (1)	Terrestrial Time
TT (2)	Transverse Traceless (gauge)
UHECR	UltraHigh Energy Cosmic Ray
ULE	Ultra-Low-Expansion (glass)
USB	Upper SideBand

UT	Universal Time
UTC	Temps Universel Coordonné
UV	UltraViolet
UVOT	UltraViolet and Optical Telescope (on Swift)
VERITAS	Very Energetic Radiation Imaging Telescope Array System
Virgo	a gravitational-wave interferometer; not an acronym
VLA	Very Large Array (New Mexico)
VLBA	Very Long Baseline Array
VLBI	Very Long Baseline Interferometry
WFC	Wide Field Camera (on ROSAT)
WFC3	Wide Field Camera 3 (on HST)
WFIRST	Wide Field InfraRed Survey Telescope
WIDAR	Wideband Interferometer Digital ARchitecture
WMAP	Wilkinson Microwave Anisotropy Probe
XMM-Newton	X-ray Multimirror Mission-Newton
XRT	X-Ray Telescope (on Swift)
YAG	Yttrium Aluminum Garnet

Appendix C
Additional reading

Chapter 1 – Astrophysical information

Rytov, S. M., Kravtsov, Yu. A., & Tatarskii, V. I. (1987). *Principles of statistical radiophysics*, **1–4**. Berlin: Springer-Verlag.

Chapter 2 – Photometry

Bessell, M. S. (2005). Standard photometric systems. *ARA&A*, **43**, 293–336.
Bradt, H. (2004). *Astronomy methods*. Cambridge: Cambridge University Press.
Kitchin, C. R. (2009). *Astrophysical techniques*, 5th edn. Boca Raton: CRC Press.
Rybicki, G. B. & Lightman, A. P. (1979). *Radiative processes in astrophysics*. New York: Wiley.

Chapter 5 – Detection systems

McLean, I. S. (2008). *Electronic imaging in astronomy: detectors and instrumentation*, 2nd edn. Berlin: Springer.

Chapter 7 – Stochastic processes & noise

van Kampen, N. G. (2007). *Stochastic processes in physics and chemistry*. Amsterdam: Elsevier.

Chapter 8 – Optics

Klein, M. V. & Furtak, T. E. (1986). *Optics*, 2nd edn. New York: Wiley.
Yoder, P. R., Jr. (1993). *Opto-mechanical systems design*, 2nd edn. New York: Marcel Decker.

Chapter 9 – Interference

Wolf, E. (2007). *Introduction to the theory of coherence and polarization of light.* Cambridge: Cambridge University Press.

Chapter 10 – Spectroscopy

Bowen, I. S. & Vaughan, A. H. Jr. (1973). "Nonobjective" gratings. *PASP*, **85**, 174–176.

Mar, D. J., Marsh, J. P., Deen, C. P. *et al.* (2009). Micromachined silicon grisms for infrared optics. *Appl. Opt.*, **48**, 1016–1029.

Traub, W. A. (1990). Constant-dispersion grism spectrometer for channeled spectra. *J. Opt. Soc. Am. A*, **7**, 1779–1791.

Chapter 11 – Ultraviolet, x-ray, and gamma ray astronomy

Atwood, W. B., Abdo, A. A., Ackermann, M. *et al.* (2009). The large area telescope on the Fermi gamma-ray space telescope mission. *ApJ*, **697**, 1071–1102.

Fishman, G. J., Meegan, C. A., Wilson, R. B. *et al.* (1994). The first BATSE gamma-ray burst catalog. *ApJS*, **92**, 229–283.

Levine, A. M., Bradt, H., Wei, C. *et al.* (1996). First results from the all-sky monitor on the Rossi x-ray timing explorer. *ApJ*, **469**, L33–L36.

Tanaka, Y., Inoue, H., & Holt, S. S. (1994). The X-ray astronomy satellite ASCA. *PASJ*, **46**, L37–L41.

Ubertini, P., Lebrun, F., Di Cocco, G. *et al.* (2003). IBIS: The imager on-board INTEGRAL. *A&A*, **411**, L131–L139.

Vedrenne, G., Roques, J.-P., Schönfelder, V. *et al.* (2003). SPI: The spectrometer aboard INTEGRAL. *A&A*, **411**, L63–L70.

Chapter 12 – Radio receivers, spectrometers, and interferometers

Mather, J. C., Fixsen, D. J., & Shafer, R. A. (1993). Design for the COBE far-infrared absolute spectrophotometer (FIRAS). *Proc. SPIE*, **2019**, 168–179.

Page, L. *et al.* (2003). The optical design and characterization of the microwave anisotropy probe. *ApJ*, **585**, 566–586.

Smoot, G. *et al.* (1990). COBE differential microwave radiometers: instrument design and implementation. *ApJ*, **360**, 685–695.

Chapter 13 – Modern statistical methods

Bretthorst, G. L. (1988). Bayesian spectrum analysis and parameter estimation. In *Lecture notes in statistics*, **48**, J. Berger, S. Fienberg, J. Gani, K. Krickeberg, & B. Singer, eds. New York: Springer-Verlag.

Feigelson, E. D. & Nelson, P. I. (1985). Statistical methods for astronomical data with upper limits. I. Univariate distributions. *ApJ*, **293**, 192–206.

Gull, S. F. & Skilling, J. (1984). Maximum entropy method in image processing. *IEE Proc.*, **131**, 646–650.

Jaynes, E. T. (2003). *Probability theory*. Cambridge: Cambridge University Press.

Jeffreys, H. (1939). *Theory of probability*, 3rd edn. Oxford: Oxford University Press.

Mathews, J. & Walker, R. L. (1970). *Mathematical methods of physics*, 2nd edn. Menlo Park: Benjamin.

Miller, R. G. Jr. (1981). *Survival analysis*. New York: Wiley.

Narayan, R. & Nityananda, R. (1986). Maximum entropy image restoration in astronomy. *ARA&A*, **24**, 127–170.

Parratt, L. G. (1961). *Probability and experimental errors in science*. New York: Wiley.

Schmitt, J. H. M. M. (1985). Statistical analysis of astronomical data containing upper bounds: General methods and examples drawn from X-ray astronomy. *ApJ*, **293**, 178–191.

Skilling, J. & Gull, S. F. (1991). Bayesian maximum entropy image reconstruction. In *Spatial statistics and imaging*, A. Possolo, ed., 341–367, Hayward: Institute of Mathematical Statistics.

Chapter 14 – Neutrino detectors

Learned, J. G. & Mannheim, K. (2000). High-energy neutrino astrophysics. *Annu. Rev. Nucl. Part. Sci.*, **50**, 679–749.

Maltoni, M., Schwetz, T., Torola, M., & Valle, J. W. F. (2004). Status of global fits to neutrino oscillations. *New J. Phys.*, **6**, 122.

Chapter 15 – Cosmic ray detectors

Abraham, J. *et al.* (Pierre Auger Collaboration) (2010). The fluorescence detector of the Pierre Auger observatory. *Nucl. Instrum. Methods Phys. Res. A*, **620**, 227–251.

Bichsel, H., Groom, D. E., & Klein, S. R. (2008). Passage of particles through matter. In *Review of particle physics*. *Phys. Lett. B*, **667**, 267–280.

Boyle, P. J. (TRACER Project) (2008). Cosmic ray composition at high energies: the TRACER project. *Adv. Space Res.*, **42**, 409–416.

Fleisher, R. L., Price, P. B., & Walker, R. M. (1975). *Nuclear tracks in solids*. Berkeley: University of California Press.

Gaisser, T. K. (1990). *Cosmic rays and particle physics*. Cambridge: Cambridge University Press.

Grieder, P. K. F. (2001). *Cosmic rays at Earth*. Amsterdam: Elsevier.

Lemoine, M. & Sigl, G. (2001). *Physics and astrophysics of ultra-high-energy cosmic rays*. Lecture Notes in Physics, **576**. Berlin: Springer.

Papini, P. *et al.* (2008). In-flight performances of the PAMELA satellite experiment. *Nucl. Instrum. Methods Phys. Res. A*, **588**, 259–266.

Rybicki, G. B. & Lightman, A. P. (1979). *Radiative processes in astrophysics*. New York: Wiley.

Seo, E.-S. *et al.* (1997). Advanced thin ionization calorimeter to measure ultrahigh energy cosmic rays. *Adv. Space Res.*, **19**, 711–718.

Stone, E. C. *et al.* (1998). The cosmic-ray isotope spectrometer for the Advanced Composition Explorer. *Space Sci. Rev.*, **86**, 285–356.

Yoshida, S. (2003). *Ultra-high energy particle astrophysics*. Hauppauge: Nova Science.

Yoshida, T. *et al.* (2004). BESS-polar experiment. *Adv. Space Res.*, **33**, 1755–1762.

Chapter 16 – Gravitational waves

Abbott, B. *et al.* (2004). Detector description and performance for the first coincidence observations between LIGO and GEO. *Nucl. Instrum. Methods Phys. Res. A*, **517**, 154–179.

Akutsu, T. *et al.* (2008). Search for continuous gravitational waves from PSR J0835-4510 using CLIO data. *Class. Quantum Grav.*, **25**, 184013.

Armano, M. *et al.* (2009). LISA pathfinder: the experiment and the route to LISA. *Class. Quantum Grav.*, **26**, 094001.

Barriga, P., Zhao, C., & Blair, D. C. (2005). Optical design of a high power mode-cleaner for AIGO. *Gen. Relativ. Gravit.*, **37**, 1609–1619.

Barsotti, L., Evans, M., & Fritschel, P. (2010). Alignment sensing and control in advanced LIGO. *Class. Quantum Grav.*, **27**, 084026.

Braginskii, V. B. (2000). Gravitational-wave astronomy: new methods of measurements. *Physics–Uspekhi*, **43**, 691–699.

Burgay, M. *et al.* (2005). The highly relativistic binary pulsar PSR J0737-3039A: discovery and implications. In *Binary Radio Pulsars*, ASP Conference Series, **328**, F. A. Rasio and I. H. Stairs, eds. 53–58.

Camp, J. B. & Cornish, N. J. (2004). Gravitational wave astronomy. *Annu. Rev. Nucl. Part. Sci.*, **54**, 525–577.

Daw, E. J., Giaime, J. A., Dormand, D., Lubinski, M., & Zweizig, J. (2004). Long-term study of the seismic environment at LIGO. *Class. Quantum Grav.*, **21**, 2255–2273.

ESA/NASA (2009). Laser Interferometer Space Antenna (LISA) Payload Preliminary Design Description, LISA-MSE-DD-0001.

Flanagan, E. E. (1993). Sensitivity of the Laser Interferometer Gravitational Wave Observatory to a stochastic background, and its dependence on the detector. *Phys. Rev. D*, **48**, 2389–2407.

Gottardi, L., de Waard, A., Usenko, O. *et al.* (2007). Sensitivity of the spherical gravitational wave detector MiniGRAIL operating at 5 K. *Phys. Rev. D*, **76**, 102005.

Kalogera, V. *et al.* (2004). The cosmic coalescence rates for double neutron star binaries. *ApJ*, **601**, L179–L182.

Lang, R. N. & Hughes, S. A. (2006). Measuring coalescing massive binary black holes with gravitational waves: the impact of spin-induced precession. *Phys. Rev. D*, **74**, 122001.

Plissi, M. V. *et al.* (2000). GEO 600 triple pendulum suspension system: seismic isolation and control. *Rev. Sci. Instrum.*, **71**, 2539–2545.

Racca, G. D. & McNamara, P. W. (2010). The LISA pathfinder mission. Tracing Einstein's geodesics in space. *Space Sci. Rev.*, **151**, 159–181.

Sigg, D. (for the LIGO Scientific Collaboration) (2008). Status of the LIGO detectors. *Class. Quantum Grav.*, **25**, 114041.

Takahashi, R. *et al.* (TAMA collaboration) (2008). Status of TAMA300. *Class. Quantum Grav.*, **21**, S403–S408.

Thorne, K. S. (1980). Multipole expansions of gravitational radiation. *Rev. Mod. Phys.*, **52**, 299–340.

Tyson. J. A. & Gifford, R. P. (1978). Gravitational-wave astronomy. *ARA&A*, **16**, 521–554.

Vahlbruch, H., Chelkowski, S., Hage, B. *et al.* (2006). Coherent control of vacuum squeezing in the gravitational-wave detection band. *Phys. Rev. Lett.*, **97**, 011101.

Vinet, J.-Y. (2008). Instruments for gravitational wave astronomy on ground and in space. In *Gravitation and experiment*, T. Damour, B. Duplantier, & V. Rivasseau, eds., 111–138. Basel: Birkhäuser Verlag.

Whitcomb, S. E. (2008). Ground-based gravitational-wave detection: now and future. *Class. Quantum Grav.*, **25**, 114013.

Willems, B. & Kalogera, V. (2004). Constraints on the formation of PSR J0737-3039: the most probable isotropic kick magnitude. *ApJ*, **603**, L101–L104.

Yamamoto, K. *et al.* (2008). Current status of the CLIO project. *J. Phys. Conf. Ser.*, **122**, 012002.

Chapter 17 – Polarimetry

Cotton, W. D. (1999). Polarization in interferometry. In *Synthesis imaging in radio astronomy II*, ASP Conference Series, **180**, G. B. Taylor, C. L. Carilli, and R. A. Perley, eds., 111–126.

Heiles, C. & Crutcher, R. (2005). Magnetic fields in diffuse H I and molecular clouds. In *Cosmic magnetic fields*, Lecture Notes in Physics, **664**, 137–182. Berlin: Springer.

Kemball, A. J. (1999). VLBI polarimetry. In *Synthesis Imaging in Radio Astronomy II*, ASP Conference Series, **180**, G. B. Taylor, C. L. Carilli, and R. A. Perley, eds., 499–511.

Landi Degl'Innocenti, E., Bagnulo, S., & Fossati, L. (2007). Polarimetric standardization. In *The future of photometric, spectrophotometric, and polarimetric standardization*, ASP Conference Series, **999**, C. Sterken, ed., arXiv:astro-ph/0610262v1.

Rao, R., Crutcher, R. M., Plambeck, R. L., & Wright, M. C. H. (1998). High-resolution millimeter-wave mapping of linearly polarized dust emission: magnetic field structure in Orion. *ApJ*, **502**, L75.

Sault, R. J. & Cornwell, T. J. (1999). The Hamaker–Bregman–Sault measurement equation. In *Synthesis imaging in radio astronomy II*, ASP Conference Series, **180**, G. B. Taylor, C. L. Carilli, & R. A. Perley, eds., 657–669.

References

Abadie, J. *et al.* (LIGO and Virgo Collaborations) (2010a). Search for gravitational-wave inspiral signals associated with short gamma-ray bursts during LIGO's fifth and Virgo's first science run. *ApJ*, **715**, 1453–1461.

Abadie, J. *et al.* (LIGO and Virgo Collaborations) (2010b). All-sky search for gravitational-wave bursts in the first joint LIGO-GEO-Virgo run. *Phys. Rev. D*, **81**, 102001.

Abadie, J. *et al.* (2010c). Calibration of the LIGO gravitational wave detectors in the fifth science run. *Nucl. Instrum. Methods Phys. Res. A*, **624**, 223–240.

Abbasi, R. U. *et al.* (High Resolution Fly's Eye Collaboration) (2005). A study of the composition of ultra-high-energy cosmic rays using the High-Resolution Fly's Eye. *ApJ*, **622**, 910–926.

Abbasi, R. U. *et al.* (High Resolution Fly's Eye Collaboration (2008a). First observation of the Greisen-Zatsepin-Kuzmin suppression. *Phys. Rev. Lett.*, **100**, 101101.

Abbasi, R. U. *et al.* (High Resolution Fly's Eye Collaboration (2008b). An upper limit on the electron-neutrino flux from the HiRes detector. *ApJ*, **684**, 790–793.

Abbasi, R. U. *et al.* (High Resolution Fly's Eye Collaboration) (2010a). Indications of proton-dominated cosmic-ray composition above 1.6 EeV. *Phys. Rev. Lett.*, **104**, 161101.

Abbasi, R. U. *et al.* (High Resolution Fly's Eye Collaboration) (2010b). Analysis of large-scale anisotropy of ultra-high energy cosmic rays in HiRes data. *ApJ*, **713**, L64–L68.

Abbott, B. *et al.* (LIGO and TAMA Collaborations) (2005). Upper limits from the LIGO and TAMA detectors on the rate of gravitational-wave bursts. *Phys. Rev. D*, **72**, 122004.

Abdurashitov, J. N. *et al.* (1994). Results from SAGE (The Russian-American gallium solar neutrino experiment). *Phys. Lett. B*, **328**, 234–248.

Abraham, J. *et al.* (Pierre Auger Collaboration) (2008a). Observation of the suppression of the flux of cosmic rays above 4×10^{19} eV. *Phys. Rev. Lett.*, **101**, 061101.

Abraham, J. *et al.* (Pierre Auger Collaboration) (2008b). Correlation of the highest-energy cosmic rays with the positions of nearby active galactic nuclei. *Astropart. Phys.*, **29**, 188–204.

Abraham, J. *et al.* (Pierre Auger Collaboration) (2010). Measurement of the depth of maximum of extensive air showers above 10^{18} eV. *Phys. Rev. Lett.*, **104**, 091101.

Abramowitz, M. & Stegun, I. A. (1970). *Handbook of mathematical functions*. Washington: National Bureau of Standards.

Accadia, T. *et al.* (2010). Status and perspectives of the Virgo gravitational wave detector. *J. Phys. Conf. Ser.*, **203**, 012074.

Achterberg, A. *et al.* (IceCube Collaboration) (2006). First year performance of the IceCube neutrino telescope. *Astropart. Phys.*, **26**, 155–173.

Achterberg, A. *et al.* (IceCube Collaboration) (2007). Detection of atmospheric muon neutrinos with the IceCube 9-string detector. *Phys. Rev. D*, **76**, 027101.

Agresti, J. (2008). Researches on non-standard optics for advanced gravitational wave interferometers. Ph.D. thesis, University of Pisa.

Aharmim, B. *et al.* (SNO Collaboration) (2005). Electron energy spectra, fluxes, and day-night asymmetries of ^{8}B solar neutrinos from measurements with NaCl dissolved in the heavy-water detector at the Sudbury Neutrino Observatory. *Phys. Rev. C*, **72**, 055502.

Ahlen, S. P. (1980). Theoretical and experimental aspects of the energy loss of relativistic heavily ionizing particles. *Rev. Mod. Phys.*, **52**, 121–173.

Ahmad, Q. R. *et al.* (SNO Collaboration) (2001). Measurement of the rate of $\nu_e + d \rightarrow p + p + e^-$ interactions produced by ^{8}B solar neutrinos at the Sudbury Neutrino Observatory. *Phys. Rev. Lett.*, **87**, 071301.

Ahmed, S. N. *et al.* (SNO Collaboration) (2004). Measurement of the total active ^{8}B solar neutrino flux at the Sudbury Neutrino Observatory with enhanced neutral current sensitivity. *Phys. Rev. Lett.*, **92**, 181301.

Ahn, H. S. *et al.* (CREAM Collaboration) (2007). The cosmic ray energetics and mass (CREAM) instrument. *Nucl. Instrum. Methods Phys. Res. A*, **579**, 1034–1053.

Ahrens, J. *et al.* (IceCube Collaboration) (2001). *IceCube preliminary design document* (revision 1.24; www.icecube.wisc.edu/science/publications/pdd/pdd.pdf).

Allen, C. W. (2001). *Allen's astrophysical quantities*, 4th edn., A. N. Cox, ed. New York: Springer.

Amblard, A., Cooray, A., & Kaplinghat, M. (2007). Search for gravitational waves in the CMB after WMAP3: foreground confusion and the optimal frequency coverage for foreground minimization. *Phys. Rev. D*, **75**, 083508.

Antonucci, R. (1993). Unified models for active galactic nuclei and quasars. *ARA&A*, **31**, 473–521.

Armstrong, J. W., Estabrook, F. B., & Tinto, M. (1999). Time-delay interferometry for space-based gravitational wave searches. *ApJ*, **527**, 814–826.

Arqueros, F., Hörandell, J. R., & Keilhauer, B. (2008). Air fluorescence relevant for cosmic-ray detection. Summary of the 5th fluorescence workshop, El Escorial, 2007. *Nucl. Instrum. Methods Phys. Res. A*, **597**, 1–22.

Ashie, Y. *et al.* (Super-Kamiokande Collaboration) (2005). Measurement of atmospheric neutrino oscillation parameters by Super-Kamiokande I. *Phys. Rev. D*, **71**, 112005.

The astronomical almanac for the year 2009 (2009). US Naval Observatory and HM Nautical Almanac Office (also http://asa.usno.navy.mil).

Bahcall, J. N., Pinsonneault, M. H., & Basu, S. (2001). Solar models: current epoch and time dependences, neutrinos, and helioseismological properties. *ApJ*, **555**, 990–1012.

Bahcall, J. N., Serenelli, A. M., & Basu, S. (2005). New solar opacities, abundances, helioseismology, and neutrino fluxes. *ApJ*, **621**, L85–L88.

Ballardin, G. *et al.* (2001). Measurement of the VIRGO superattenuator performance for seismic noise suppression. *Rev. Sci. Instrum.*, **72**, 3643–3652.

Beatty, J. J. & Westerhoff, S. (2009). The highest-energy cosmic rays. *Annu. Rev. Nucl. Part. Sci.*, **59**, 319–345.

Bennett, C. L. *et al.* (2003). The microwave anisotropy probe mission. *ApJ*, **583**, 1–23.

Bessell, M. S. & Brett, J. M. (1988). JHKLM photometry: standard systems, passbands, and intrinsic colors. *PASP*, **100**, 1134–1151.

Bessell, M. S., Castelli, F., & Plez, B. (1998). Model atmospheres broad-band colors, bolometric corrections and temperature calibrations for O-M stars. *A&A*, **333**, 231–250.

Bevington, P. R. & Robinson, D. K. (2003). *Data reduction and error analysis for the physical sciences*, 3rd edn. New York: McGraw-Hill.

Bildsten, L. (1998). Gravitational radiation and rotation of accreting neutron stars. *ApJ*, **501**, L89–L93.

Blair, D. G., ed. (1991). *The detection of gravitational waves*. Cambridge: Cambridge University Press.

Born, M. & Wolf, E. (1999). *Principles of optics*, 7th edn. Cambridge: Cambridge University Press.

Boyer, J. H., Knapp, B. C., Mannel, E. J., & Seman, M. (2002). FADC-based DAQ for HiRes Fly's Eye. *Nucl. Instrum. Methods Phys. Res. A*, **482**, 457–474.

Braccini, S. *et al.* (WG2 Suspension group) (2009). *Superattenuator seismic isolation measurements by Virgo interferometer: a comparison with the future generation antenna.* Einstein Telescope scientific note ET-025-09.

Bracewell, R. N. (2000). *The Fourier transform and its applications*, 3rd edn. New York: McGraw-Hill.

Burke, B. E., Mountain, R. W., Harrison, D. C. *et al.* (1991). An abuttable CCD imager for visible and X-ray focal plane arrays. *IEEE Trans. Electron Dev.*, **38**, 1069–1076.

Callen, H. & Welton, T. (1951). Irreversibility and generalized noise. *Phys. Rev.*, **83**, 34–40.

Caves, C. M. (1980a). Quantum-mechanical radiation-pressure fluctuations in an interferometer. *Phys. Rev. Lett.*, **45**, 75–79.

Caves, C. M., Thorne, K. S., Drever, R. W. P., Sandberg, V. D., & Zimmermann, M. (1980b). On the measurement of a weak classical force coupled to a quantum-mechanical oscillator. I. Issues of principle. *Rev. Mod. Phys.*, **52**, 341–392.

Caves, C. M. (1981). Quantum-mechanical noise in an interferometer. *Phys. Rev. D*, **23**, 1693–1708.

Cesarsky, C. J. (1980). Cosmic-ray confinement in the galaxy. *ARA&A*, **18**, 289–319.

Chang, F.-Y., Chen, P., Lin, G.-L., Noble, R., & Sydora, R. (2009). Magnetowave induced plasma wakefield acceleration for ultrahigh energy cosmic rays. *Phys. Rev. Lett.*, **102**, 111101.

Cherry, M. L., Hartmann, G., Müller, D., & Prince, T. A. (1974). Transition radiation from relativistic electrons in periodic radiators. *Phys. Rev. D*, **10**, 3594–3607.

Chiang, H. C. *et al.* (2010). Measurement of cosmic microwave background polarization power spectra from two years of BICEP data. *ApJ*, **711**, 1123–1140.

Cleveland, B. T. *et al.* (1998). Measurement of the solar electron neutrino flux with the Homestake chlorine detector. *ApJ*, **496**, 505–526.

Corbitt, T., Chen, Y., Khalili, F. *et al.* (2006). Squeezed-state source using radiation-pressure-induced rigidity. *Phys. Rev. A*, **73**, 023801.

Cox, R. T. (1946). Probability, frequency, and reasonable expectation. *Am. J. Phys.*, **14**, 1–13.

Crutcher, R. M., Troland, T. H., Goodman, A. A. *et al.* (1993). OH Zeeman observations of dark clouds. *ApJ*, **407**, 175–184.

Crutcher, R. M., Troland, T. H., Lazareff, B., Paubert, G., & Kazès, I. (1999). Detection of the CN Zeeman effect in molecular clouds. *ApJ*, **514**, L121–L124.

Cutler, C. & Thorne, K. S. (2002). An overview of gravitational-wave sources. In *Proceedings of 16th international conference on general relativity and gravitation* (GR16), N. Bishop & S. D. Maharaj, eds., 72–111, Singapore: World Scientific.

Cuttaia, F. *et al.* (2004). Analysis of the pseudocorrelation radiometers for the low frequency instrument onboard the PLANCK satellite. *Proc. SPIE*, **5498**, 756–767.

Davis, L. Jr. & Greenstein, J. L. (1951). The polarization of starlight by aligned dust grains. *ApJ*, **114**, 206–240.

den Herder, J. W. *et al.* (2001). The reflection grating spectrometer on board XMM-Newton. *A&A*, **365**, L7–L17.

Diaconis, P. & Efron, B. (1983). Computer-intensive methods in statistics. *Sci. Am.*, **248**, 116–130.

Efron, B. (1981). Censored data and the bootstrap. *J. Am. Stat. Assoc.*, **76**, 312–319.

ESA/NASA (2009a). Laser Interferometer Space Antenna (LISA) Mission Concept, LISA-PRJ-RP-0001.

ESA/NASA (2009b). Laser Interferometer Space Antenna (LISA) Measurement Requirements Flowdown Guide, LISA-MSE-TN-0001.

Esposito, J. A. *et al.* (1999). In-flight calibration of EGRET on the Compton gamma-ray observatory. *ApJS*, **123**, 203–217.

Estabrook, F. B. & Wahlquist, H. D. (1975). Response of Doppler spacecraft tracking to gravitational radiation. *Gen. Relativ. Gravit.*, **6**, 439–447.

Estabrook, F. B., Tinto, M., & Armstrong, J. W. (2000). Time-delay analysis of LISA gravitational wave data: elimination of spacecraft motion effects. *Phys. Rev. D*, **62**, 042002.

Falta, D., Fisher, R., & Khanna, G. (2010). Gravitational wave emission from the single-degenerate channel of type Ia supernovae. arXiv:1011.6387v1.

Forward, R. L. (1978). Wideband laser-interferometer gravitational-radiation experiment. *Phys. Rev. D*, **17**, 379–390.

Fryer, C. & Kalogera, V. (1997). Double neutron star systems and natal neutron star kicks. *ApJ*, **489**, 244–253.

Fukuda, Y. *et al.* (1998). Evidence for oscillation of atmospheric neutrinos. *Phys. Rev. Lett.*, **81**, 1562–1567.

Fukuda, S. *et al.* (2001). Solar ^{8}B and hep neutrino measurements from 1258 days of Super-Kamiokande data. *Phys. Rev. Lett.*, **86**, 5651–5655.

Gahbauer, F., Hermann, G., Hörandel, J. R., Müller, D., & Radu, A. A. (2004). A new measurement of the intensities of the heavy primary cosmic-ray nuclei around 1 TeV amu^{-1}. *ApJ*, **607**, 333–341.

Gaisser, T. & Stanev, T. (2008). Cosmic rays. In *Review of particle physics*. *Phys. Lett. B*, **667**, 254–260.

Gelman, A., Roberts, G. O., & Gilks, W. R. (1996). Efficient Metropolis jumping rules. In *Bayesian statistics*, Vol. 5, J. Bernardo, J. Berger, A. Dawid, & A. Smith, eds., 599–607, Oxford: Oxford University Press.

Gelman, A., Carlin, J. B., Stern, H. S., & Rubin, D. B. (2004). *Bayesian data analysis*, 2nd edn. Boca Raton: CRC Press.

Genzel, R. & Karas, V. (2007). The galactic center. *Proc. Int. Astron. Union Symp.*, **238**, 173–180.

Ghez, A. M., Klein, B. L., Morris, M., & Becklin, E. E. (1998). High proper-motion stars in the vicinity of Sagittarius A*: evidence for a supermassive black hole at the center of our galaxy. *ApJ*, **509**, 678–686.

Gillespie, A. & Raab, F. (1993). Thermal noise in the test mass suspensions of a laser interferometer gravitational-wave detector prototype. *Phys. Lett. A*, **178**, 357–363.

Ginzburg, V. L. & Tsytovich, V. N. (1979). Several problems of the theory of transition radiation and transition scattering. *Phys. Rep.*, **49**, 1–89.

Giunti, C. & Kim, C. W. (2007). *Fundamentals of neutrino physics and astrophysics.* Oxford: Oxford University Press.

Goda, K., Mikhailov, E. E., Miyakawa, O. *et al.* (2008). Generation of a stable low-frequency squeezed vacuum field with periodically poled $KTiOPO_4$ at 1064 nm. *Opt. Lett.*, **33**, 92–94.

Goldreich, R. & Kylafis, N. D. (1981). On mapping the magnetic field direction in molecular clouds by polarization measurements. *ApJ*, **243**, L75–L78.

Gorbunov, D. S., Tinyakov, P. G., Tkachev, I. I., & Troitsky, S. V. (2004). Testing the correlations between ultrahigh-energy cosmic rays and BL Lac-type objects with HiRes stereoscopic data. *JETP Lett.*, **80**, 145–148.

Goßler, S. *et al.* (2003). Mode-cleaning and injection optics of the gravitational-wave detector GEO600. *Rev. Sci. Instrum.*, **74**, 3787–3795.

Gradshteyn, I. S. & Ryzhik, I. M. (1980). *Table of integrals, series, and products* (corrrected and enlarged edition), A. Jeffrey, ed. New York: Academic Press.

Gross, E. P. (1955). Shape of collision-broadened spectral lines. *Phys. Rev.*, **97**, 395–403.

Hamaker, J. P. & Bregman, J. D. (1996). Understanding radio polarimetry. III. Interpreting the IAU/IEEE definitions of the Stokes parameters. *A&AS*, **117**, 161–165.

Hamaker, J. P., Bregman, J. D., & Sault, R. J. (1996). Understanding radio polarimetry. I. Mathematical foundations. *A&AS*, **117**, 137–147.

Hamamatsu Photonics (2006). *Photomultiplier tubes: basics and applications*, 3rd edn.

Hampel, W. *et al.* (GALLEX Collaboration) (1999). GALLEX solar neutrino observations: results for GALLEX IV. *Phys. Lett. B*, **447**, 127–133.

Hanbury Brown, R., Jennison, R. C., & Das Gupta, M. K. (1952). Apparent angular sizes of discrete radio sources: observations at Jodrell Bank, Manchester. *Nature*, **170**, 1061–1063.

Hecht, E. (2002). *Optics*, 4th edn. Reading: Addison-Wesley.

Helstrom, C. W. (1991). *Probability and stochastic processes for engineers*, 2nd edn. New York: Macmillan.

Hild, S., Chelkowski, S., Freise, A. *et al.* (2010). A xylophone configuration for a third generation gravitational wave detector. *Class. Quantum Grav.*, **27**, 015003 (www.iop.org/EJ/abstract/0264-9381/27/1/015003/).

Hillas, A. M. (1984). The origin of ultra-high-energy cosmic rays. *ARA&A*, **22**, 425–444.

Hillas, A. M. (1996). Differences between gamma-ray and hadronic showers. *Space Sci. Rev.*, **75**, 17–30.

Hopkins, A. M. & Beacom, J. F. (2006). On the normalization of the cosmic star formation history. *ApJ*, **651**, 142–154.

Horiuchi, S., Beacom, J. F., & Dwek, E. (2009). Diffuse supernova neutrino background is detectable in Super-Kamiokande. *Phys. Rev. D*, **79**, 083013.

Hu, W. & White, M. (1997). CMB anisotropies: total angular momentum method. *Phys. Rev. D*, **56**, 596–615.

Hughes, S. A. (2001). Evolution of circular, nonequatorial orbits of Kerr black holes due to gravitational-wave emission. II. Inspiral trajectories and gravitational waveforms. *Phys. Rev. D*, **64**, 064004.

Hulse, R. A. & Taylor, J. H. (1975). Discovery of a pulsar in a binary system. *ApJ*, **195**, L51–L53.

IAU (1974). Polarization definitions (Commission 40). *Trans. Int. Astron. Union*, **15B**, 166.

IEEE (1969). IEEE standard #211: definitions of terms for radio wave propagation. *IEEE Trans. Antennas Propag.*, **AP-17**, 270–275.

In't Zand, J. J. M. (1992). A coded-mask imager as monitor of galactic X-ray sources. Ph.D. thesis, University of Utrecht.

Irwin, K. D. & Hilton, G. C. (2005). Transition-edge sensors. In *Cryogenic particle detection*, C. Enss, ed. Topics in Applied Physics, **99**, 63–149, Berlin: Springer.

Jackson, J. D. (1998). *Classical electrodynamics*, 3rd edn. New York: Wiley.

Jahoda, K., Markwardt, C. B., Radeva, Y. *et al.* (2006). Calibration of the Rossi x-ray timing explorer proportional counter array. *ApJS*, **163**, 401–423.

Jansen, R. A. (2006). *Astronomy with charged coupled devices* (e-book: www.public. asu.edu/~rjansen/ast598/ast598_jansen2006.pdf).

Jansen, F., Lumb, D., Altieri, B. *et al.* (2001). XMM-Newton observatory. I. The spacecraft and operations. *A&A*, **365**, L1–L6.

Johnson, H. L. (1966). Astronomical measurements in the infrared. *ARA&A*, **4**, 193–206.

Johnson, H. L. & Morgan, W. W. (1953). Fundamental stellar photometry for standards of spectral type on the revised system of the Yerkes spectral atlas. *ApJ*, **117**, 313–352.

Kitchin, C. R. (2009). *Astrophysical techniques*, 5th edn. Boca Raton: CRC Press.

Kliger, D. S., Lewis, J. W., & Randall, C. E. (1990). *Polarized light in optics and spectroscopy*. Boston: Academic Press.

Kuroda, K. *et al.* (LCGT Collaboration) (2010). Status of LCGT. *Class. Quantum Grav.*, **27**, 084004.

Lamarre, J. M. *et al.* (2003). The Planck high frequency instrument, a third generation CMB experiment, and a full sky submillimeter survey. *New Astron. Rev.*, **47**, 1017–1024.

Larson, D. *et al.* (2011). Seven-year Wilkinson microwave anisotropy probe (WMAP) observations: power spectra and WMAP-derived parameters. *ApJS*, **192**, 16.

Lazarian, A. & Cho, J. (2005). Grain alignment in molecular clouds. In *Astronomical polarimetry: current status and future directions*, ASP Conference Series, **343** A. Adamson, C. Aspin, C. J. Davis, & T. Fujiyoshi, eds. 333–345.

Lazarian, A., Goodman, A. A., & Myers, P. C. (1997). On the efficiency of grain alignment in dark clouds. *ApJ*, **490**, 273–280.

Lèna, P., Lebrun, F., & Mignard, F. (1998). *Observational astrophysics*, 2nd edn. Berlin: Springer.

Loredo, T. J. (1992). Promise of Bayesian inference for astrophysics. In *Statistical challenges in modern astronomy*, E. D. Feigelson & G. J. Babu, eds., 275–306. New York: Springer-Verlag.

Lyons, L. (1991). *A practical guide to data analysis for physical science students*. Cambridge: Cambridge University Press.

MacKay, D. J. C. (2003). *Information theory, inference, and learning algorithms*. Cambridge: Cambridge University Press.

Mathews, J. (1962). Gravitational multipole radiation. *J. Soc. Indust. Appl. Math.*, **10**, 768–780.

Mathews, J. & Walker, R. L. (1970). *Mathematical methods of physics*. Menlo Park: Benjamin.

Matthews, J. N. (2010). Overview of the high resolution fly's eye: some results from the HiRes experiment. In *Proc. 2009 Snowbird particle astrophysics and cosmology workshop*, ASP Conference Series, **426**, D. B. Kieda & P. Gondolo, eds., 3–10.

McKenzie, K., Grosse, N., Bowen, W. P. *et al.* (2004). Squeezing in the audio gravitational-wave detection band. *Phys. Rev. Lett.*, **93**, 161105.

Meers, B. J. (1988). Recycling in laser-interferometric gravitational-wave detectors. *Phys. Rev. D*, **38**, 2317–2326.

Mie, G. (1908). Beiträge zur optik trüber medien, speziell kolloidaler Metallösungen. *Ann. Physik*, **330**, 377–445.

Misner, C. W., Thorne, K. S., & Wheeler, M. A. (1973). *Gravitation*. San Francisco: Freeman.

Nyquist, H. (1928). Thermal agitation of electric charge in conductors. *Phys. Rev.*, **32**, 110–113.

Ogliore, R. (2007). The sulfur, argon, and calcium isotopic composition of the galactic cosmic ray source. Ph.D. thesis, Caltech.

Oke, J. B. (1964). Photoelectric spectrophotometry of stars suitable for standards. *ApJ*, **140**, 689–693.

Oppenheimer, B. R. & Hinkley, S. (2009). High-contrast observations in optical and infrared astronomy. *ARA&A*, **47**, 253–289.

Papoulis, A. (1991). *Probability, random variables, and stochastic processes*, 3rd edn. New York: McGraw-Hill.

Perryman, M. A. C. *et al.* (1997). The HIPPARCOS catalogue. *A&A*, **323**, L49–L52.

Peters, P. C. & Mathews, J. (1963). Gravitational radiation from point masses in a Keplerian orbit. *Phys. Rev.*, **131**, 435–440.

Press, W. H. (1997). Understanding data better with Bayesian and global statistical methods. In *Unsolved problems in astrophysics*, J. N. Bahcall & J. P. Ostriker, eds., pp. 49–60, Princeton: Princeton University Press.

Press, W. H., Teukolsky, S. A., Vetterling, W. T., & Flannery, B. P. (2007). *Numerical recipes: the art of scientific computing*, 3rd edn. Cambridge: Cambridge University Press.

Price, P. B. & Fleisher, R. L. (1971). Identification of energetic heavy nuclei with solid dielectric track detectors: applications to astrophysical and planetary studies. *Annu. Rev. Nucl. Sci.*, **21**, 295–334.

Prior, G. (for SNO Collaboration) (2009). Results from the Sudbury Neutrino Observatory phase III. *Nucl. Phys. B (Proc. Suppl.)*, **188**, 96–100.

Raffelt, G. (1996). *Stars as laboratories for fundamental physics*. Chicago: University of Chicago Press.

Ramsey, N. F. (1949). A new molecular beam resonance method. *Phys. Rev.*, **76**, 996.

Reitz, J. R., Milford, F. J., & Christy, R. W. (1979). *Foundations of electromagnetic theory*, 3rd edn. Reading: Addison-Wesley.

Rieke, G. H. (2002). *Detection of light: from the ultraviolet to the submillimeter*, 2nd edn. Cambridge: Cambridge University Press.

Roberts, G. O., Gelman, A., & Gilks, W. R. (1997). Weak convergence and optimal scaling of random walk Metropolis algorithms. *Annu. Appl. Prob.*, **7**, 110–120.

Röser, S. & Bastian, U., eds. (1991). *PPM star catalogue*. Heidelberg: Spektrum Akademischer Verlag.

Saikia, D. J. & Salter, C. J. (1988). Polarization properties of extragalactic radio sources. *ARA&A*, **26**, 93–144.

Sault, R. J. Hamaker, J. P., & Bregman, J. D. (1996). Understanding radio polarimetry. II. Instrumental calibration of an interferometer array. *A&AS*, **117**, 149–159.

Schönfelder, V. *et al.* (1993). Instrument description and performance of the imaging gamma-ray telescope COMPTEL aboard the Compton gamma-ray observatory. *ApJS*, **86**, 657–692.

Schroeder, D. J. (2000). *Astronomical optics*, 2nd edn. San Diego: Academic Press.

Seidelmann, P. K. (2006). *Explanatory supplement to the astronomical almanac*, rev. edn. Mill Valley: University Science Books.

Sivia, D. S. (1996). *Data analysis: a Bayesian tutorial*. Oxford: Clarendon Press.

Soffitta, P. *et al.* (2003). Techniques and detectors for polarimetry in X-ray astronomy. *Nucl. Instrum. Meth. A*, **510**, 170–175.

Stone, E. C. *et al.* (1998). The solar isotope spectrometer for the Advanced Composition Explorer. *Space Sci. Rev.*, **86**, 357–408.

Streetman, B. G. & Banerjee, S. K. (2005). *Solid state electronic devices*, 6th edn. Englewood Cliffs: Prentice Hall.

Strömgren, B. (1966). Spectral classification through photoelectric narrow-band photometry. *ARA&A*, **4**, 433–472.

Strüder, L., Briel, U., Dennerl, K. *et al.* (2001). The European photon imaging camera on XMM-Newton: the pn-CCD camera. *A&A*, **365**, L18–L26.

Sutton, E. C. & Wandelt, B. D. (2006). Optimal image reconstruction in radio interferometry. *ApJS*, **162**, 401–416.

Sze, S. M. & Ng, Kwok K. (2006). *Physics of semiconductor devices*, 3rd edn. Hoboken: Wiley.

Tatarskii, V. I. (1971). *The effects of the turbulent atmosphere on wave propagation.* Jerusalem: Israel Program for Scientific Translations.

Thompson, A. R., Moran, J. M., & Swenson, G. W. (2001). *Interferometry and synthesis in radio astronomy*, 2nd edn. New York: Wiley.

Thorne. K. S. (1987). Gravitational radiation. In *Three hundred years of gravitation*, S. W. Hawking & W. Israel, eds., 330–458. Cambridge: Cambridge University Press.

Timothy, J. G. (1983). Optical detectors for spectroscopy. *PASP*, **95**, 810–834.

Tinbergen, J. (1996). *Astronomical polarimetry*. Cambridge: Cambridge University Press.

Tinto, M. & Armstrong, J. W. (1999). Cancellation of laser noise in an unequal-arm interferometer detector of gravitational radiation. *Phys. Rev. D*, **59**, 102003.

Tinto, M. & Dhurandhar, S. V. (2005). Time-delay interferometry. *Living Rev. Relativity*, **8**, 4 (http://www.livingreviews.org/lrr-2005-4).

Tretyakov, M. Yu., Koshelev, M. A., Dorovskikh, V. V., Makarov, D. S., and Rosenkranz, P. W. (2005). 60-GHz oxygen band: precise broadening and central frequencies of fine-structure lines, absolute absorption profile at atmospheric pressure, and revision of mixing coefficients. *J. Mol. Spectrosc.*, **231**, 1–14.

Turner, M. J. L., Abbey, A., Arnaud, M., *et al.* (2001). The European photon imaging camera on XMM-Newton: the MOS cameras. *A&A*, **365**, L27–L35.

Vahlbruch, H., Khalaidovski, A., Lastzka, N. *et al.* (2010). The GEO 600 squeezed light source. *Class. Quantum Grav.* **27**, 084027.

Vallerga, J. V., Kaplan, G. C., Siegmund, O. H. W. *et al.* (1989). Imaging characteristics of the extreme ultraviolet explorer microchannel plate detectors. *IEEE Trans. Nucl. Sci.*, **36**, 881–886.

van de Hulst, H. C. (1957). *Light scattering by small particles*. New York: Dover.

Virtue, C. J. (SNO Collaboration) (2001). SNO and supernovae. *Nucl. Phys. B Proc. Suppl.*, **100**, 326–331.

Wakely, S. P. (2002). Precision x-ray transition radiation detection. *Astropart. Phys.*, **18**, 67–87.

Wang, L. & Wheeler, J. C. (2008). Spectropolarimetry of supernovae. *ARA&A*, **46**, 433–474.

Weber, J. (1966). Observation of the thermal fluctuations of a gravitational-wave detector. *Phys. Rev. Lett.*, **17**, 1228–1230.

Weisberg, J. M. & Taylor, J. H. (2005). The relativistic binary pulsar B1913+16: thirty years of observations and analysis. In *Binary radio pulsars*, ASP Conference Series, **328**, F. A. Razio & I. H. Stairs, eds., 25–31.

Wilson, R. N. (1996). *Reflecting telescope optics I*. Berlin: Springer.

Wilson, R. N. & Delabre, B. (1995). New optical solutions for very large telescopes using a spherical primary. *A&A*, **294**, 322–338.

Wolf, E. (2007). *Introduction to the theory of coherence and polarization of light.* Cambridge: Cambridge University Press.

Zuber, K. (2004). *Neutrino physics.* Bristol: Institute of Physics Publishing.

Index

Index

Printed in the United States
By Bookmasters